The Restoration of Rivers and Streams

The Restoration of Rivers and Streams

Theories and Experience

Edited by

James A. Gore

Faculty of Natural Sciences
University of Tulsa

BUTTERWORTH PUBLISHERS

Boston · London

Sydney · Wellington · Durban · Toronto

An Ann Arbor Science Book

Ann Arbor Science is an imprint of Butterworth Publishers.

Library of Congress Cataloging in Publication Data
Main entry under title:

The restoration of rivers and streams.

Includes bibliographies and index.
Contents: Introduction / James A. Gore—The use of meander parameters in restoring hydrologic balance to reclaimed stream beds / Victor R. Hasfurther—Water quality restoration and protection in streams and rivers / Edwin E. Herricks and Lewis L. Osborne—[etc.]
1. Stream conservation—Addresses, essays, lectures.
I. Gore, James A.
QH75.R47 1985 627'.12 84-16965
ISBN 0-250-40505-9

Butterworth Publishers
80 Montvale Avenue
Stoneham, MA 02180

10 9 8 7 6 5 4 3 2 1

Printed in the United States of America

CONTENTS

Introduction

James A. Gore

Streams and rivers are members of a small group of geologically enduring features on the surface of the earth. Indeed, these running water systems have been major factors in the formation of many dominant landforms. As enduring environments, streams and rivers are the primary habitats for uniquely adapted vertebrates, invertebrates, and plants. Many of these groups, some of which are geologically quite old and primative, survive only in lotic environments. Because running water systems have the potential to provide nourishment, energy, and transportation to humans, overuse and misuse of streams and rivers have resulted in severe damage to these ecosystems as well.

With human energy demands exceeding the readily obtainable sources of oils and gas, alluvial valley floors have provided the easiest entry to deposits of alternative energy sources such as coal and uranium. In addition to providing power for electricity generation, the erosive forces of streams and rivers have cut into the *overburden* (the material covering the deposit) to make mechanical access to the deposits much cheaper via surface mining techniques. In the northern Great Plains, for example, the erosive powers of the Powder and the Tongue rivers have decreased overburdens to less than one meter in some places. After removal of this thin layer of topsoil and deposited gravels, rich seams of low-sulfur coal have been exposed (Gore and Johnson 1981). Surface coal mines are located as close as possible or upon these shallow overburden areas. Figures I.1 through I.3 illustrate the size of a typical mine and the erosive forces of an adjacent river. Until passage of the Surface Mine Reclamation and Control Act of 1977, relocation of stream and river channels into previously mined pits was often the common practice. Vinikour (1980) documented the total loss of the running water community in these new lake-like aquatic ecosystems. Cairns et al. (1977) have also documented similar impacts from surface mine operations in the eastern United States.

Even before the time that American pioneers followed the valleys of the Missouri, Platte, and Wind rivers to reach Oregon, humankind constructed its major transportation routes along river flood plains. Increased construction in recent years has "necessitated" the relocation of streams to accommodate four-

Figure I.1 A typical coal surface mine operation in the western United States. This mine is operating or has operated on alluvial plains. Photo by author.

lane highway bridges or new rails and roadbed to transport industrial and domestic goods. When the relocated and/or restored channels were not properly structured to enhance biological productivity, the results have been obvious and notable. Chapman and Knudsen (1980) reported a decline in total habitat in altered areas used in livestock and flood control programs. The effect of this channelization was reduced salmonid standing crops. Barton (1977) noticed a distinct change in faunal composition of benthic communities impacted by highway construction. Even the effects of building a trench and pipeline across a stream can have significant impacts without proper restoration (Tsui and McCart 1981).

Alteration of stream and river habitat that requires some mitigating action can often take more subtle forms. Alteration and restoration of water quality after point and nonpoint source impacts must be considered. The need to restore stream water quality after acid mine drainage impacts is obvious to most managers. But, without erosion control, actions like clear-cut logging (Newbold et al. 1980) and even urban runoff (Judy, Chapter 9, this book) can effectively alter biota in receiving streams. In some cases the misuse or "overuse" of a recreational resource may cause the degradation of a stream ecosystem so that enhancement of stream habitat is required to maintain stability and production (Burgess and Bides 1980).

I became interested in the restoration or reclamation of rivers when I was involved in a project to restore a portion of the Tongue River, in Wyoming, which

Figure I.2 It is economically advantageous to mine or construct bridges or roads near riverine environments where the erosive power of the river aids construction activity. The forested area of this picture is occupying the opposite side of the high wall shown in Figure I.1. Photo by author.

had been diverted and relocated several times to facilitate further coal removal. With the advice and help of mine engineers and reclamation specialists, a new channel had been constructed and observed for attainment of a stable "restored" ecosystem (Gore and Johnson 1981; Gore 1979). Although I was able to model invertebrate colonization trends and to predict use of available habitat by chosen fish species, an essential truth about the definition of *river restoration* was discovered. Hynes (1970), in his classic text on stream and river ecology, stated that streams have an inherent ability to cleanse themselves. Cairns et al. (1977) also showed that rivers exhibit a clear natural recovery process. What, then, is *reclamation* or *restoration* of a river ecosystem? Unlike terrestrial reclamation projects in which a restored area is often revegetated and new animals introduced, most river restoration projects entail the design and emplacement of suitable habitat structures to attract invading pioneer and colonizing biota. In essence, *river restoration* is the process of *recovery enhancement.* Recovery enhancement enables the river or stream ecosystem to stabilize (some sort of trophic balance) at a much faster rate than through the natural physical and biological processes of habitat development and coloniza-

Figure I.3 After mining or construction activity, restoration can begin. This newly cut and reconstructed post-mining river channel yet requires revegetation and structures to enhance biotic activity. Photo by author.

tion. Recovery enhancement should establish a return to an ecosystem which closely resembles unstressed surrounding areas. Since a stream ecosystem and its energy dynamics are a product of external and internal actions (Hynes 1975; Vannote et al. 1980), restoration must include considerations of in-stream habitat improvement *and* hydrologic stability, riparian restoration, and water quality improvement. Indeed, and as some of the authors in this book suggest, trophic relationships in streams also depend upon stability and structure of riparian fauna (Hynes 1970), often overlooked in mitigation procedures.

A stream manager cannot address the issues of restoration without identifying what restoration means and what its various elements are. No one person can establish all criteria and standards for successful restoration projects. It requires a group effort based on comprehensive field work and background of ecological theory and experience. With the encouragement of several of my colleagues, some pushing by graduate students, and my own desire to broaden my knowledge of theory and practices, I have attempted to collect works by biologists, hydrologists, engineers, and other stream managers working in the field. The body of the material presented in this book represents the theory and experience of academicians and stream managers who have attempted to establish criteria and standards for a great variety of restoration projects.

This book is intended to display theories, experiences, and techniques that have proven to be of good use in enhancing the recovery of damaged running water ecosystems. A wieldy volume of information cannot necessarily be comprehensive. Most of the case studies presented here emphasize large impacts from surface mining. Yet many of the same techniques can easily be applied to mitigation of impacts from highway and bridge construction and agricultural channelization. The stream manager using these works must be able to wisely employ needed measures on a site-specific basis.

There are a number of publications on structures and restoration practices (see, for example, White and Brynildson 1967; U.S. Forest Service 1969; Nelson et al. 1978; Stream Enhancement Research Committee 1980). Unfortunately, many of these works are not always accessible or even known to nonfederal and/or state agency researchers and managers. Most preexisting publications seem to concentrate on a single aspect of restoration, often fish habitat improvement (i.e., cover and spawning structures). In this book, I hope to present a review of some previous efforts and provide a selection of restoration ideas and alternatives to the stream manager. Either through case histories or specific techniques, chapters will attempt to address these restoration questions:

1. What have historical practices and successes been?
2. What new or additional techniques are being used in recent restoration projects?
3. Are new techniques more successful than the "old stand-by?"
4. How can all of the techniques be integrated into an overall recovery enhancement project?

Reference sections for each chapter provide an extensive library of additional works for consultation.

There is always need for improvement of techniques and refinement of management programs. I hope that this book will provide the impetus for these advances. A tool is good only if one learns proper use and need. The restoration of damaged or impacted running water ecosystems can provide important tests of ecological theory as well as the more practical conservation of a geologically old and unique environment.

It seems inevitable that the demands of an energy hungry, mobile nation will outweigh the protests of the environmentalists, conservationists, and ecologists. The constituencies with legislative power reside in the ten or so largest cities in the United States. In these cities, energy and transportation demands are high while views of pristine streams and rivers are few. Often the need for moral and economic support of restoration projects is not apparent. It behooves us, as biologists, hydrologists, engineers, and resource users to establish the best possible restoration techniques and to develop an integrated approach to recovery enhancement that will accommodate the impacts of resource development *and* the preservation of a unique ecosystem. There is no guarantee that these studies and projects will be inexpensive or require a minimum of field work. The ultimate benefits of stream and river integrity to local and visiting consumers, perhaps not always measurable in economic units, are nevertheless of great value.

REFERENCES

Barton, B.A. 1977. Short-term effects of highway construction on the limnology of a small stream in southern Ontario. *Freshw. Biol.* 7:99–108.

Burgess, S.A., and J.R. Bides. 1980. Effects of stream habitat improvements on invertebrates, trout populations, and mink activity. *J. Wildl. Manage.* 44(4):87–880.

Cairns, J., Jr., K.L. Dickson, and E.E. Herricks. 1977. *Recovery and Restoration of Damaged Ecosystems.* Charlottesville: University Press of Virginia.

Chapman, D.W., and E. Knudsen. 1980. Channelization and livestock inpacts on salmonid habitat and biomass in western Washington. *Trans. Amer. Fish. Soc.* 109:357–63.

Gore, J.A. 1979. Patterns of initial benthic recolonization of a reclaimed coal strip-mined river channel. *Can. J. Zool.* 12:2429–39.

Gore, J.A., and Lora S. Johnson. 1981. Strip-mined river restoration. *Water Spectrum* 13:31–38.

Hynes, H.B.N. 1970. *The Ecology of Running Waters.* Toronto, Ontario: Univ. of Toronto Press.

———. 1975. Edgardo Baldi Memorial Lecture: The stream and its valley. *Verh. Internat. Verein. Limnol.* 19:1–15.

Judy, R.D., Jr. 1985. Enhancement of urban water quality through control of nonpoint source pollution: Denver, Colorado. In *The Restoration of Rivers and Streams,* edited by J.A. Gore. Stoneham, Mass.: Butterworth Publishers, pp. 247–276.

Nelson, R.W., G.C. Horak, and J.E. Olson, eds. 1978. *Western Reservoir and Stream Habitat Improvements Handbook.* Western Energy and Land Use Team, Fort Collins, CO: U.S. Fish and Wildlife Service. FWS/OBS-78-56.

Newbold, J.D., D.C. Erman, and K.B. Roby. 1980. Effects of logging on macroinvertebrates in streams with and without buffer strips. *Can. J. Fish. Aquat. Sci.* 37:1076–85.

Stream Enhancement Research Committee. 1980. *Stream Enhancement Guide.* Province of British Columbia Vancouver, B.C.: Government of Canada, Ministry of Environment.

Tsui, P.T.P., and P.J. McCart. 1981. Effects of stream-crossing by a pipeline on the benthic macroinvertebrate communities of a small mountain stream. *Hydrobiologia* 79:271–76.

U.S. Forest Service. 1969. *Wildlife Habitat Improvement Handbook.* Washington, D.C.: Cat. No. FSH 2609:11.

Vannote, R.L., G.W. Minshall, K.W. Cummins, J.R. Sedell, and C.E. Cushing. 1980. The river continuum concept. *Can. J. Fish. Env. Sci.* 37:130–37.

Vinikour, W.S. 1980. Biological consequences of stream routing through a final-cut strip-mine pit: benthic macroinvertebrates. *Hydrobiologia* 75:33–43.

White, R.J., and O.M. Brynildson. *Guidelines for Management of Trout Stream Habitat in Wisconsin.* Tech. Bull. No. 39. Madison, Wis.: Wisconsin Dept. of Nat. Res.

CHAPTER 1

Water Quality Restoration and Protection in Streams and Rivers

Edwin E. Herricks and Lewis L. Osborne*

Department of Civil Engineering
University of Illinois Urbana-Champaign
208 North Romine Street
Urbana, Illinois 61801

From the Greek philosopher Thales originated a philosophy and science which asserted water was the origin of all things. Aristotle also included water in his listing of the elements (earth, air,. fire, and water). Today we recognize that water is essential for ecosystem maintenance. The quality of water affects the development of society, and is essential for the health and well-being of all people.

The restoration and protection of water quality has grown in importance and significance as our understanding of the complex processes involved in both the maintenance of healthy, functioning ecosystems and our understanding of mechanisms of degradation has improved. We have come to recognize the connection between ecosystem integrity and water resources issues and the need for restoration as well as protection of water resources, particularly where human impact has been severe. Examples such as the restoration of the Thames estuary (Gameson and Wheeler 1977), the improvement of Lake Washington (Edmondson 1977), and the recovery of streams following spills or chronic stress (Cairns and Dickson 1977; Herricks 1977) demonstrate the potential for recovery in systems with severely degraded water quality. The restoration of lakes and estuaries points the way toward sound management practices while studies of spills illustrate the opportunities provided by the fundamental resiliency of streams. International interest and legislation are an indication that as development proceeds we should not lose sight of the importance of water resources (in particular water quality) in both developed and undeveloped countries.

The following discussion of restoration and protection of water quality in streams and rivers recognizes the role water plays in ecosystem processes. Water

* Present address: Department of Urban and Regional Planning, University of Illinois Urbana-Champaign, 1003 West Nevada, Urbana, Illinois 61801.

quality cannot easily be discussed from a single disciplinary perspective because issues span a number of disciplines and relate to physical, chemical, and biological components of the ecosystem. In addition, the separation of purely technical issues from economic, political, and social factors is impossible when constraints on restoration or protection efforts are considered.

In organizing our discussion of the restoration and protection of water quality in streams, we felt it important to first review the context in which restoration and protection of water quality is viewed, identify uses and impacts, and then discuss the general approaches to restoration and protection which are available to water quality managers.

PERSPECTIVES ON RESTORATION AND PROTECTION

Restoration and protection are concepts basic to the formulation of water quality regulation in the United States. The declaration of goals and policy in Title I, Section 101 of the Clean Water Act (33 U.S.C. 466 et seq.) reads, "The objective of this Act is to restore and maintain the chemical, physical, and biological integrity of the Nation's waters." The need for restoration arises from the public's perception, supported by empirical data, that water quality is degraded from some idealized condition and maintenance of the status quo is not acceptable. This has led to the establishment of regulations to protect high quality water resources from degradation and to the improvement and restoration of existing water quality. The justification for water quality regulations is protection of public health and safety as well as the prevention of further environmental degradation.

To initiate a discussion of the restoration of stream water quality, several conceptual and technical issues associated with restoration must be evaluated. The issues include what restoration is, how it is assessed, the relationship of impact to restoration, and the integration of water quality in watershed restoration and protection.

Restoration

The definition of restoration and an assessment of success or failure of restoration efforts are related. Implicit in the concept of water quality restoration is some knowledge of the undisturbed or natural state of the stream system. Restoration of water quality can be defined as returning the concentrations of substances to values typical of undisturbed conditions. In the absence of data on undisturbed states, restoration is often assessed on the basis of a derived standard which may or may not represent actual undisturbed conditions. A standard, often in the absence of a real understanding of its derivation, is accepted as an endpoint to restoration efforts.

The assessment of restoration has depended on two assumptions. The first assumption is that all streams and rivers should meet an idealized standard, quantified in terms of presently undisturbed streams and rivers. Undisturbed streams are uncommon and those that are available for comparison are either small streams or large, high gradient, cold water rivers. Both of these stream types are unlike the low gradient, warm water streams or rivers that bear the brunt of man's activities. Using small streams or wild and scenic rivers as a model for water quality regulation may protect the greatest number of streams (small low-order streams dominate stream numbers and length categories) or meet aesthetic standards, but regulation based on these systems unrealistically limits management which is based on assimilative capacity. The definition of restoration in streams and rivers heavily used by man is either impossible due to the absence of suitable comparative systems or inaccurate because the systems available for comparison are inappropriate.

The second assumption is that in the absence of suitable comparative systems, a water quality standard, and thus a definition of restoration, can be established through extrapolation of laboratory testing results. Standards are based on criteria. Criteria are established based on identified uses of the stream and the level of protection desired. The form of a standard, either numerical or narrative, is established by legislative action. A detailed review of criteria identification and standard development is provided in later discussions on use and criteria. The critical issue that must be addressed here is the validity of the use of standards to define restoration. If restoration is defined based on an acceptable concentration of a substance, the standard, that concentration must not only identify direct effects, but also account for the range of indirect effects possible in a complex ecosystem. Few existing standards can meet such a rigorous test of validity. The definition of restoration based on existing water quality standards is inappropriate, but as standard development incorporates both direct and indirect effects, standards will be useful for restoration assessment.

Protection

The next issue that must be addressed as a part of this review of water quality perspectives is the implication of the word *maintain* in the phrase "to restore and maintain the chemical, physical, and biological integrity of the Nation's waters." This wording suggests that maintenance and protection are largely synonymous when used in the context of water quality management. When maintenance of water quality is the goal, by definition, a specified quality level will be kept or defended providing protection with maintenance. It is possible to maintain any water quality condition, but we must protect water quality from degradation. Protection then better describes the process, not just the goal.

The process of protection in water quality management must meet variable goals depending on the existing condition of the watercourse in question. If the water quality of a stream or river is already degraded, the process of protection requires no further degradation of quality—a nondegradation policy. Although nondegradation applies even to pristine waters, a more standard definition of protection, shielding from injury or loss, is applied when quality is not to be changed. For most streams and rivers maintenance of water quality at least meets the nondegradation policy, but water quality may be changed over a defined reach (typically the mixing zone) taking advantage of stream assimilative capacity. In reality then, protection is a compromise between the desire for pristine conditions and the practical realization that the presence of humans precludes achieving this goal. Thus, protection in a practical sense implies application of available technology to assure the nondegradation of water quality. The same problems inherent in the definition of restoration (e.g., dependence on established criteria or standards) apply to protection.

If the recent history of protection of water quality is reviewed, the selection of parameters protected (or, conversely, identification of what is causing damage) is directly related to the capabilities and quality of measurement of the concentrations of substances and compounds in water. For example early developments in bacteriological and virological analyses brought into focus the potential for infection through water supplies. The earliest standards were drinking water standards. General utilization of gas chromatography and mass spectrometry has led to the identification of new compounds and produced new areas of environmental concern. With each analytical improvement the protection of water quality requires a new assessment of risk or hazard, and new criteria and standards.

To assure protection, water quality management now requires the use of sophisticated monitoring procedures and analytical techniques. Each technological development requires a new assessment of risk or hazard. As often as technological discoveries are made, review and evaluation, redefinition of conditions, and honest scientific disagreement lead to conflicts in approaches to protection. When even a well-supported result is continually quoted, it takes on the appearance of "gospel," and the entire field of water quality management is at risk. The risk arises from a loss of credibility if the gospel is successfully challenged and protection for valuable resources is lost.

Ecological Considerations

Discussion of water quality restoration and protection is difficult without understanding the intricate relationships between the biotic and abiotic elements of the stream ecosystem. When considering water quality management, holistic approach finds support in both recent legal interpretation of regulations and accepted ecological principles. Only by following the development and integration of both biotic and abiotic factors is it possible to evaluate restoration and protection procedures from a sound ecological position. Abiotic factors include flow regime, habitat structure, and water quality (see Fig. 1.1). The primary biotic factor is fixed carbon or food resources. Additional factors which control the presence and abundance of organisms include tolerance, acclimation, adaptation, and life history characteristics. Both abiotic and biotic factors may vary seasonally and geographically.

An understanding of the attributes that distinguish individual and population responses is imperative in the formulation of restoration and protection methods

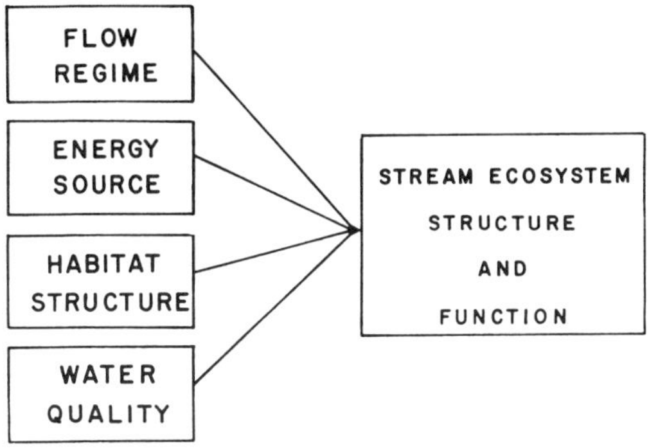

Figure 1.1 General environmental factors affecting the structure and function of stream ecosystems. Modified from Karr and Dudley 1980.

and procedures. The attributes useful in population analysis include population size, size distribution within the population, spatial distribution, life history patterns, genetic composition of individuals in a geographically defined population, as well as genetic differences in isolated populations and evidence of evolutionary trends in geographically separated populations. The metrics of these parameters are determined based on total population assessments and are often expressed as mean values with known standard deviation or range.

After attributes are quantified, determination of tolerance to an environmental alteration is possible using toxicity tests. The ability of an individual to successfully maintain itself is a function of its genetic composition and ability to acclimate to a new set of conditions. Acclimation can occur only within tolerance limits, which are set by the genetic composition of the individual. Once conditions are outside the individual's tolerance limits, the organism will be excluded from the habitat. A population, on the other hand, may be subjected to the same set of conditions, but survival is not based on the limited genetic information of the individual, but rather on the gene pool of the population. The population also has the capacity to change tolerance limits through adaptive processes. Thus, the ability of a population to maintain itself when environmental conditions change is a function of the gene pool (defining the capabilities for acclimation) and the population's ability to adapt to new environmental conditions.

When considering the ecological effect of water quality, any analysis of ecological response will depend on the level of ecological organization specified (individual, population, community, etc.). Adaptation is a population attribute, and acclimation is an attribute of the individual. An organism's expression of preference in a habitat may not always indicate the habitat that is optimal for the population. The preference demonstrated may be constrained by the habitat available, requiring acclimation to existing conditions. For example, an individual may tolerate degraded water quality to take advantage of a set of physical habitat conditions. Thus, the use of organism presence as an indication of the expected population response to a water quality condition may be interpreted differently depending on whether it is considered an individual or a population attribute.

When analyzing both biotic and abiotic components of an ecosystem a basic tenet of environmental chemistry should be considered. This tenet is that the properties of a chemical in the environment interact with the properties of the environment to affect the movement, persistence, and fate of the chemical. A similar statement can be made about biological systems where interaction of an organism with its environment alters the environment. Direct and indirect effects must be considered. Norris (1981) refers to the direct effects of chemicals on organisms as "those that result from the direct contact of a specific organism with a chemical agent." Indirect effects are those which "do not require a direct interaction between the chemical and organism." Such indirect effects are most often observed when a link within the trophic organization of a food web is broken or interrupted (Paine 1980).

PREVAILING CONDITIONS

The importance of restoration efforts and the applicability of any of a number of restorative techniques is dependent on the prevailing conditions which exist in the stream. If a stream is to be restored and protected, stream water quality cannot be dissociated from existing conditions in the watershed (Hynes 1975). In fact, restoration of water quality is so dependent on watershed conditions that any water quality restoration or protection activity must include remedial action on the watershed. The following discussions of prevailing watershed and water quality conditions set the stage for later discussions of methods to effect restoration of stream water quality.

Water quality may be degraded by natural changes to a watershed (fire, flooding, etc.) but these natural changes are often limited in extent and/or duration. The effects of human activities on watersheds include disturbances which create nonpoint sources of water quality degradation (pollution), while industrial development and population growth produce point source pollutants that are highly concentrated. When discussing restoration and protection both nonpoint and point sources of pollution must be considered. McElroy et al. (1975) reported that of the 916 million hectares of land in the United States 97% was rural in nature, and they concluded that all of the rural land was a potential source of nonpoint pollution. Sixty-four percent of the total land area in the United States was dedicated to either agriculture or silviculture (Table 1.1), while mining and construction occupied 0.6% (McElroy et al. 1975). Polls and Lanyon (1980) have stated that "It is becoming increasingly evident that the water quality goals established by the Clean Water Act . . . cannot be attained by regulating and controlling only point source pollution . . . (for) in many areas of the country pollutants emanating from nonpoint sources comprise the major contribution to water quality degradation." Likens et al. (1978) have demonstrated that removal or cutting of hardwood forests sets into motion a variety of effects including: increasing the concentrations of dissolved chemicals, accelerating erosive processes and transport of particulate matter, and decreasing the pH of drainage water.

Agricultural nonpoint effects on water quality may be discounted because

Table 1.1 Data on land use in the United States

Land Use Category	%	*Millions of Hectares*
Farmland in grass	18	218
Cropland, plus farmsteads and roads	14	167
Construction (annual)	<1	0.59
Commercial forest (includes farm woodlands and forests)	17	202
Annual harvest of forests (growing stock)	<1	4.45
Subsurface mines	<1	2.8
Surface mines	<1	1.2
Active surface mines	<1	0.14
Mineral waste storage	50	592.79
Total	100	1188.97

Source: Modified from McElroy et al. 1975.

the amount of change in concentration of any substance may be small; however, the land area involved in agriculture points to the magnitude of the problem. Agricultural modifications in the watershed and along stream channels alter the substrate through increased siltation rates; increase the amount of solar radiation received, which increases water temperatures, also creating oxygen imbalances; and alter seasonal flow patterns, especially when irrigation is employed or riparian vegetation is removed.

Increased sediment loads and rates of sedimentation have a pronounced effect on water quality (Meyer 1979) and have been shown to have adverse effects on the stream biota by altering spawning times and behavior (Muncy et al. 1979) or suffocating eggs (Johnson 1961).

Seasonal flow patterns, particularly fluctuating water levels, are an integral component of stream ecosystems. Stream aquatic life has evolved either behaviorally or physiologically to accommodate natural periodicities and fluctuations in flow regimes. Modification of the land surface by agriculture, however, has resulted in higher flood peaks and longer low-flow periods that occur more frequently (Karr and Dudley 1981). Borman et al. (1969) have demonstrated that flood events in natural watersheds have a dampened hydrograph, while those in agricultural watersheds tend toward sharp and extreme peaks. Furthermore, in agricultural areas, late summer low-flow periods are often extended (Karr and Dudley 1981). Such patterns affect water quality, degrading conditions to significantly affect the reproductive success of many organisms. Alteration of current velocities which are critical to the survivability of most lotic organisms (see Hynes 1970) alters flushing rates and potential and changes the residence time for any contaminant in the stream.

In a study of factors which control water quality in agricultural watersheds, Schlosser and Karr (1980) reported that during runoff events, water quality is primarily governed by hydrological processes. They also found that stream organisms were limited by watershed conditions (such as riparian vegetation) which are often controlled or modified by hydrological processes. The resulting ecological relationships involved complex interactions between watershed topography, channel morphology, and magnitude of the runoff event. The role of water quality in the maintenance of stream ecosystem conditions is equally complex. Schlosser and Karr (1980) observed that seasonal changes in amounts of precipitation have a substantial effect on the influx of nutrients. Alteration of nutrient levels affects primary production, particularly if riparian vegetation is removed. Water quality as well as physical habitat are modified and stream ecosystems are degraded.

These results demonstrate the importance of various hydrologic, geomorphic, and edaphic processes in regulating water quality in agricultural watersheds. Schlosser and Karr (1980) have warned that modeling efforts to predict water quality or management efforts to enhance water resources that are based solely on erosive potential in the watershed while ignoring near- and inchannel processes will not be widely applicable and successful. These authors suggest that during low or base flow conditions, "emphasis should be placed on linking the hydrological theory of transport of inorganic material (Stall and Yang 1972; Yang and Stall 1974) to the biological theory of production and transport of organic materials (Cummins 1974)." During runoff events, "emphasis should be placed on linking the agricultural theory of erosion prediction (Wischmeier and Smith 1965) to the geomorphological theory of stream equilibrium and sediment transport (Leopold et al. 1964)." Thus, restoration and protection of water quality in agricultural areas requires an understanding of and capacity to merge many disciplines.

PERSPECTIVES ON IMPACT

Within the context of the above discussion and implicit within a discussion of restoration is the existence of an impact. Essentially an impact (unless otherwise noted) refers specifically to artificial alterations and can be viewed from three overlapping perspectives. Clarification of each perspective is essential if one is to examine the potential success of restoration and protection procedures.

The *ecological perspective* in impact analysis recognizes the importance of any alteration in the structure, composition, or functional capabilities of an ecosystem that would not occur in the absence of human activities. From a regulatory standpoint, the action following from such a perspective is determination of the concentration of a substance, if any, that can be released into an aquatic system without altering any component of the natural framework. Although heuristically desirable, such a perspective is socially, economically, and technically impractical. The fact that all organisms are subjected to some degree of natural stress (due to prevailing environmental conditions being less than optimal along at least one niche axis) makes it difficult to distinguish human-caused impacts from natural stress. From a practical standpoint one will only be able to assess impact within the capabilities of available techniques, methodologies, and resources. Therefore, restoration must be limited by the same factors. The value of the ecological perspective is found in the requirement that we increase our understanding of the response of ecosystems to various impacts, allowing the development of new and more sensitive methods of impact analysis. The ecological perspective should not be discarded or ignored by politically motivated actions.

The two other perspectives on impact are the *criterion perspective* and the *impairment of designated use perspective.* Both incorporate components of the ecological perspective but have an economic, social, or political emphasis. An impact, from a criterion perspective, involves exceeding an established criterion value; while the designated use perspective identifies impact when a stream or stream reach no longer supports a previous use or a higher and better use is unattainable because of pollution. Thus, the goal of restoration within the latter two impact perspectives may be more clearly defined politically and operationally relative to the heuristic ecological perspective. Attainment of defined goals is, nonetheless, difficult. Furthermore, the goals themselves are only as valid as the data base on which either the criterion or designated use was established, thus the need for quantitative ecological studies (Herricks and Cairns 1982).

USE AND CRITERIA

Although the total miles of streams and rivers that are affected by agriculture and silviculture may be greater than the miles of streams that are affected by industrial and municipal effluents, the intensity of point source effects as well as the potential for treatment and improvement of defined effluents has focused regulatory attention on wastewater treatment. When considering present water quality in relation to existing use, it is helpful to differentiate between small streams and larger rivers and to identify the uses defined for reaches in the same watercourse.

A review of methods for restoration of stream water quality must deal with at least two general classes of restorative techniques. The class of techniques for smaller streams and rivers can be reasonably well defined because research has concentrated on low-order streams. The class of techniques for larger rivers suffers from a lack of basic information and often becomes clouded with nontechnical issues raised when uses are defined by social, political, or economic constraints. In both classes of techniques, restoration is typically judged on the basis of meeting some set of quality criteria. In the United States these criteria have a basis of regulatory support in the Clean Water Act. That act defines acceptable uses for the nation's waters, and regulations establishing water quality standards are based on criteria that identify concentrations of water quality parameters that will allow maintenance of the specified use. Use designations include ". . . public water supply, propagation of fish and wildlife, recreational purposes, and agricultural, industrial, and other purposes, and also taking into consideration their use and value for navigation" (Sec. 303.c.2 of the Clean Water Act).

A compilation of the actual designated use categories employed by forty-eight states and the District of Columbia (U.S.EPA 1980) reveals that there are a minimum of fifteen general categories of stream designated uses presently utilized

(Table 1.2). The legal requirements and complexity of each category and the number of individual categories employed by each state vary from single use designations to a detailed and comprehensive multiple use specification found in California. Three principal categories dominate the list numerically with only one category, fish and wildlife propagation, listed for all states examined (henceforth including the District of Columbia). The fish and wildlife category was further subdivided by approximately 35% of the states to distinguish between warm and cold water habitats. Besides fish and wildlife propagation, maintenance of stream quality for the purpose of primary recreation (i.e., swimmable) and as a public water supply (i.e., drinkable) are the most common designated uses. The recognized importance of these three use categories demonstrates the states' resolve for maintaining and/or restoring the integrity of a number of aquatic systems and simultaneously protecting the health of the public.

The economic reality of the designated uses is represented by numerous other use categories that require water quality less suitable for aquatic life (Table 1.2). The types and number of specific use categories employed by each state are basically a function of the most common and profitable industrial, agricultural, and/or commercial activity within each state.

The concentrations of various substances that support a designated use are the basis of water quality protection efforts. The criteria published by the U.S.

Table 1.2 Categories of designated stream uses and the number of states (from a total of 48 states and the District of Columbia) legislatively possessing such designated uses.

Designated Stream Use	*Number of States*
Fish and wildlife propagation	49
Cold water habitats	18
Warm water habitats	17
Public water supply	47
Primary recreation	47
Secondary recreation	40
Agricultural/industrial water supply	39
Industrial operations	23
Growth of shellfish	19
Navigation	18
Aesthetics	13
Fish migration	10
Preservation areas	9
Hydropower generation	7
Fish spawning	4
Multiple use*	3
Groundwater recharge	3

Source: based upon information available in U.S.EPA 1980.

* Denotes more than one group of uses for a stream, in which case the highest use designation in the category must be maintained.

Environmental Protection Agency are designed to support aquatic life maintenance and propagation, which is generally considered the most stringent as well as the highest and best use for the stream or river. This defined ecological connection indicates why aquatic life criteria are most commonly used to assess water quality restoration and protection.

A review of the history of the development of criteria in water quality regulation provides an opportunity to evaluate the technical support for the assessment of restoration in both large and small streams. The development of criteria and the standards they support is highlighted by controversy, often based on differing opinion or scientific methodology.

The first water quality criteria employed physical tests for taste and odor and were improved by measures of temperature, color, and turbidity. The earliest quantification of criteria was noted in 1784 when waters were assessed as drinkable if they would "dissolve soap without forming lumps . . . and deposit nothing or very little by tests" (Hinman 1920). As water-borne diseases were recognized other criteria were applied. Disease-related criteria typically dealt with residues, ammonia, and bacteria per volume of water (McKee and Wolf 1963). The controversy associated with criteria included parameter selection, measurement methods, and concentration limits. Often ad hoc criteria were established based on minimal scientific input. For example, in the early part of this century Morse and Wolman (1918) argued that methods for establishing a criterion or standard must be adequately evaluated and this evaluation should antedate the establishment of limiting values.

Water quality regulation requires a standard suitable for legal enforcement but that will also protect environmental integrity. If a standard originates in a narrowly focused criteria, few environmental benefits may result. Enforcement requirements may unnecessarily limit impact assessment (particularly from the ecological perspective) and emphasize only politically or economically expedient regulatory approaches.

From their establishment, standards have fallen on hard times. Sedgwick has been quoted (Schroepfer 1942) as describing standards as "devices to save lazy minds the trouble of thinking . . . standards are often the guess of one worker, easily seized upon, quoted and requoted until they assume the semblance of authority." This is best emphasized in Sprague's (1976) examination of available data identifying "safe" levels of toxicants. In his analysis Sprague employed the rationale that without levels of truly safe concentrations of toxicants, there can be no acceptable strategy for water pollution control.

The tabulated results of Sprague's investigation (Table 1.3) of recommendations for "safe" levels of toxic pollutants were taken from the *Water Quality Criteria* bluebook (1972). Sprague found that almost half of the recommendations had no numerical value or support, were arbitrary (including the use of arbitrary application factors), while 13% were essentially lethal concentrations. The remaining 35 criteria on sublethal or chronic data constituted less than half of the total number of recommendations examined. Of this latter group, one third of those concerned with invertebrate organisms employed *Daphnia* or *Gammarus* as test organisms, neither of which are common in most streams. It is readily apparent from Sprague's

Table 1.3 Tabulated apparent bases for recommendations given in the sections on freshwater and marine aquatic life of U.S. government's *Water Quality Criteria 1972* (National Academy of Sciences/National Academy of Engineering 1974).

Criteria Used	*Cases*	%
1. No clear basis	10	13.3
2. No specific value or an arbitrary level (including application factor)	20	26.7
3. Laboratory lethal level alone	10	13.3
4. Reproduction:		
Fish	7	9.3
Invertebrates	5	6.7
Algae	1	1.3
5. Laboratory physiological tests (fish, invertebrates, or algae)	10	13.3
6. Tissue residue	≈7	9.3
7. Field Observations	≈5	6.7
Total	75	99.9

Source: modified from Sprague 1976.

analysis that as of 1972 we were still far from achieving a good *experimental* basis for water quality criteria. This review suggests that future criteria should be based upon experimental studies that scientifically, not haphazardly, evaluate the long-term and sublethal effects of pollutants. Initiation and completion of such studies are particularly important when dealing with pollutants that have an effective life in the environment measured in years or decades, or that are transported through the food chain, or affect only specific life stages or age classes or organisms. The following studies provide an example of experimental analysis useful in interpreting the effectiveness of experimental studies is setting criteria.

Menendez (1976) reported that the numbers of viable brook trout (*Salvelinus fontinalis*) eggs were significantly reduced when eggs were exposed to waters with a pH value of 5.0 and below. Similarly embryo hatchability was significantly less at all pH levels below 6.5, while adult brook trout were able to survive in waters with pH values of 5.0 and above (maximum pH tested 7.1). These results were similar to those reported by Mount (1973) who found that reproduction in fathead minnows (*Pimephales promelas*) was affected at substantially higher pH values than those levels that induced adult mortality.

Numerous investigators have presented evidence that immature or early instar aquatic macroinvertebrates are more sensitive to toxicants than are older classes (e.g., Maki et al. 1975; Clubb et al. 1975). Heit and Fingerman (1977) demonstrated that the ability of crayfish to cope with inorganic mercury was dependent on temperature, and the size and sex of the organism, while Lee and Buikema (1979) have demonstrated that *Daphnia pulex* undergoes cyclic changes in susceptibility to chromate which appears to correspond with the organism's molting pattern. Thus, it is essential to closely examine a criterion that is based on limited age groups of test organisms prior to setting the goal for restoration of an individual parameter.

The same can be said for those chemicals that bioaccumulate. Sodergren (1971; 1973) studied the transport, distribution, and degradation of a number of compounds including p,p'DDT, p,p'DDE, lindane, and Clophen A 50 (proprietary name for mixture of polychlorinated biphenyls) in a laboratory microcosm. Through a series of classical experimental manipulations Sodergren confirmed that the test substances were transported through the food chain. He further demonstrated differential degradation rates for the various compounds as they passed through the food chain. Moser et al. (1972) studied the effects of PCBs and DDT on the composition of mixed cultures of algae and similarly concluded that "exposure to chlorinated hydrocarbon compounds is probably greater than the concentrations in natural waters would indicate, as these substances are rapidly absorbed from the water by organisms." These authors further demonstrated that exposure to PCBs and DDT significantly diminished the competitive success of the normally more successful (i.e., competitively dominant) algae *Thalassiosira pseudonana,* and increased that of *Dunaliella tertiolecta* at all concentrations tested.

Maki et al. (1973) reported sublethal concentrations of the organo-phosphate Dibrom significantly altered oxygen consumption and reduced the tolerance of two stream insects (*Hydroperla crosbyi* and *Corydalus cornutus*) and the golden shiner (*Notemigonus crysoleucas*). These authors concluded that "although selected fish and fish food organisms may be able to survive in oxygen poor environments, when oxygen depletion is combined with a sublethal presence of pesticide concentration, these two factors may interact to cause massive fish kills."

The above studies demonstrate that establishment of water quality criteria for protection and restoration of aquatic ecosystems can be of serious consequence to future environmental quality and restoration success if based upon apparently sublethal concentrations found in the water (especially those prone to bioaccumulation), or upon studies that have employed only adult organisms. Failure to consider potential synergistic reactions with other water quality parameters can cause analysts to miss continued degradation of water quality. These studies also demonstrate the potential for structural changes within the community, even at sublethal concentrations, due to slight alterations in competitive capabilities. Similar conditions are likely to exist in various predator-prey relationships (see Hammons 1981 for review).

In general, past emphasis has been on bioassay and toxicity testing typically with a single species and using pure compounds in a controlled laboratory environment. The result has been promulgation of criteria based on extensive testing, but the controversies continue. Issues such as multiple species interactions, laboratory-to-field extrapolations, complex effluent characterization and testing, and water quality based standards now dominate criteria and standard development. If a criterion or standard is used for the assessment of restoration, in the final analysis a major judgmental or subjective component still exists. A generally decreasing level of certainty about a successful management outcome pertains to stream size increase. This is due to the expanding complexity of streams as watershed area increases. This complexity is coupled with a reduced technical understanding of ecosystem components and function in larger rivers.

ELEMENTS OF RESTORATION AND PROTECTION

The primary element of any restoration approach is elimination of the activity of a substance which degrades water quality within a stream or river. Note that the emphasis is on the activity of a substance, not its presence. From a practical viewpoint, elimination of trace amounts of a substance from dynamic systems such as streams is impossible. However, it is possible to identify mechanisms that inactivate a substance. This inactivation is defined as the reduction in concentration to a level that does not degrade the capacity of the system to support aquatic life.

Again, it is necessary to consider use designations and their attendant criteria and standards as the indicators of water quality degradation and corresponding restoration requirements. Specifications for navigation use may have broad limits for substance concentration. Little concern is directed to indirect effects in most use categories. When aquatic life maintenance and propagation is considered as a use, inactivation of a substance may be very difficult. Although ambient concentrations may be low, accumulation of the substance by acquatic organisms and eventual concentration through food chains can result in ecosystem degradation. The presence of indirect effects in ecosystems requires more stringent concentration limits. Once introduced into a system the continued presence of a substance or compound may preclude actual restoration.

Departing from concerns about the variable definition of degradation (and corresponding assessment of restoration), the identification of elements of water quality restoration and discussion of mechanisms of implementation are possible. The primary elements of restoration are isolation, removal, transfer, and dilution through space and time.

Isolation

Isolation of a substance requires a permanent or semi-permanent limitation on movement and transfer within a stream or river. For example, the addition of a heavy metal to a stream may result in precipitation of the metal or absorption to sediment particles on the stream bottom. If, for example, the introduction of the metal is episodic, it is possible that precipitated or sedimented material can be buried and isolated. Isolation has been the foundation of water quality management in lake systems. The characteristic precipitation of phosphorus with iron and manganese under aerobic conditions has been used as a mechanism to isolate excess phosphorus with the goal of controlling nuisance aquatic growth. In stream systems the isolation approach is subject to severe limitations. The dynamic nature of the stream maintains the potential for change in the conditions which provided initial isolation. Flooding, with increased bottom scouring, may serve to expose and move isolated materials. The result is often "slug" loading in downstream areas; the concentration of the isolated substance may be high and local impact may be quite severe. Isolation thus tends to be of limited value in streams.

Removal

Effecting water quality restoration by removal eliminates the potential for future degradation if the substance is only isolated. Removal can be effected either prior to introduction of a substance to a stream or directly from the watercourse. Removal of a substance at its source (prior to its discharge) is the preferred approach and is the thrust of existing water quality regulation. Before discharge, the substance is generally concentrated and removal efficiencies will be relatively high while costs will be minimized.

Typically though, stream water quality restoration requires removal of a substance following an unintentional spill or after the hazard of a discharge is recognized following some extended period of release. Unfortunately, the process involved in removal of a substance from a stream or river may exacerbate any existing degradation and create severe localized impacts. For example, dredging material from a stream bottom can be expected to affect water column suspended solids and dissolved oxygen. In addition, the disturbance of the bottom may release nontarget substances that may create both short- and long-term water quality degradation. The substance is often present in low concentrations and treatment efficiencies are low. In reality it may be impossible to remove all of the substance because of significant incorporation into food chains.

Thus in water quality restoration efforts, the removal of a substance must be considered a remedial action unless removal can take place when high efficiency and effective treatment can be employed, prior to discharge into a stream or river. The eventual disposal of concentrated material that has been removed must be considered. As in isolation difficulties, if the potential for entrance back into the stream or river exists, removal will be fruitless. The disposal process must put the substance out of reach of any watershed mechanisms that could result in reintroduction to the stream system.

Transfer

Restoration of water quality by transfer or transport mechanisms has been the time-honored method of water quality improvement. Streams and rivers flow along a gradient, carrying materials downstream. The primary transfer-based restoration technique takes advantage of the potential of stream systems to remove a substance from an area of concern. In considering substance transfer, no transformation of the substance is assumed, only movement from one place to another (this is the typical definition of a conservative pollutant). As might be expected, transfer of a substance only solves the problem in the reach that loses the substance; however, there will be a reach of the stream that will be subject to additional loading and possible accumulation to dangerous levels.

Transfer is a time-honored tradition because the problems created by a discharge are soon out of sight and mind; and, more fundamentally, transfer may result in assimilation of the substance within the stream, thus reducing or eliminating the potential for degradation. Transfer methods also take advantage of another characteristic of stream systems, dilution. The translocation of a substance or compound downstream will often result in a decreasing concentration due to the inflow of tributaries or groundwater. If that inflow is not contaminated by the substance, concentrations of the target substance may drop below dangerous levels and no degradation will be noted downstream. Unless transfer is accompanied by corresponding changes in concentration, the degradation problem will only be transferred, not solved.

Dilution through Space and Time

As just discussed, the potential benefits of transfer lie in the opportunity for dilution. Substance concentration can change along the length of the stream (through space) or may change through time in response to a number of mechanisms. Dilution along the length of the stream follows from the concept of a stream as a continuum constant, often gradual, with change in characteristics as it progresses downstream. Addition of dilution water volumes identifies only one mechanism of potential water quality restoration. Along with dilution water volumes, all other physical characteristics of the stream or river also change. Wetted perimeter increases, and sediment size decreases, producing significantly increased surface area for absorption and possibly precipitation (primary isolation mechanisms). The potential exists for significant dilution of a substance within the food chain. As the river grows in size, biomass and ecosystem complexity increase, providing the opportunity for both dilution and removal (through prey removal, adult emergence, etc.) of a substance incorporated in biological tissues.

The dilution of a substance through time is also possible if addition of the substance to the system is halted. Even conservative substances may be modified, isolating or removing them from the ecosystem. Classic examples are available for heavy metals. Sulfide precipitates are often insoluble. Through time, metals precipitate under certain environmental conditions and may actually be removed from the stream as channels change. Similarly, concentrations of substances which may bioaccumulate may be diluted through time due to depuration or modification of the substances to forms that present little potential for water quality degradation. The fact that streams and rivers are dynamic systems suggests that the ecosystem is capable of assimilating a wide variety of insults and maintaining "integrity." The difficulty arises when insufficient space or time is provided to allow these cleansing processes to occur. The obvious solution is implementation of management strategies that provide a mix of isolation, removal, transfer, and dilution to restore and maintain water quality.

Elements of Protection

As previously defined, protection can be related to the application of control technology designed to reduce or eliminate the concentration of a substance at its source. In the United States effluent treatment is technology based, depending on the best practical or best available technology for removal or control of substance concentration. Recognition that treatment technologies typically only control substance concentrations is important. The 100% removal of a substance is often an elusive goal. In a practical sense, the incremental costs for greater removal of a substance at a high rate places complete protection at a cost that is limiting to most industries and municipalities. In addition to prohibitive costs, treatment technologies often produce unwanted side effects. Disposal of substances concentrated by treatment is a significant problem. Also, treatment effectiveness is often judged solely on the capacity for removal of a substance. The typical industrial or municipal effluent is a complex mixture of substances. Treatment that reduces or removes a substance from a waste stream may actually produce a final effluent that will have greater impact on the receiving system. Thus, the design of technology-based protection measures must include provisions for substance reduction or elimination and receiving system use maintenance.

Protection from Nonpoint Sources of Pollution

When nonpoint sources of water quality degradation are considered, protection can involve two approaches. The first requires implementation of best management practices at the source to reduce or eliminate eventual contributions of a substance to a stream or river. Any disturbance on a watershed can lead to water quality degradation, but through minimizing watershed disturbance control of contaminant input to streams and rivers is possible. Maintenance of riparian vegetation or stream buffer strips and reduction of erosion lowers the potential for substance movement by surface runoff, thereby reducing the potential for water quality degradation. The protection methods for nonprint sources of degradation are typically watershed-oriented management practices. The second option for nonpoint source control is collection and treatment. In some nonpoint sources, collection of channel flow to a point where technology-based treatment can be applied is possible. The effectiveness of this approach is mixed, usually subject to treatment designs that can accommodate widely varying flow conditions. Costs are often prohibitive because substance concentrations are low and associated treatment costs high.

SUMMARY

Restoration and protection of stream quality are concepts basic to the formulation of water quality regulation in the United States. In practice, the goal of both restoration and protection is the return to, or maintenance of some preconceived notion of an undisturbed state. Since few undisturbed streams and rivers exist, arbitrary measures of restoration effectiveness are often based on readily accepted criteria and standards of water quality. The success of restoration and protection efforts and the applicability of any technique or methodology that restores water quality or protects existing uses is dependent on physical, chemical, and biological characteristics of the stream ecosystem and prevailing use and disturbance in each watershed. Assessment of restoration and protection is dependent on the scientific validity of the criterion value used as an endpoint. Protection and restoration of stream water quality requires a knowledge, appreciation, and proper juxtaposition of several fields of science.

The primary methods of restoration are isolation, removal, transfer, and dilution through space and time of substances which degrade water quality or affect ecosystem structure and function. Protection of stream water quality is often technology based. The application of treatment technologies typically meets protection requirements where point sources of effluents, containing high concentrations of substances that degrade water quality, are encountered. The potential for water quality degradation from nonpoint sources is greater (entire watersheds may be involved), but substance concentrations are generally less than point source effluents. Protection of stream water quality affected by nonpoint sources of pollution is dependent on the implementation of best management practices that control substance entry into stream systems.

REFERENCES

Borman, F.H., G.E. Likens, and J.S. Eaton. 1969. Biotic regulation of particulate and solution losses from a forest ecosystem. *BioScience* 19:600–10.

Cairns, J., Jr., and K.L. Dickson. 1977. Recovery of streams from spills. In *Recovery and Restoration of Damaged Ecosystems,* edited by J. Cairns, Jr., K.L. Dickson, and E.E. Herricks. Charlottesville: University of Virginia Press.

Clubb, R.W., A.R. Gaufin, and J. Lords. 1975. Acute cadmium toxicity studies upon nine species of aquatic insects. *Environ. Res.* 9:332–41.

Cummins, K.W. 1974. Structure and function of stream ecosystems. *BioScience* 24:631–41.

Edmondson, W.T. 1977. Recovery of Lake Washington from eutrophication. In *Recovery and Restoration of Damaged Ecosystems,* edited by J. Cairns, Jr., K.L. Dickson, and E.E. Herricks. Charlottesville: University of Virginia Press.

Gameson, A.H.L., and A. Wheeler. 1977. Restoration and recovery of the Thames Estuary. In *Recovery and Restoration of Damaged Ecosystems,* edited by J. Cairns, Jr., K.L. Dickson, and E.E. Herricks. Charlottesville: University of Virginia Press.

Hammons, A. 1981. *Methods for Ecological Toxicology* Ann Arbor, Mich: Ann Arbor, Science Publ.

Heit, M., and M. Fingerman. 1977. The influence of size, sex, and temperature on the toxicity of mercury to two species of crayfishes. *Bull. Environ. Contam. Toxicol.* 18:572–80.

Herricks, E.E. 1977. Recovery of streams from chronic pollutional stress—acid mine drainage. In *Recovery and Restoration of Damaged Ecosystems,* edited by J. Cairns, Jr., K.L. Dickson, and E.E. Herricks. Charlottesville: University of Virginia Press.

Herricks, E.E., and J. Cairns, Jr. 1982. Biological monitoring Part III—Receiving system methodology based on community structure. *Water. Res.* 16:141–53.

Hinmann, J.J. 1920. Standards of water quality. *Amer. Wat. Works Assoc.* 7:821–29 (1920).

Hynes, H.B.N. 1970. *The Ecology of Running Waters.* Liverpool, England: Liverpool Univ. Press.

______. 1975. The stream and its valley. *Verh. Internat. Verein. Limnol.* 19:1–15.

Johnson, F.H. 1961. Walleye egg survival during incubation on several types of bottoms in Lake Winnibigoshish, Minnesota, and connecting waters. *Trans. Am. Fish. Soc.* 90:312–22.

Karr, J.R., and D.R. Dudley. 1981. Ecological perspective on water quality goals. *Environ. Manag.* 5:55–68.

Lee, D.R., and A.L. Buikema. 1979. Molting-related sensitivity of *Daphnia pulex* in toxicity testing. *J. Fish. Res. Board Can.* 36:1129–32.

Leopold, L.B., M.G. Wolman, and J.P. Miller. 1964. *Fluvial Processes in Geomorphology.* San Francisco: W.H. Freeman and Co.

Likens, G.E., F.H. Bormann, R.S. Pierce, and W.A. Reiner. 1978. Recovery of a deforested ecosystem. *Science* 199:492–96

McElroy, A.D., S.Y. Chiu, J.W. Nebgen, A. Aleti, and A.E. Vandergrift. 1975. Water pollution from nonpoint sources. *Wat. Res.* 9:675–81.

McKee, J.E., and H.W. Wolf. 1963. *Water Quality Criteria.* 2d ed. Publ. No. 3A. Sacramento, CA: California State Water Quality Control Board.

Maki, A.W., K.W. Stewart, and J.K.S. Silvey. 1973. The effects of Dibrom on respiratory activity of the stonefly, *Hydroperla crosbyi,* hellgramite, *Corydalus cornutus,* and the golden shiner, *Notemigonus crysoleucas. Trans. Am. Fish. Soc.* 4:806–15.

Maki, A.W., L. Geissel, and H.E. Johnson, 1975. Comparative toxicity of larval lampricide (TFM: 3-trifluoromethyl-4 nitrophenol) to selected benthic macroinvertebrates. *J. Fish. Res. Board Can.* 32:1455–59.

Menendez, R. 1976. Chronic effects of reduced pH on brook trout (*Salvelinus fontinalis*). *J. Fish. Res. Board Can.* 33:118–23.

Meyer, F.P. 1979. Seasonal fluctuation in the incidence of disease on fish farms. In *A Symposium on Diseases of Fishes and Shellfishes,* edited by S.F. Snieszko. Am. Fish. Soc. Spec. Publ. No. 5.

Morse, B.B., and A. Wolman. 1918. The practicability of adapting standards of quality for water supplies. *J. Am. Wat. Works Assoc.* 5:198–228.

Moser, J.L., N.S. Fisher, and C.F. Wurster, 1972. Polychlorinated biphenyls and DDT alter species composition in mixed cultures of algae. *Science* 176:533–35.

Mount, D.I. 1973. Chronic effects of low pH on fathead minnow survival, growth, and reproduction. *Wat. Res.* 7:987–93.

Muncy, R.J., G.J. Atchinson, R.W. Bulkley, B.W. Menzel, L.G. Perry, R.C. Summerfelt. 1979. *Effects of Suspended Solids and Sediments on Reproduction and Early Life of Warmwater Fish: A review.* EPA–600/3–79–042. Washington D.C.: U.S. Environmental Protection Agency.

Norris, L.A. 1981. Ecotoxicology at the watershed level. In *Testing for Effects of Chemicals in Ecosystems,* edited by Nat. Res. Council, Comm. on Nat. Resources. Washington, D.C.: National Academy of Science Press.

Paine, R.T. 1980. Food webs–linkage, interaction, strength, and community structure. The Third Tansley Lecture. *J. Anim. Ecol.* 49:667–685.

Polls, I. and R. Lanyon. 1980. Pollutant concentrations from homogenous land uses. *J. Environ. Engineer. Div. EE.* 1:69–80.

Schlosser, I.J., and J.R. Karr. 1980. *Determinants of Water Quality in Agricultural Watersheds.* UILU-WRC-80-0147. Urbana, IL.: University of Illinois, Water Resources Center.

Schroepfer, G. 1942. An analysis of stream pollution and stream standards. *Sewage and Ind. Wastes* 14:1030–43.

Sodergren, A. 1971. Accumulation and distribution of chlorinated hydrocarbons in cultures of *Chlorella pyrenoidosa* (Chlorophycae). *Oikos* 22:215–20.

———. 1973. Transport, distribution, and degradation of chlorinated hydrocarbon residue in aquatic model ecosystems. *Oikos* 24:30–41.

Sprague, J.B. 1976. Current status of sublethal tests of pollutants on aquatic organisms. *J. Fish. Res. Board Can.* 33:1988–92.

Stall, J.B., and T.C. Yang. 1972. Hydraulic geometry and low stream flow regime. Res. Rept. No. 54. Urbana, IL.: University of Illinois, Water Resources Center.

U.S. Environmental Protection Agency. 1980. *Water Quality Standards Criteria Summaries: A Compilation of State and Federal Criteria.* Vol. I. Washington, D.C.: Office of Water Standards and Regulations.

Wischmeier, W.H., and D.D. Smith. 1965. *Predicting rainfall-erosion losses from cropland east of the Rocky Mountains.* Agriculture Handbook No. 282. Washington, D.C.: U.S. Dept. of Agriculture.

Yang, C.T., and J.B. Stall. 1974. *Unit Stream Power for Sediment Transport in Alluvial Rivers.* Res. Rept. 88. Urbana, IL.: University of Illinois, Water Resources Center.

CHAPTER 2

The Use of Meander Parameters in Restoring Hydrologic Balance to Reclaimed Stream Beds

Victor R. Hasfurther
Civil Engineering Department
University of Wyoming
Laramie, WY 82071

A river or stream is dynamic through time. Change is one of the most common features associated with river and stream channels. In general, this change is very slow, however, and only over long periods of time is it actually noticeable to most individuals. As a result, engineers, ecologists, and others involved with the hydrologic balance of a stream many times treat the stream system as static (i.e. unchanging in shape, slope, or pattern). One only has to observe a favorite fishing stream or pleasure area along a stream to notice that the stream channel itself probably looks slightly different now than it did two or three years previous.

In general, most streams are continually changing position and shape as a consequence of hydraulic forces acting on their beds and banks. This stress is mainly a result of climatic changes from year to year in the amount of water flow variation that occurs in the stream. Over time, the stream system adjusts to this natural variation in flow and develops a pattern which puts the system in a quasi-steady, yet dynamic, situation where only unusual climatic events or human activities may cause rapid changes in the nature of the stream channel morphology (dimensions, shape, or pattern). It should be noted, however, that streams are, in fact, the most actively changing of all geomorphic forms, especially alluvial stream systems. It is the rule rather than the exception that banks will erode, sediments will be deposited, and floodplains, islands, etc., will undergo change with time (Richardson et al. 1975). The problem in all this comes when humans induce change upon the system without taking the necessary steps to restore the quasi-steady situation and thus set in motion a response by the stream system to adjust to this change, which results in the propagated response along great distances from the human-induced action.

In planning and designing stream channel restoration and stream system balance, it is critically important to avoid the geometric stress thresholds of a stream at which dramatic and significant undesirable landscape modifications occur. It is desirable to approximate a range of appropriate stream channel features that will cause the stream system to respond to its natural inclinations (the stream pattern which would exist under normal conditions) as if no human action had occurred. This intervention will result in controlled sediment production and produce a channel similar to the existing stream channel. The end result will be a system where stream habitat should be equivalent to predisturbance and, hopefully, a slight enhancement of habitat and form.

This chapter discusses methods and techniques for restoring a stream channel to its natural inclinations after a human-induced change such as surface mining, road construction, etc. The main emphasis will be on meander parameters and their importance in stream channel stability.

STREAM SYSTEM FACTORS

In spite of the complexity of a stream system, the same basic factors govern the delicate balance of all streams. It is important that stream managers understand and work with these basic natural factors that govern the stream system. These natural factors are (1) geologic, (2) hydrologic, (3) hydraulic, and (4) geometric. Together these factors interact to develop the stream system.

Geologic Factors

Geologic factors influence the nature and amount of sediment production and the development of meanders due to topography and soil conditions. Topography determines overall slope of the area and can be a limiting factor in meander formation as a result of the location of large relief areas (hills) which will automatically change the direction of the stream channel. The abruptness and amount of change will be a characteristic of the soil material. The amount of sediment production will also depend upon the type of soil and general slope of the stream channel.

Hydrologic Factors

Hydrologic factors will influence the variations in flow and runoff and thus the type of meander system developed by the stream. Long-term climatic fluctuations can cause variations in runoff which can cause major changes in a stream's morphology. Along with the soil conditions, the amount and type of vegetation on the landscape will have a great influence on runoff and associated infiltration characteristics. The hydrologic effects of changes in land use can result in major modifications in runoff characteristics and thus the channel morphology. Land use changes could have a dramatic effect on meander characteristics and on morphology in general if not addressed in the reclamation of a disturbed area.

Hydraulic Factors

Hydraulic factors include depth, slope, and velocity of a stream. These factors are the characteristics which directly produce bank cutting, sediment transport, and the like. Hydraulic factors tend to change channel cross-sectional shape, pool and riffle formation, and meander shape.

The hydraulics of flow in streams is complex. Some of the major complications are (a) the large number of interrelated variables (depth, slope, and velocity) in describing the response of natural or imposed changes to the stream system and (b) the continual change of stream patterns and channel geometry with changes in flow and sediment discharge. By changing the slope of a stream, it is possible to change a stream from a fairly stable situation (meandering stream) that has fairly tranquil flow to an unstable situation (braided stream, very dynamic) that has high velocities and carries large quantities of sediment.

Geometric Factors

Geometric factors consist of the channel cross-sectional shape, stream pattern (straight, meandering, or braided) and the pool-riffle pattern that may exist on smaller streams. On many alluvial type streams, significantly different channel dimensions, shapes, and patterns are associated with amount of discharge and sediment load, indicating that changes in these variables can cause significant adjustments to the geometric factors. Perhaps the most exceptional example was the Cimarron River located in southwestern Kansas which was a stream approximately 15 m wide in the late 1800s and into the early 1900s. During the 1930s a series of floods widened the channel to almost 370 m occupying the greater part of the flood plain. By the 1960s, the channel had receded to a width of approximately 150 m (Schumm and Lichty 1965).

The example of the Cimarron River indicates how important it is to understand the many factors involved in the mechanics of a stream system. Artificial changes in a stream, by disturbance of the stream channel through mining or placement of flood control structures, can have far-reaching impacts on the stream system for many kilometers upstream and downstream of the disturbance. It is important to study all the factors whenever artificial disturbance of the stream is to take place and to avoid the critical stream system stress threshold factors which produce dramatic and significant channel modifications.

MEANDERS

Stream patterns can be broadly classified as straight, meandering, braided or some combination of these classifications (Leopold and Wolman 1957). Straight and meandering stream sections are considered reasonably stable. A combination of discharge, slope, and suspended sediment load generally determines the type of stream pattern (Leopold and Wolman 1957, Chitale 1970, Skinner 1971, Schumm

1977). As pointed out by Chitale (1970), straight and meandering refer to direction changes while braiding refers to multiple channels. Thus, streams are more appropriately grouped into single or multiple channels with single channels further divided into straight or meandering. Multiple channels are comprised of braided channels and alluvial fans. Braided channels are generally straight unless constrained to a winding path by valley walls. The categories are not unique, for a stream can possess characteristics of straight, meandering, and braided in different reaches of its course. The division between meandering (nonstraight) and straight is arbitrary

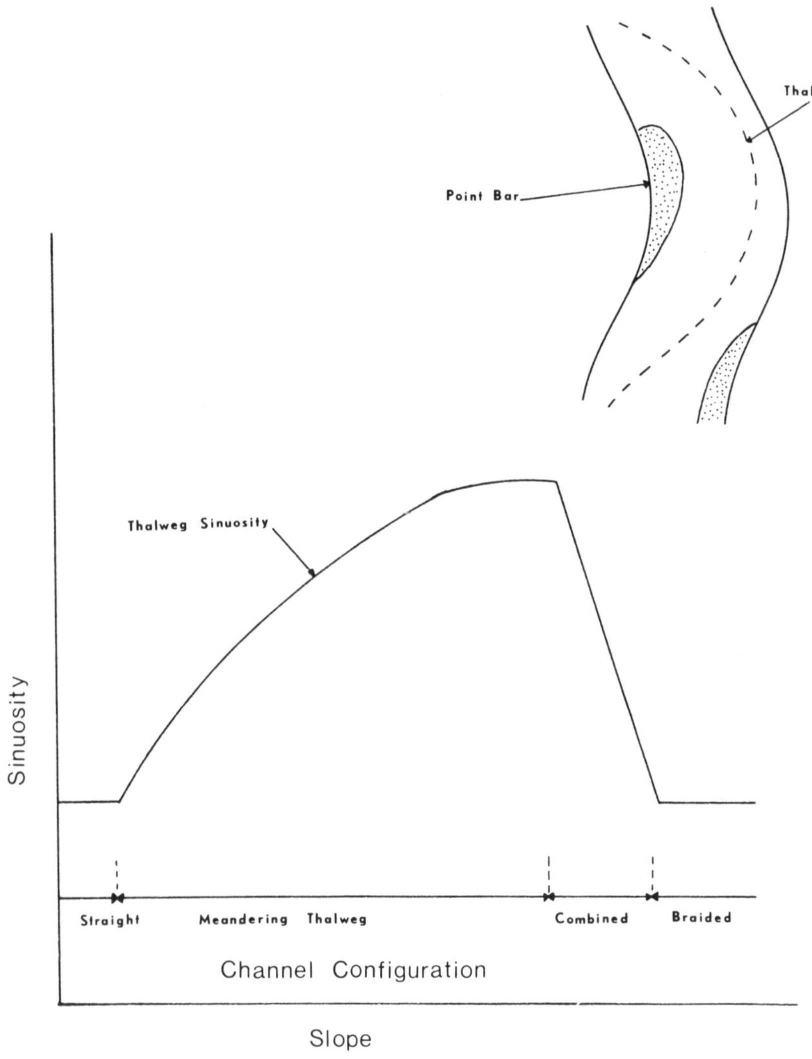

Figure 2.1 Sinuosity vs. slope with constant discharge. After Richardson et al. 1975.

but Leopold et al. (1964) designated single channels with sinuosities, a term to be defined later, greater than 1.50 as meandering where an absolutely straight channel would have a sinuosity of 1.00. Figure 2.1 illustrates differences between meandering and straight channels. Langbein and Leopold (1966) found 10 times the stream width to be the maximum length a natural stream will adopt a straight course.

The term *meander* comes from the word *miandras,* the name of a tortuous stream in Turkey which is known today as Menderes (Langbein and Leopold 1966). The term is generally used to cover all nonstraight, single channels although Matthes (1941) restricted the definition to regular S-shaped waveforms. Lane restricted the definition to include geologically uncontrolled waveforms in unconsolidated alluvium (Skinner 1971). The more general definition of meandering as nonstraight single channels, modifying the word with appropriate adjectives as necessary will be used herein.

The adjective *free* refers to meanders occurring in unconsolidated alluvium (water-deposited material) free to migrate and develop waveforms without constraints from valley walls (geologic factors), terrain, or significant distortion from heterogeneous alluvium (Carlston 1965). Ideal uniform meanders established in flume experiments approximate a natural free meander. A related term is *alluvial river* defined by Schumm (1977) as a river free to adjust its river pattern, hydraulic dimensions, and slope flowing in a channel composed of material presently carried by the river (designated as an alluvial channel). Care must be exercised in using the term *alluvial channel* because of the present turmoil associated with the similar sounding term *alluvial valley floor,* which has a statutory definition.

Other modifiers for categorizing meanders are regular or irregular, simple or compound, and meander bends (curves) that are acute (hairpin) or flat (Chitale 1970) (see Fig. 2.2). *Regular* meanders are composed of bends with uniform curvature and spectral wavelengths (defined later) and if spectrally analyzed as a time series, the meandering would possess a single frequency. Langbein and Leopold (1966) observed that the appearance of meander regularity depends upon the constancy of the ratio of wavelength to radius of curvature. *Irregular* meanders are deformed in shape and may have a varying meander belt width (defined later) and/or wavelength. Terrain, nonhomogeneous alluvium, variable discharges from tributaries, or water loss to permeable strata (stream system factors) may be responsible for stream meander irregularity.

Simple and regular meanders are similar. However, the term *simple* is a more appropriate antonym for compound. *Simple* meanders have one dominant meander belt width and wavelength. *Compound* meanders may originate on streams with more than one dominant discharge. A similar situation occurs with misfit streams. There are cases of meandering valleys containing meandering streams with shorter wavelengths than the valley. The stream is described as *underfit* by geomorphologists with the implication that: (1) there is a definite physical (not merely statistical) relationship between stream discharge and meander wavelength, (2) the valley meanders were formed by a former layer stream, and (3) something caused a significant stream size reduction. The cause for the discharge change is

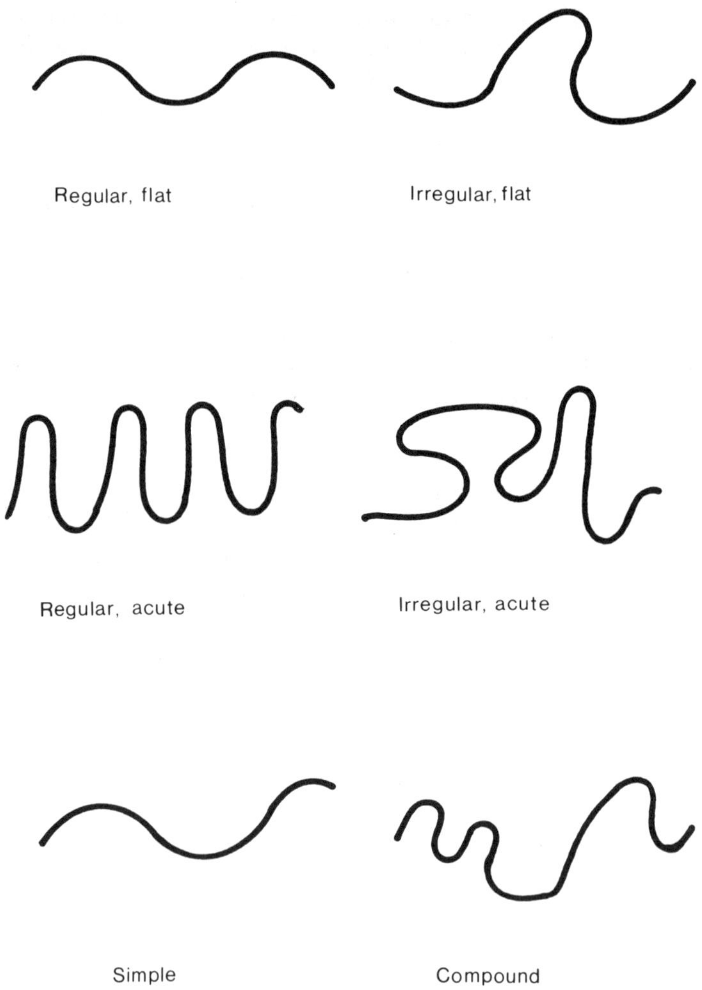

Figure 2.2 Basic meander patterns. Adapted from Chitale 1970.

most likely climatic though stream capture is possible (Dury 1965). Consequently, a compound meander would be a complex shape requiring two or more ideal constant discharges to form. By definition, the *dominant discharge* is the appropriate constant discharge that would be equivalent in developing the present channel shape which was formed from the variable flow of a natural river (Henderson 1966). Determining the most appropriate dominant discharge value for the type of stream under study is difficult. The flow may be hypothetical (such as a two-year recurrence flood) or actual (such as bankfull discharge, the discharge at which a stream first overflows onto its flood plain).

Meander Parameters

In selecting parameters to quantify the meander shapes, wavelength, sinuosity, radius of curvature, and peak to peak amplitude or meander belt width are often used. Figure 2.3 is a definition sketch of the stream parameters most commonly used to describe meanders.

Wavelength has been described as: (1) twice the linear distance between successive inflection points (Leopold and Wolman 1957; Dury 1965), (2) twice the arc distance between two successive inflection points, and (3) the reciprocal of the dominant frequency from spectral analysis (Speight 1965; Ferguson 1975). In this chapter, the first definition will be called *linear wavelength* (λ_1), definition two, the *meander length* (M_1) or band length when referring to one-half of the meander length, and the third definition, *spectral wavelength* (λ_s). When unmodified, wavelength refers to the general concept.

There are advantages and disadvantages to each definition. Comparing linear wavelength (λ_1) and meander length (M_1), the meander length may be intuitively more meaningful a parameter for relating to the actual flow hydraulics since the fluid actually traverses this distance. For linear wavelength, the orientation of the measured line between two successive inflection points may differ significantly

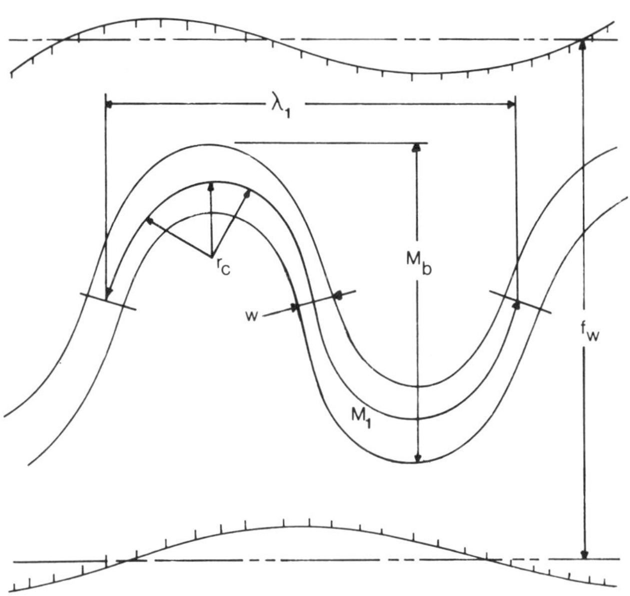

Figure 2.3 Definition sketch of meander terms. λ_1 = linear wavelength, M_1 = meander length, W = channel width measured at crossing (inflection point), M_b = meander belt width, f_w = flood plain width (belt of meandering), r_c = radius of curvature (= ρ, in text).

from the local or regional stream flow direction (Ferguson 1975; Speight 1965).

Meander pattern irregularity complicates the estimation of linear wavelength (λ_l) and meander length (M_l). The investigator must locate a representative, fairly symmetrical meander loop or average over many meander loops. The spectral wavelength (λ_s) is determined using all meander loops in an entire reach and thus eliminates some inherent disadvantages of linear wavelength and meander length. Meanders smaller or larger than a subjectively chosen representative meander are not assumed to be atypical nor does a fairly normal distribution of meander loops have to be assumed for averaging to represent the dominant wavelength.

Spectral wavelength (λ_s) also has the advantage of repeatability. While a skilled researcher could be consistent and reliable in the linear wavelength or meander length determination, two individuals may differ in technique.

Sinuosity or tortuosity (P) is a parameter that also has several definitions. Leopold and Wolman (1957) defined sinuosity as the ratio of stream length or thalweg (lowest thread or deepest part along the flow channel) length to valley length (equivalent to the ratio of valley slope to stream slope). Friedken (1945) defined sinuosity as the ratio of thalweg distance to arc distance. Leopold and Wolman (1966) also used the ratio of arc distance (meander length, M_l) to linear wavelength. For a uniform meander, the definitions are equivalent but irregularity introduces complications. In this chapter, the sinuosity ratio (P) will be as defined by Friedkin because it eliminates the subjectiveness in obtaining valley length or visual wavelength measurements and is easier to program for the computer.

Goodman (1974) defined the *radius of curvature* (ρ) as:

$$\rho = \frac{\left(1 + \left(\frac{dy}{dx}\right)^2\right)^{3/2}}{\frac{d^2y}{dx^2}} \tag{2.1}$$

Equation 2.1 cannot be used for several numerical reasons. First, a derivative requires a single valued function which is not always obtainable for a naturally meandering stream. However, a complex computer program could possibly break the stream into single valued segments. Next, the derivatives must be calculated by numerical differentiation, an inaccurate process. These initial errors are compounded after squaring, cubing, and taking the square root of the first derivative and then dividing by the second derivative. Consequently, the radius of curvature at each point along the stream can be approximated by the radius of a circle which passes through one point and two nearby points. This approximation is similar to that employed by Leighly (1936) and Brice (1973) when circles of various radii were visually fit to mapped meander loops.

The final parameter, *meander belt width* (M_b), is defined as the normal distance between tangents drawn on the convex sides of successive bends (Fig. 1.3). The definition of a free meander stipulates that the width of the confining terrain or consolidated strata be greater than the meander belt width. Meander belt width differs from the peak to peak amplitude of a waveform by the channel width. The technical difference is insignificant and the terms are often used interchangeably. While *meander belt* connotes a region, the term is often used as a synonym of *meander belt width* by geomorphologists. A related term is the *flood plain width* (f_w) or *belt of meandering* which is the approximate width of the stream's valley.

Other Parameters

A few hydrologic and hydraulic parameters are sometimes used to correlate with meander parameters. The main parameters to be used are drainage area (A_{drain}), stream bankfull width (W), discharge (Q), sediment load index (M), stream gradient (S_{ch}), and depth of flow (d). Drainage area and discharge are values which are easily understood and defined.

Stream bankfull width (W) (also defined as indicated on Figure 2.3) is the width where the maximum change in slope of the channel cross section occurs or where the first significant break in slope occurs.

Sediment load index (M) is determined by bed and bank soil samples. The sampling depth should be less than 10 inches and a sample obtained from the bed, bank, and flood plain or first terrace at each cross section considered in a given stream reach. For the minus 200 fraction of soil in the sample, a sediment load index value is determined using Schumm's (1960) equation:

$$M = \frac{C_{bed}\,(W) + C_{bank}(2d)}{W + 2d} \tag{2.2}$$

where

C_{bank} = silt and clay percentages in channel banks
C_{bed} = bed silt and clay percentages

For M to be a valid index, the stream must be stable and alluvial such that the exposed channel consists of material currently being transported.

Stream gradient (S_{ch}) is the change in elevation divided by the channel length between two particular points making up the stream reach. Stable drainages generally exhibit a decreasing slope in the downstream direction. The rate of decrease is most profound near the headwaters of the stream.

The *channel depth* (d) is defined as the maximum depth occurring in the cross-section of the channel. In most streams, this value is slightly greater than the hydraulic depth which is the cross-sectional area divided by the surface width.

Meander Parameter Relationships

Lane (1955), Leopold and Maddock (1953), Santos-Cayudo and Simons (1973), Schumm (1971), and Rechard and Hasfurther (1980) found a number of general relationships between meander parameters, hydrologic parameters, and hydraulic parameters in streams. Some of these relationships are:

1. Depth is directly proportional to discharge and inversely proportional to the bed material discharge.
2. Channel width is directly proportional to discharge and to sediment load.
3. Channel shape (width depth ratio) is directly related to sediment load. This is not true of ephemeral streams, however (Rechard and Hasfurther 1980).
4. Meander wavelength is directly proportional to discharge and to sediment load.
5. Gradient is inversely proportional to discharge and directly proportional to sediment load and grain size.
6. Sinuosity is proportional to valley slope and inversely proportional to sediment load.

These qualitative relationships should give some idea of the response a stream would have to changes imposed upon it. Richardson et al. (1975) have developed a table (Table 2.1) which indicates the response of alluvial channels to change in magnitude of different hydrologic and hydraulic factors.

Lane (1955) suggested that for a stream channel to be stable, water discharge (Q) and slope (S_{ch}) must be proportional to sediment load (Q_s) and bed material size (d_s):

$$QS_{ch} \ \alpha \ Q_s d_s \tag{2.3}$$

Assume, now, that mining occurred through a stream channel and that the reconstructed stream had an increased slope due to removal of a number of meanders. If water discharge and bed material size are more or less the same after reconstruction, then equilibrium status will have to be achieved by increased sediment load. Once this sediment load is delivered downstream of the mine area, a similar adjustment would have to result in the receiving area in terms of deposition. Equation 2.3 is a true relationship and should be considered in all cases of reclamation design. Meanders must be designed into the reconstructed stream channel so that slope along with discharge and bed material size are not altered significantly from preimpact conditions.

A number of studies have been conducted in order to develop general relationships between measurable hydraulic and hydrologic conditions and meander parameters. Leopold and Wolman (1957) developed relationships for meander length (M_1) and radius of curvature (ρ) in terms of channel width:

$$\rho = 2.42\ W \tag{2.4}$$

and

$$M_1 = 10.9\ W^{1.01} \tag{2.5}$$

Table 2.1 Qualitative response of alluvial channels.

Variable		*Change in Magnitude of Variable*	*Effect on*						
			Regime of Flow	*River Form*	*Resistance to Flow*	*Energy Slope*	*Stability of Channel*	*Area*	*Stage*
Discharge	(a)	+	+	M → B	±	−	−	+	+
	(b)	−	−	B → M	∓	+	+	−	−
Bed-Material Size	(a)	+	−	M → B	+	+	±	+	+
	(b)	−	+	B → M	−	−	±	−	−
Bed-Material Load	(a)	+	+	B → M	−	−	+	−	−
	(b)	−	−	M → B	+	+	−	+	+
Wash Load	(a)	+	+		−	−	±	−	−
	(b)	−	−		+	+	±	+	+
Viscosity	(a)	+	+		−	−	±	−	−
	(b)	−	−		+	+	±	+	+
Seepage force	(a)	Outflow	−	B → M	+	−	+	+	+
	(b)	Inflow	+	M → B	−	+	−	−	−
Vegetation	(a)	+	−	B → M	+	−	+	+	+
	(b)	−	+	M → B	−	+	−	−	−
Wind	(a)	Downstream	+	M → B	−	+	−	−	−
	(b)	Upstream	−	B → M	+	−	−	+	+

Source: Adapted from Richardson et al. 1975

where W is measured in feet. Leopold and Miller (1956) found that meander length could also be correlated to drainage area. Based on studies in New Mexico, the relationship is:

$$M_1 \; \alpha \; A_{drain}^{0.395} \tag{2.5}$$

where A_{drain} is measured in square miles. Dury (1964) found that meander length could also be expressed as:

$$M_1 = \frac{Q_m^{0.48}}{M^{0.74}} \tag{2.7}$$

where Q_m is the mean annual flood (in cubic feet per second) and M is given by Equation 1.2. Divis (1982) developed another equation for meander length for ephemeral streams, in the Powder River Basin of Wyoming, which included factors of drainage area, channel slope, and elevation difference.

Schumm (1977) related sinuosity (P) to sediment index for alluvial rivers of the Great Plains as:

$$P = 1.38 \; M^{0.17} \tag{2.8}$$

Rechard and Hasfurther (1980) found that, for ephemeral streams in Wyoming:

$$P = 0.94 \; M^{0.25} \tag{2.9}$$

Rechard and Hasfurther also found a relationship for radius of curvature for ephemeral streams. Bhowmik and Stall (1979) developed a relationship between sinuosity and drainage area for the Kaskaskia River in Illinois.

The above studies indicate that relationships do exist between meander parameters and other hydrologic and hydraulic parameters. However, a consistent set of relationships which could be used in the design of reclaimed streams does not exist for large numbers of streams and widely varying conditions. The studies also indicate that meander patterns are important to stream stability and, as a result, should be included in design of reclaimed or disturbed channels.

Studies by Leopold and Wolman (1966), Rechard and Hasfurther (1980), and Divis (1982) suggest that meanders have a fundamental wavelength masked by secondary wavelengths resulting from geologic and hydrologic influences. Divis (1982) proposed that these fundamental wavelengths tend to be related by integers of 2 progressively replaced by a higher multiple downstream. A determination of the fundamental wavelength, then, will be important in design of reclaimed channels.

MEANDER DESIGN

Fluvial morphologists have identified meandering as a primary means of dissipating excess stream energy (Leopold and Wolman 1957; Schumm 1977). Therefore, meandering is a potential design technique to stabilize channels. One proposed method of recreating an appropriate meander channel pattern is to replace the meander exactly as found before disturbance, *the carbon copy technique.* While the technique does have its merits, it is based upon several assumptions which are not always applicable. First, the stream pattern before disturbances is assumed stable and appropriate. Second, the factors affecting stream patterns are assumed to have identical values after stream restoration. However, not all potentially disturbed stream reaches are currently stable nor do all influential factors such as bed and bank material remain constant.

A second method involves the use of *empirical relationships* which are generally not exacting and apply to small geographic regions. The extrapolation of these equations is questionable and potentially misleading. For example, when examining the hydraulic geometry of ephemeral channels in the eastern Powder River Basin, Apley (1976) found very little correlation of bed and bank material to the width/depth ratio in contrast to Schumm's (1977) findings in the Great Plains. While Apley's study did not examine meanders, it did suggest different factors were important to Powder River Basin streams than were found by Schumm for the Great Plains streams. The "regime" equations of Lacey (1930), Blench (1957), and Simons and Albertson (1960) also fall into the empirical category and are strictly engineering-oriented for artificial channels, more than for natural channels. Bhowmik (1981) gave a variation of the regime theory in trying to consider geomorphic principles.

A third method could be classified as a *natural* approach. A valley is created with the reclaimed material and the intent is to allow natural processes to take over and form their own channel and drainage basin morphology. The disequilibrium associated with this approach would cause many more problems than it could possibly solve.

A fourth approach (and the one suggested for use) is a *systems* approach which includes meander analysis and an evaluation of the geomorphology of the disturbed area and its effect on the surrounding undisturbed areas. Lidstone (1982) suggested a similar approach in designing stream channels.

MEANDER ANALYSIS

A natural stream channel is generally constrained to some extent by the geology of the area in terms of relief, slope, and soil material. When the valley area is disturbed and the soil material replaced, a somewhat different relief will exist and the disturbed material will become relatively homogeneous in character. Many of the minor geologic controls, which cause small irregularities in the meander

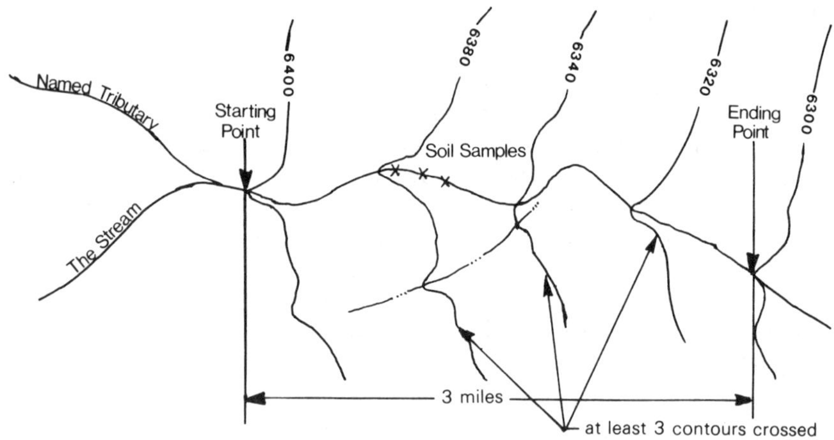

Figure 2.4 A typical stream meander reach and measurements required for Fourier analysis.

pattern, will be removed. The stream channel in the reclaimed area will tend more towards a fairly regular meander pattern which should be characteristic of the fundamental meander wavelength existing before disturbance. Langbein and Leopold (1966) and Rechard and Hasfurther (1980) have demonstrated the existence of these fundamental meander wavelengths. To reclaim the "carbon copy" channel after disturbance would have little purpose since the minor controls have been removed and some new controls created.

The methodology proposed here is to perform a basinwide analysis of the stream channel to determine the fundamental wavelength, mean radius of curvature, and meander belt width in areas determined to be reasonably free of geologic control. These same areas should have measured sinuosity, channel width and depth, and soil bed and bank analysis. Leopold and Wolman (1957), Divis (1982), and Rechard and Hasfurther (1980) have suggested a method of analysis using Fourier analysis of angular departures of a direction of angle transform. This transform involves determining the angular displacement of a given channel segment from the mean valley direction of the channel or another appropriate direction. The angular displacement is plotted as a function of distance along the channel and results in a plot which is appropriate for Fourier transform analysis. Rechard and Hasfurther (1980) and Divis (1982) have developed digitization techniques which allow for quick computation and plotting of the appropriate meander parameters. U.S. Geological Survey 7.5 minute quadrangle maps are appropriate for use with the digitization procedure. At least three stream reaches should be analyzed.

Figure 2.4 displays a typical stream reach while Figure 2.5 shows the analysis of a given reach with the results of the Fourier analysis. The radius of curvature used should be the mean value for the entire plot. Divis (1982) has interpreted the analysis in the following manner:

> The dominant peak of the amplitude plot does not necessarily coincide with the dominant visual wavelength as determined in map view. This effect is readily explained

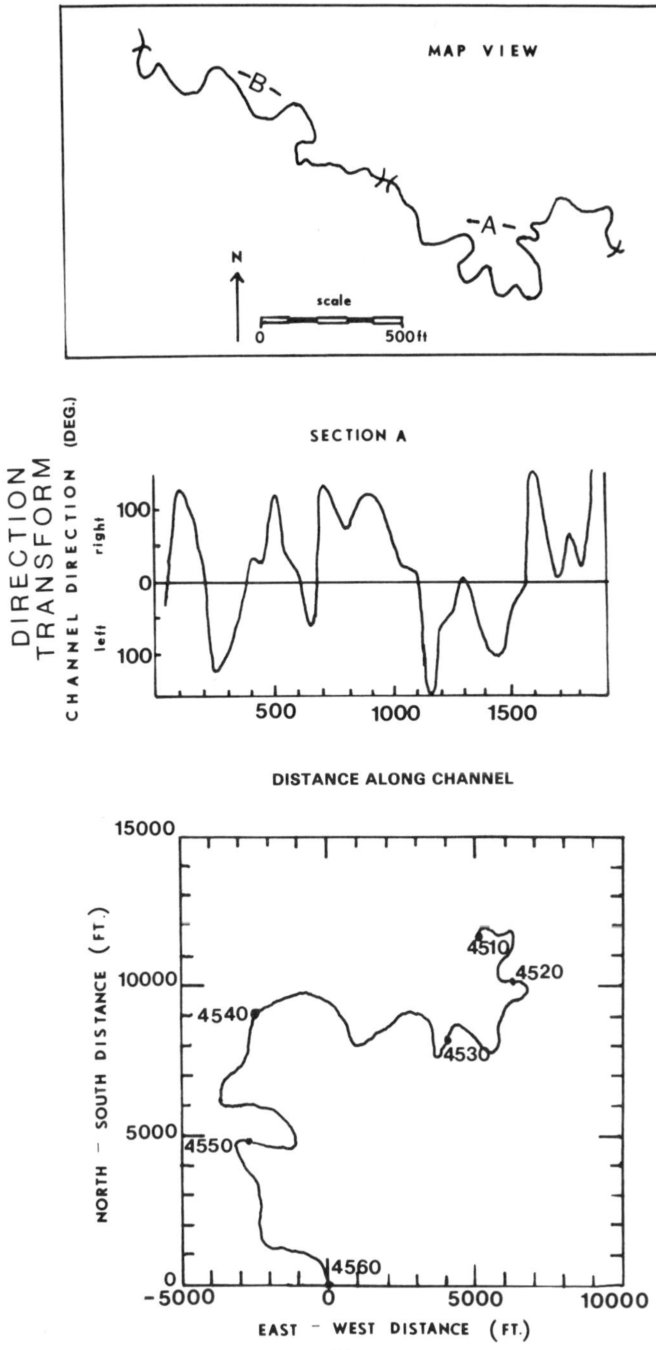

(a)

Figure 2.5 Typical Fourier analysis. From Divis 1982.

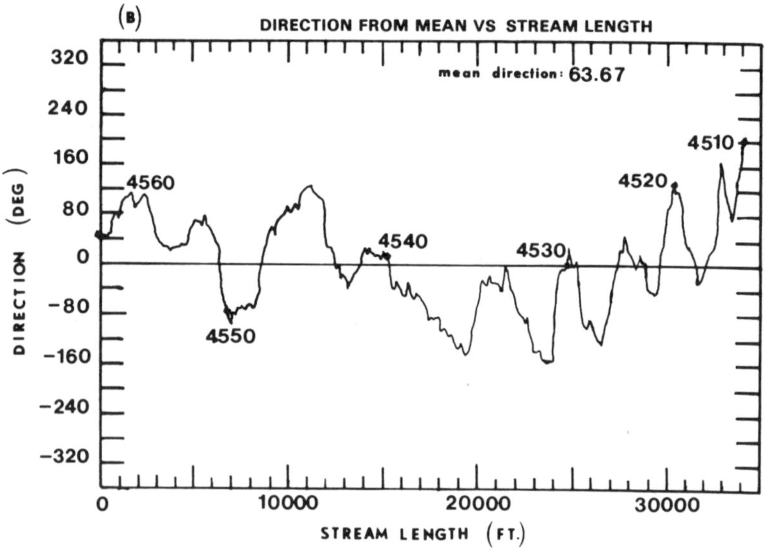

(b)

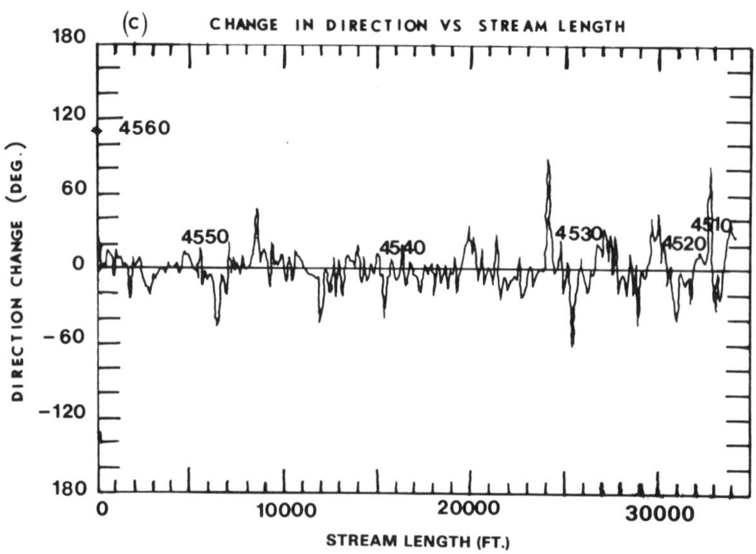

(c)

Figure 2.5 (*Continued*)

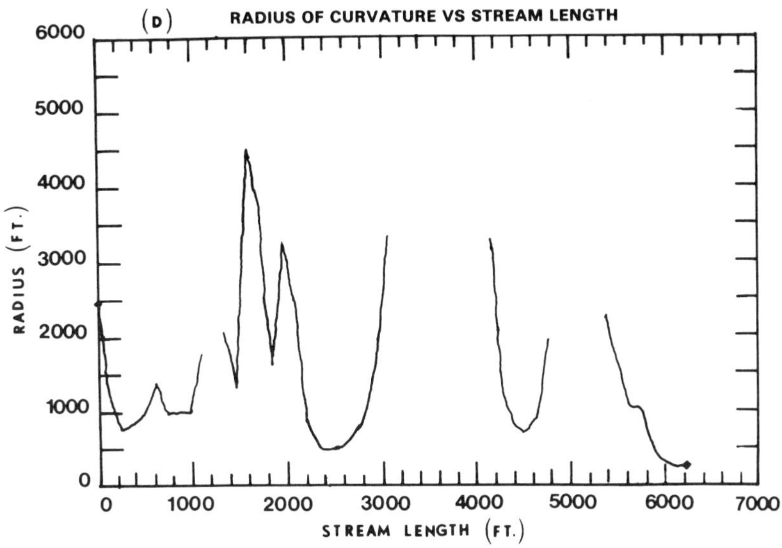

(d)

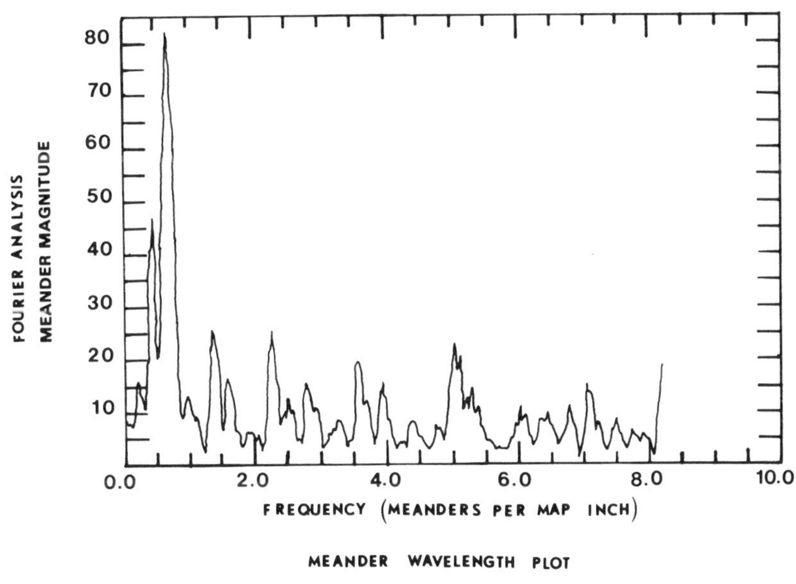

Figure 2.5 (*Concluded*)

> by an examination of the operation of the Fourier transform function. In a combination of several superimposed wavelengths the lower wavelength, or lower frequency, tends to dominate the spectral plot because a greater portion of the data set contains a portion of the long wavelength signal thus, it would not be unusual to expect a fundamental wavelength which is also the dominant visual wavelength to be masked in part by lower and higher order harmonics, particularly lower order harmonics. This effect may be resolved by comparison of several different analyzed sections of a channel and intercomparison of transform angle data and wavelength and amplitude frequency plots. Although some care is necessary interpreting the products of the Fourier analysis it is generally possible to locate the dominant meander wavelength.

If a small regional analysis by the Fourier analysis has been performed, the regional relationships should give values close to the individual basin analysis. This should help in those instances where the design methodology and equipment are not available to the individual designer.

Design Considerations

The use of the fundamental wavelength, mean radius of curvature, gradient restoration (to the same slope as preceded by the disturbance), width, sinuosity, and meander belt width should be used together to develop a stable meander pattern for the entire reach if soil conditions are considered to be approximately the same as predisturbance. If this is not the case, then some adjustment to the pattern should be made with Equation 2.3 as a guide to the action to be taken. Short-term characteristics such as vegetation establishment and slightly higher runoff directly after disturbance should not be considered in the main design. However, short-term solutions (sediment traps or small ponds on tributaries) to handle increased sediment load should be considered until vegetation can be reestablished on the disturbed watershed. The result should be a stable reclaimed section after a short period of time with little effect on the undisturbed portion of the stream and watershed.

Divis (1982) recommended that an inner or pilot channel be constructed with characteristics to hold the mean annual flood. Pool and riffle patterns with an optimum spacing of six times the channel width are suggested for this pilot channel (Leopold et al. 1964). A flood plain should be provided upstream of the pilot channel.

It cannot be overemphasized that the reclamation of a stream is a very delicate process which is complicated by a large number of variables. All stream system factors need to be examined before and after reclamation. If the reclaimed stream channel design has zones of instability, these should be analyzed and corrective measures taken as soon as possible. The newly designed and constructed stream channel will have a period of self-adjustment early which should display only local effects. Gore and Johnson (1980) observed such local effects on reclaimed coal-surface-mined rivers in Wyoming. The ultimate effect, however, will be a hydrologically stabilized channel with controlled sediment deposition and transport, and flow characteristics adequate for the establishment of habitat enhancement structures for aquatic biota.

REFERENCES

Apley, T.E. 1976. "The hydraulic geometry of the ephemeral channel of the eastern Powder River basin." Master's thesis, University of Wyoming, Laramie.

Bhowmik, N.G. 1981. Hydraulic considerations in the alteration and design of diversion channels in and around surface mined areas. In *Nat. Symp. on Surface Mining, Hydrology, Sedimentology, and Reclamation,* edited by D.H. Graves, 97–104. Lexington, KY: University of Kentucky.

Bhowmik, N.G., and J.B. Stall. 1979. *Hydraulic Geometry and Carrying Capacity of Floodplains.* Urbana, IL: University of Illinois, Water Res. Ctr. Res. Rpt. No. 145.

Blench, T. 1957. *Regime Behavior of Canals and Rivers.* London: Butterworth Sci. Publ.

Brice, J. 1973. Meandering pattern of the White River in Indiana—an analysis. In *Fluvial Geomorphology.* Binghamton: State University of New York.

Carlston, C.W. 1965. Flow and channel characteristics of free meander geometry to stream discharge and its geomorphic implications. *Amer. J. Sci.* 263:864–85.

Chitale, S.V. 1970. River channel patterns. *J. Hydraul. Div. Proc. Am. Soc. Civil Eng.,* Vol. 7038, No. HY1:201–21.

Divis, A.F. 1982. Numerical analysis—applications to surface mine reclamation. In *Hydrology Symp. on Surface Coal Mines in Powder River Basin,* edited by R.R. Stowe, 191–217. Gillette, WY: Gillette Area Groundwater Monitoring Organization.

Dury, G.H. 1964. *Principles of Underfit Streams.* U.S. Geol. Surv. Prof. Paper 452-A, pp. 1-A67

______. 1965. Theoretical Implications of Underfit Streams. U.S. Geol. Surv. Prof. Paper 452-C.

Ferguson, R.I. 1975. Meander irregularity and wavelength estimation. *J. Hydrology* 26:315–33.

Friedken, J.F. 1945. A laboratory study of the meandering of alluvial rivers. In *Fluvial Geomorphology,* edited by S.A. Schumm, 237–82. Stroudsburg, Pa: Dowden, Hutchinson, and Ross, Inc.

Goodman, A.W. 1974. *Analytic Geometry and the Calculus.* New York: MacMillan Publ. Co.

Gore, J.A., and L.S. Johnson. 1980. Establishment of biotic and hydrologic stability in a reclaimed coal strip-mined river channel. Laramie, WY: Inst. Energy and Environ., Univ. Wyoming.

Henderson, F.M. 1966. *Open Channel Flow.* New York: Macmillan Publ. Co.

Lacey, G. 1930. Stable Channels in alluvium. *Proc. Inst. of Civil Engineers* 229:259–384

Lane, E.W. 1955. The importance of fluvial morphology in hydraulic engineering. *Am. Soc. Civil Engin. Proc.* 81(745):1–17.

Langbein, W.B., and L.B. Leopold. 1966. *River Meanders—Theory of Minimum Variance.* U.S. Geol. Surv. Prof. Paper 422-H, pp. H1–H45.

Leighly, J. 1936. Meandering arroyos of the dry southwest. *Geograph. Rev.* 26:270–82.

Leopold, L.B., and T. Maddock, Jr. 1953. *The Hydraulic Geometry of Stream Channels and Some Physiographic Implications.* U.S. Geol. Surv. Prof. Paper 252, pp.1–57.

Leopold, L.B., and J.P. Miller. 1956. Ephemeral Streams–Hydraulic Factors and Their Relation to the Drainage Net. Prof. Paper 282-A. U.S. Geological Survey, Reston, VA.

Leopold, L.B., and M.G. Wolman. 1957. *River Channel Patterns: Braided, Meandering, and Straight.* USGS, Reston, VA: U.S. Geol. Surv. Prof. Paper 282-B, pp. 39–45.

Leopold, L.B., and M.G. Wolman. 1966. River meanders. *Bull. Geol. Soc. Am.* 71:769–94.

Leopold, L.B., M.G. Wolman, and J.P. Miller. 1964. *Fluvial Processes in Geomorphology.* San Francisco: W.H. Freeman and Co.

Lidstone, C.D. 1982. Stream channel reconstruction and drainage basin stability. In *Hydrology Symp. on Surface Coal Mines in Powder River Basin,* edited by R.R. Stowe. Gillete, WY: Gillette Area Groundwater Monitoriting Organization. p. 43–57.

Matthes, G.H. 1941. Basic aspects of stream meanders. *Trans. Amer. Geophys. Union,* pp. 632–38.

Rechard, R.P., and V.R. Hasfurther. 1980. The use of meander parameters in the restoration of mined stream beds in the eastern Powder River basin. Laramie, WY: Rocky Mtn. Inst. Energy and Environ., University of Wyoming.

Richardson, E.V., D.B. Simons, S. Karaki, K. Mahmood, and M.A. Stevens. 1975. *Hydraulic and Environmental Design Considerations.* Fort Collins, CO: Colorado State Univ.

Santos-Cayudo, J., and D.B. Simons. 1973. River response. In *Environmental Impact of Rivers,* edited by H.W. Shen. Fort Collins, CO: Water Resources Publ.

Schumm, S.A. 1960. *The Shape of Alluvial Channels in Relation to Sediment Type.* U.S. Geol. Surv. Prof. Paper 352-B, pp. 17–30.

———. 1971. Fluvial geomorphology—the historical perspective. In *River Mechanics,* edited by H.W. Shen. Fort Collins, CO: Water Res. Publ.

———. 1977. *The Fluvial System.* New York: John Wiley & Sons.

Schumm, S.A., and R.W. Lichty. 1965. Time, space and causality in geomorphology. *Am. J. Sci.* 263:110–19.

Simons, D.B., and M.L. Albertson. 1960. Uniform water conveyance channels in alluvial material. *Proc. Am. Soc. of Civil Engrs.* 86(H75):33.

Skinner, M.M. 1971. "Free meander pattern in intermontane rivers." Master's thesis, Colorado State University, Fort Collins.

Speight, J.G. 1965. Meander spectra of the Angabunga River, Papua. *J. Hydrol.* 3:1–15.

CHAPTER 3

Riparian Revegetation as a Mitigating Process in Stream and River Restoration

Bertin W. Anderson and Robert D. Ohmart

Center for Environmental Studies
Arizona State University
Tempe, Arizona 85287

Native riparian habitats are rapidly disappearing in the arid Southwest (Lowe 1964; Phillips et al. 1964; Carothers et al. 1974; Ohmart et al. 1977). Wise management of the remaining riparian habitats or replacement of these communities is extremely important because they support the highest species richness and densities of wildlife of any other desert habitat (Johnson and Simpson 1971; Carothers et al. 1974; Brown et al. 1977; Hubbard 1977; Stevens et al. 1977; Wauer 1977).

As precipitation collects and drains from the desert floor, its action cuts drainageways. Consequently, the desert regions are topographically characterized by water courses, termed washes or arroyos, that drain the adjacent desert uplands and eventually converge to form larger transport systems, which empty into primary or permanently flowing rivers. These rivers have their headwaters located in high mountain areas where they also drain high elevational watersheds.

Desert riparian ecosystems are composed of these drainages, their attendant vegetation, and the fauna supported by these riparian assemblages. The drainage system itself may have permanently flowing water, an intermittent flow, or water that seldom (if ever) flows. Nevertheless, the available soil moisture is higher in these alluvial flood plains than that in the adjacent desert uplands and supports a flora distinctly different than in the adjacent desert. A working definition is: "A riparian association of any kind is one which occurs in or adjacent to drainageways and/or their floodplains and which is further characterized by species and/or life-forms different from that of the immediately surrounding non-riparian climax" (Lowe 1964, 62).

This chapter discusses only the streamside vegetation and excludes discussion of submerged and emergent aquatic vegetation. Riparian vegetation is frequently termed *phreatophytic*, denoting a collective group of plant species that have their roots located in perennial groundwater or in the capillary fringe above the water table. The term has a negative connotation among water managers and refers to those plant species which transpire large quantities of water from the water table. Consequently, phreatophytes are frequently perceived as undesirable and their removal has been viewed as positive because it constitutes water salvage or a reduction in water loss from underground aquifers.

In general, the amount and type of vegetational ground cover, the areal extent of the watershed, and the slope of the terrain are directly related to the percentage of water that will enter the drainage system as surface flow, or as percolated water. Good watersheds have a high roughness coefficient which implies a good covering of perennial grasses. The force of falling raindrops is reduced before hitting the soil, and the vegetation retards the flow of the surface water, allowing more time for the water to penetrate the soil. This slower, decreased surface water flow reduces the erosion of topsoil and mitigates the severity of flooding.

This chapter summarizes the results of our field studies of riparian habitats on the lower Colorado River and our efforts to develop from field-collected data plant community designs that would house as many vertebrate species as possible and support high densities of wildlife. The ultimate challenge was to implement the design (plant and grow the vegetation). We were also to monitor the growth of the community and quantify the wildlife it supported so that the empirical wildlife data could be compared to the predicted values derived prior to revegetation. If our efforts were successful and costs were reasonable, our designs and methodologies could be used in habitat improvement, mitigation, and operational enhancement in totally managed river systems where native revegetation has been curtailed or stopped.

BACKGROUND

In 1973 we began studying riparian vegetation under a contract with the regional office of the U.S. Bureau of Reclamation. The area we studied was along the lower Colorado River from Davis Dam, Nevada-Arizona border, to the Mexican-American international boundary south of Yuma, Arizona (Fig. 3.1). Initially the Bureau wanted answers to two questions: Can natural habitats be improved for wildlife without increasing evaporative water loss from the vegetation? and Can new habitats be designed and built that have higher intrinsic wildlife values than existing natural communities but with lower evaporative water losses than existing natural habitats? Subsequently, a third question was asked: Can high-value wildlife communities be designed and built that have less restriction to flood flows than current riparian communities? Intuitively, it seemed that reintroducing vegetation and wildlife enhancement were mutually exclusive of efforts to reduce evapotranspiration and clearing to permit flow of floodwaters.

Prior to any attempt at plant community manipulations we had to determine baseline wildlife values for each of the major plant community types in the 450-

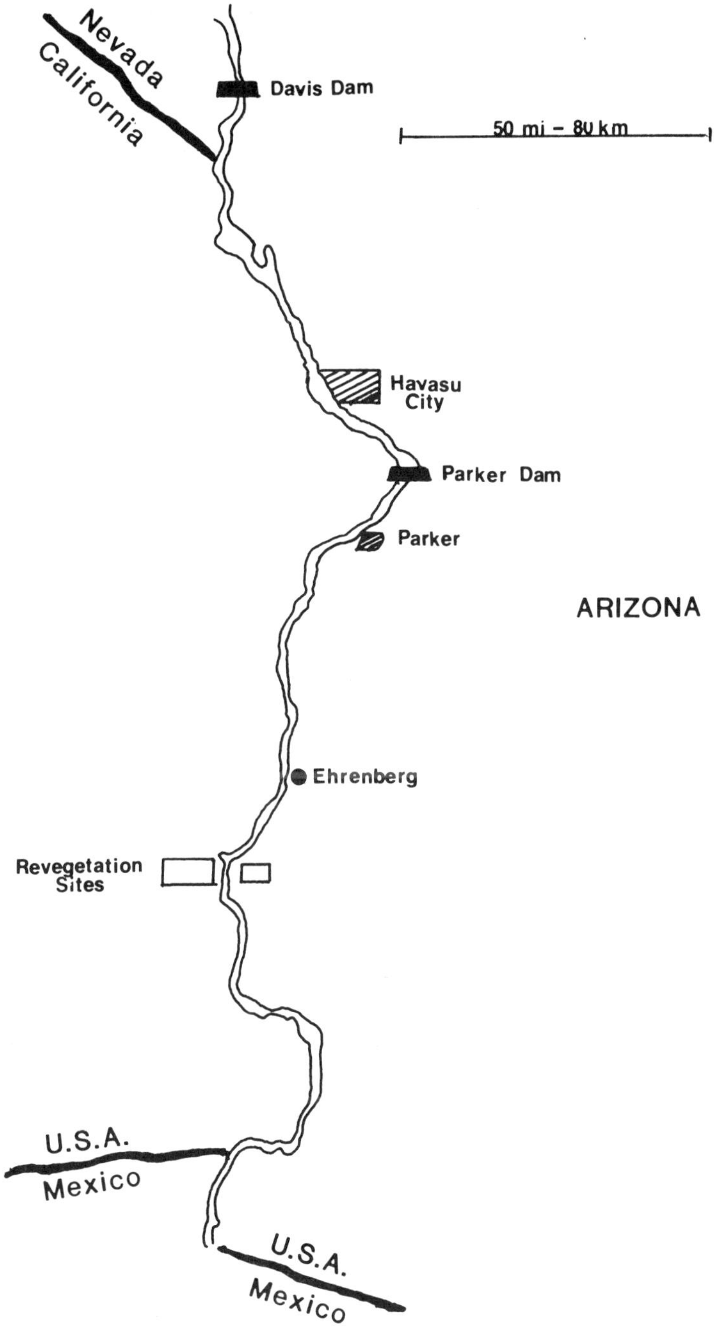

Figure 3.1 Map of the lower Colorado River showing location of revegetation plots. The initial investigations were conducted from Davis Dam to the boundary between Mexico and the United States.

km-long study area encompassing 120,000 ha. Over 100 sampling sites, varying from 20 to 40 ha, were established in the major riparian communities. Each site was sampled three times a month for birds and every six to ten weeks for small mammals. Numerous vegetation parameters were quantified so that ultimately we could examine plant-animal relationships in order to develop a plant community design that would contain all the vegetational features that were important in one or more seasons to various species of wildlife throughout their annual cycle. Plant-animal relationship data were collected monthly at all sample sites in all major community types for at least five years. This time span ensured collection of annual as well as seasonal variation in animal numbers using each community. Annual climatic changes allowed us to gain an understanding of how this important variable affects and shapes animal communities. Our study period contained one of the coldest winters recorded in the weather records of the lower Colorado River valley as well as a mild wet winter and a mild dry winter. The summers were virtually the same, with temperature maxima always exceeding 40°C and frequently approaching 48°C.

The exotic salt cedar (*Tamarix chinensis*) is the dominant plant species along the lower Colorado River. It has dense foliage and transpires large amounts of water. The density of salt cedar, combined with its high foliage volume, renders it a significant impediment to floodwaters. Native plant species, such as quail bush (*Atriplex lentiformis*), honey mesquite (*Prosopis glandulosa*), and blue palo verde (*Cercidium floridum*), all use and transpire less water per unit of occupied space than does salt cedar.

Intensive investigation was necessary to determine how evapotranspiration might be reduced and how flood flow capacity might be increased. First, replacing salt cedar with an equivalent number of native plants could lead to water salvage because of reduced evapotranspiration by native vegetation per unit of occupied space. Second, if wildlife was found to be better adapted to native vegetation, populations of birds and animals might be enhanced by removing the exotic salt cedar and replacing it with a smaller total foliage volume of native plants. Native trees such as cottonwood (*Populus fremontii*) and willow (*Salix gooddingii*), formerly numerically dominant species in the Colorado River valley (Ohmart et al. 1977) might be so attractive to wildlife that salt cedar could be removed and replaced with significantly reduced densities of these native plant species.

Preliminary Work

We first classified the vegetation in the valley and quantified the areal extent of each vegetation type (Anderson and Ohmart 1976) and the wildlife associated with it. With this information we hoped to predict what effect vegetational changes would have on wildlife in the different communities and thus we could prescribe enhancement measures that would help to increase densities of various wildlife species.

Since virtually nothing was known about growing native vegetation, we had to determine the autecological requirements for maximal growth and survival of each native plant species under natural conditions. Of even more fundamental importance, however, was the need to determine an effective way to eliminate salt cedar. Burning was ineffective and blade-clearing by bulldozer did not eliminate this tenacious species. Application of herbicides seemed to be effective, but within a few years much of the treated vegetation had recovered. In solving these problems we also had to collect cost data. Finally, we had to determine if revegetated areas would actually attract a large and diverse fauna.

Because there is significant seasonal and annual variation in wildlife populations associated with vegetation types (Anderson et al. 1981), management recommendations needed to be supported by a substantial data base and a thorough analysis of the data. Data collection was followed by three years of experimentation (1979–1981) involving manipulations of vegetation as a means of lending credence and adding clarity to conclusions drawn from the analysis of the riparian data base. The study described here involved 84 months of continuous data collection.

Revegetation Projects

Our revegetation work was initiated in 1977 on a 30-ha dredge-spoil area which had been nearly devoid of vegetation for over 15 years. It was desirable to establish vegetation on such an area, if possible, where lack of success would result in no damage to wildlife. A second site, including 20 ha, was located on the Cibola National Wildlife Refuge, about 40 km south of Blythe, California. This site was originally vegetated with salt cedar and a few widely scattered clumps of willow trees. Clearing was selective; all of the salt cedar was removed, but native willows were left intact.

RESULTS OF PRELIMINARY STUDY

Value of Salt Cedar to Birds

The value of salt cedar to birds was found to be minimal when compared with native vegetation (Anderson et al. 1977; Cohan et al. 1978). Bird species richness and densities were significantly lower in salt cedar than in other riparian communities (Tables 3.1 and 3.2). This was true for all seasons but was not pronounced in summer and late summer.

Vegetation Analysis

In order to decide which vegetation variables attracted birds and nocturnal rodents, we subjected four years of riparian vegetation data to a principal components

Table 3.1 Number of bird species present and number preferring and avoiding salt cedar by season along the lower Colorado River.

	Total species		
Season	*Recorded*	*Showing preference for or not avoiding salt cedar*	*Avoiding salt cedar*
Winter	41	1	36
Spring	45	1	38
Summer	41	10	31
Late summer	38	10	26
Fall	32	4	22

analysis to reveal the major independent trends in the vegetation data (Fig. 3.2). This analysis allowed for the evaluation of five groups of variables associated with derived variables or principal components: (1) foliage density and diversity and number of cottonwood and willow trees; (2) number of honey mesquite trees and shrubs per ha; (3) number of salt cedar trees; (4) foliage density and diversity at 0.0–0.6 m; and (5) number of screwbean mesquite (*Prosopis pubescens*) trees per ha.

Avian Habitat Analysis

Avian variables included total species richness, species richness and densities of visiting and permanent resident insectivores, and densities of doves, Gambel Quail (*Callipepla gambelii*), frugivores, and granivores. Associations between these groups and the vegetation PC's were determined with stepwise multiple regression. The results pooled for all years, seasons, and avian variables showed that foliage density and diversity and number of cottonwood and willow trees were significantly positively associated with the avian variable more frequently than was any other vegeta-

Table 3.2 Sum of avian densities in 160 ha including 40 ha each of honey mesquite, cottonwood-willow, salt cedar–honey mesquite mixes, and salt cedar. The expected is based on an even distribution among these vegetation types.

Season	*Total density in 160 ha*	*Density in salt cedar*	*Expected*	x^2	P *that deviation from expected is due to chance*
Winter	616	4	154	195	<0.001
Spring	368	27	92	70	<0.001
Summer	852	179	213	32	<0.001
Late summer	412	95	103	7	<0.001
Fall	621	31	155	11	<0.001

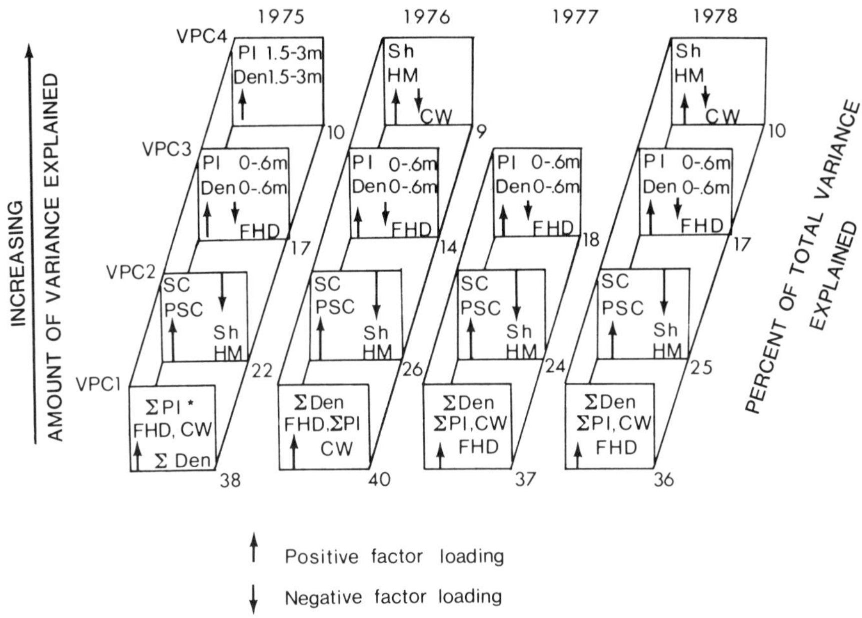

Figure 3.2 Vegetation principal components resulting from analysis of riparian vegetation in the lower Colorado River valley, 1975–1978. Variables are listed in the order of their relative contribution to the factor. *Den* = foliage density, *PI* = patchiness index, *SC* = salt cedar, *PSC* = proportion of total trees which are salt cedar, *HM* = honey mesquite, *Sh* = shrubs, *FHD* = foliage height diversity, *CW* = mixed cottonwood and willow.

tion variable (Fig. 3.3; Anderson and Ohmart 1984). Salt cedar was seldom associated positively with avian variables, but it was associated negatively 50% of the time when it was a step in a significant regression. Screwbean mesquite was positively or curvilinearly associated with the avian variables 95% of the time when it made a significant contribution. However, it made a significant contribution less frequently than did any other variable. Species richness and insectivore densities were all greater where foliage density and diversity and number of cottonwood and willow trees were greater than average (Fig. 3.4). Doves were most frequently associated with mesquite. Doves were also numerous in salt cedar, but salt cedar and foliage density and diversity seldom contributed to an explanation of dove densities. Salt cedar was associated with densities of visiting insectivores more frequently than with the other avian variables.

Frugivorous birds were associated with mistletoe (*Phoradendron californicum;* Anderson and Ohmart 1978), which, in turn, parasitized honey mesquite with far greater frequency than any other plant species in the valley. Since shrubs and honey mesquite were intercorrelated, it was not easy to determine the way other avian species reacted toward shrubs. This determination was made by varying the amount and type of shrubs while keeping other variables constant and comparing

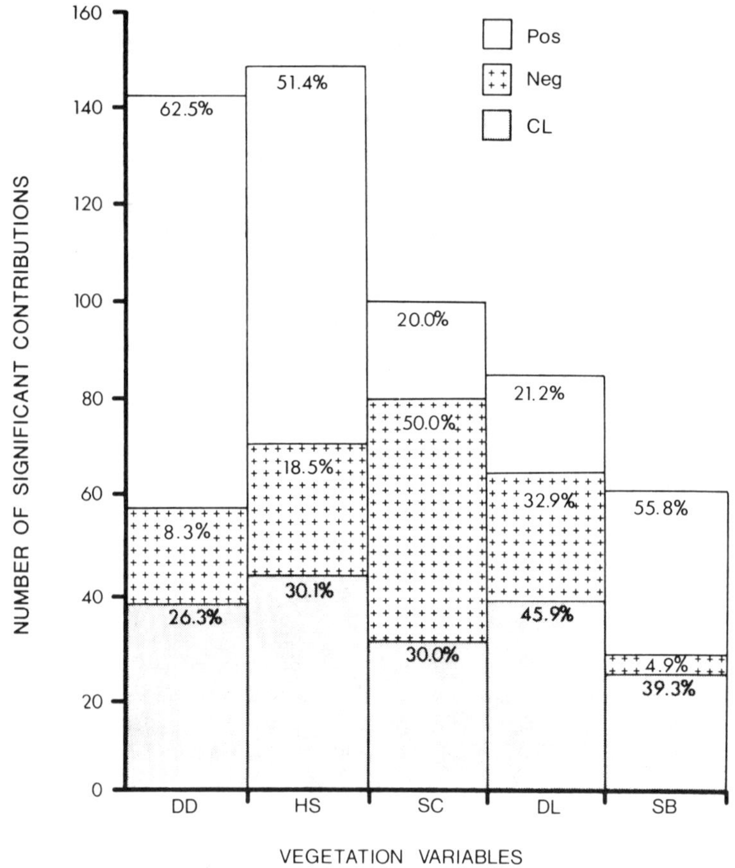

Figure 3.3 Number of times various vegetation variables were included as a step in a significant multiple linear regression equation (*Y*-axis) for all bird variables considered. The maximum number possible was 200. Percentages indicate the proportion of each type of relationship. *Pos* = positive linear relationship, *Neg* = negative linear relationship, *CL* = positive or negative curvilinear relationship. *DD* = total foliage density and diversity and the number of cottonwood and willow trees; *HS* = densities of honey mesquite trees, mistletoe, and shrubs; *SC* = salt cedar; *DL* = foliage density and diversity 0.0–0.6 m; *SB* = density of screwbean mesquite trees.

the avian variables to the average for all riparian vegetation. This revealed (Fig. 3.5) that avian species richness and insectivore, granivore, and Gambel Quail densities were all markedly above average when quail bush was present in dense stands (Anderson et al. 1978; Anderson and Ohmart in press). When annuals such as Russian thistle (*Salsola iberica*) and smotherweed (*Bassia hyssopifolia*) were present in dense patches, avian species richness was high, but the density of permanent resident insectivores was below the average for riparian vegetation, and the density

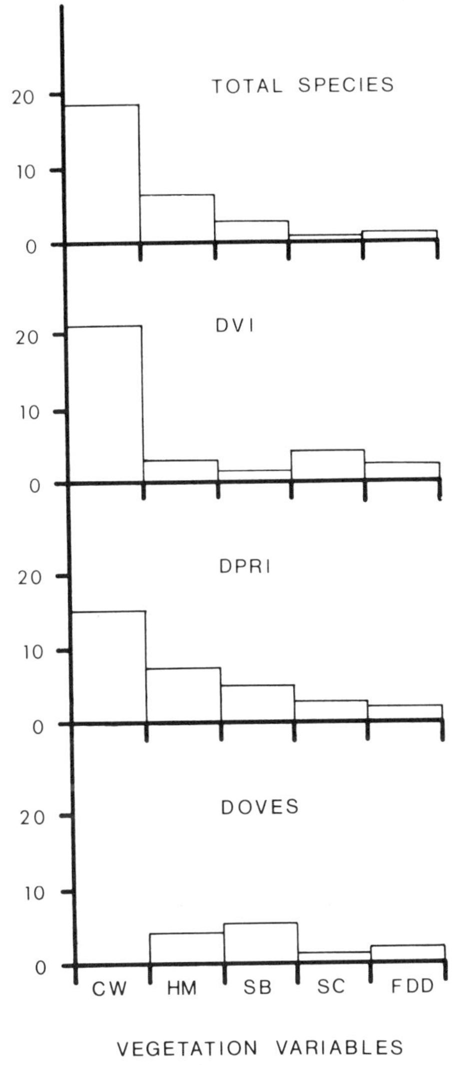

Figure 3.4 Number of times various vegetation variables were included as a step in a significant multiple linear regression equation (Y-axis). Maximum value possible for Y-axis is 25 (5 seasons for 5 years). *DVI* = density of visiting insectivores; *DPRI* = density of permanent resident insectivores; *CW* = number of cottonwood-willow trees; *HM* = honey mesquite; *SB* = screwbean mesquite; *SC* = salt cedar per 0.4 ha; and *FDD* = foliage density and diversity at 0.0–0.6 m.

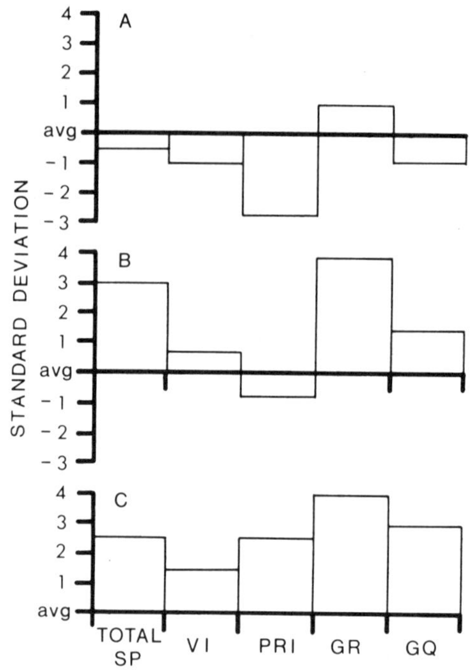

Figure 3.5 Populations of various avian groups associated with various kinds and densities of shrubs, other variables being equal. A. Very low density. B. Shrubs abundant, primarily Russian thistle and smotherweed. C. Shrubs abundant, primarily quail bush and inkweed. Avian densities are expressed in standard deviation units where the mean was for all upland riparian habitat types found in the lower Colorado River valley. *TOTAL SP* = total species; *VI* = density of visiting insectivores; *PRI* = density of permanent resident insectivores; *GR* = density of passerine granivores; and *GQ* = density of Gambel Quail.

of visiting insectivores was slightly above average. Whatever attracts insectivores (presumably insects) is not as abundant in annual plants as in quail bush. Gambel Quail densities were also substantially high when the low-level vegetation was quail bush. Therefore the presence of perennial shrubs, especially quail bush, significantly increased avian densities and diversities in an area.

Rodent Habitat Relationships

Habitats in which the greatest number of each rodent species was caught varied considerably (Anderson and Ohmart 1984). Assuming that capture rates represent rodent habitat preferences, these species can be separated into three major groups (Fig. 3.6). The species of one group, including the cactus mouse (*Peromyscus eremicus*), white-throated woodrat (*Neotoma albigula*), and western harvest mouse (*Reithrodontomys megalotis*), reached greatest densities in areas with high foliage

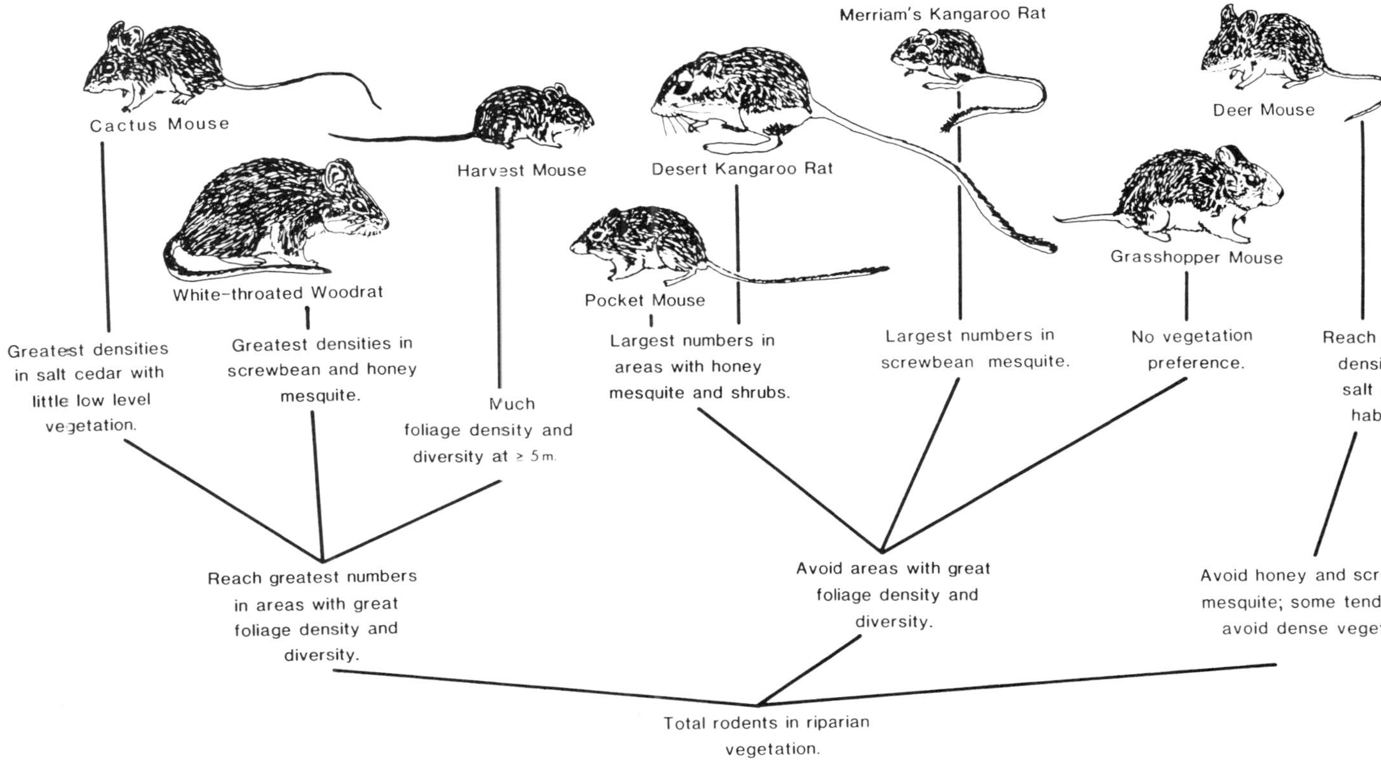

Figure 3.6 Summary of the vegetation relationships of eight species of rodents found in riparian vegetation along the lower Colorado River.

density and diversity. However, this was only a general similarity in habitat preference. The cactus mouse reached peak abundance in dense salt cedar; the white-throated woodrat was most common in mesquite vegetation; and the western harvest mouse reached greatest numbers in areas with a large amount of vegetation $\geq$5 m.

The second group, which included the desert pocket mouse (*Perognathus penicillatus*), desert kangaroo rat (*Dipodomys deserti*), Merriam's kangaroo rat (*D. merriami*), and southern grasshopper mouse (*Onychomys torridus*), was characterized by low numbers in areas with dense vegetation. This presumably indicates avoidance of these areas. The desert pocket mouse and desert kangaroo rat were primarily associated with fairly open stands of honey mesquite where shrubs were at least moderately abundant. The greatest numbers of Merriam's kangaroo rats were caught in stands of moderately dense screwbean mesquite, although the species occurred elsewhere in reduced numbers. The southern grasshopper mouse occurred only in moderately dense vegetation and seemed not to prefer any specific vegetation type.

Finally, the deer mouse (*Peromyscus maniculatus*) avoided mesquite and also to some extent very dense vegetation. It was always a numerically dominant species in disturbed areas such as burns or cleared areas.

Obviously there is no simple way to manage for all of these rodents on a single plot. The recommendation likely to benefit most rodent species would be to create an area that is horizontally diverse. If salt cedar is absent from an area, the cactus mouse will probably not attain maximum densities. This species is, however, nearly ubiquitous and reached moderate densities in several vegetation types. We concluded that an area with cottonwood and willow trees, honey mesquite trees, and shrubs should attract at least a moderate population of this species. Desert kangaroo rats are present only if the area is sandy. If sand is present, proximity of the area to the nearest desert kangaroo rat colony would determine the likelihood of the species moving into the area. If pioneering from a proximal colony is not possible, live trapping of desert kangaroo rats in one area and releasing them in a revegetated area might lead to establishment of a colony.

General Predictions

Enhancement of avian and rodent populations is most likely to be achieved if cottonwood and willow trees are present. These tree species attract, or are correlated with vegetational factors which attract, the greatest density and diversity of insectivorous birds. If frugivorous birds are to be included in the enhancement plan, mistletoe should be present. Since mistletoe parasitizes honey mesquite more extensively than any other native tree, honey mesquite should be planted with the hope that it will eventually be parasitized by mistletoe naturally or that mistletoe can be introduced artificially. Doves were also associated with honey mesquite. Gambel Quail and small granivorous birds, as well as certain rodents, were associated with shrubs. Detailed studies showed that quail bush was an excellent shrub for wildlife. Finally, since foliage density and diversity were among the vegetation variables most frequently associated with high avian densities and diversities, trees and shrubs should be planted densely and in such a pattern that foliage diversity in the vertical and horizontal dimension will be high. The design we followed is shown in Figure 3.7.

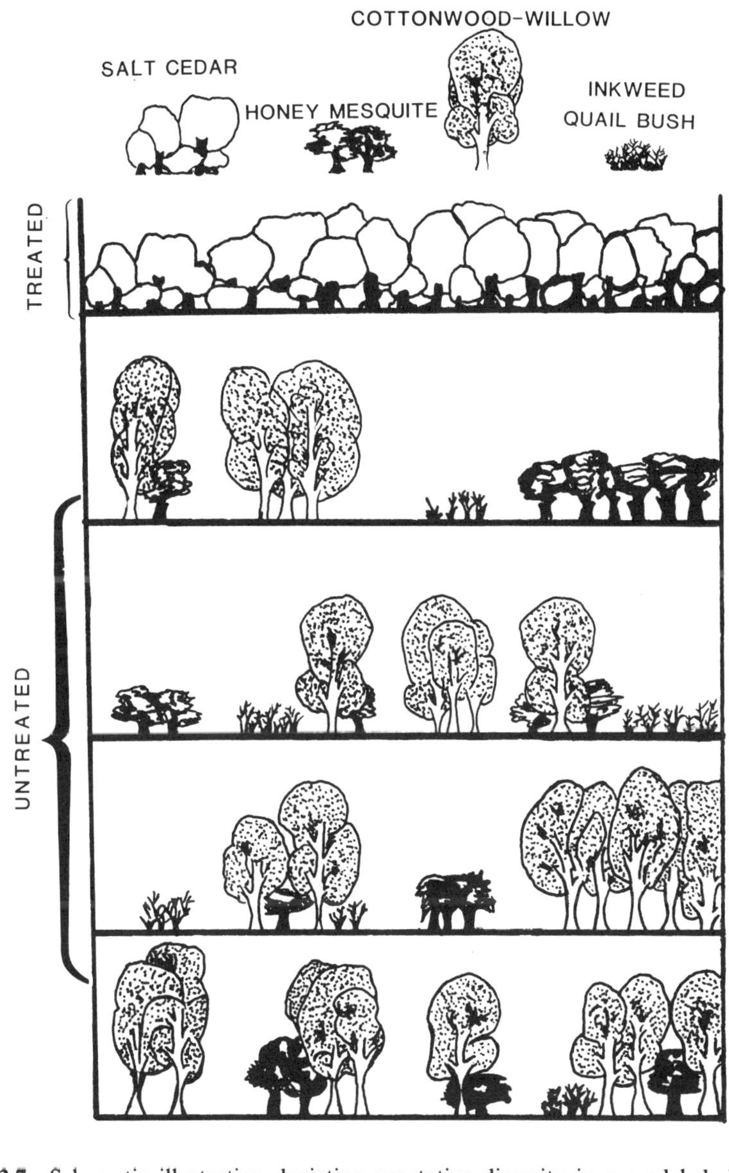

Figure 3.7 Schematic illustration depicting vegetative diversity in a model designed for enhancement of wildlife.

FACTORS AFFECTING GROWTH OF RIPARIAN VEGETATION

Importance of Tillage

Tillage is defined as the breaking up and mixing of the soil. Cottonwood and willow trees planted in sandy soil grew an average of 5 m within two growing seasons if the saplings were provided tillage to 3 m (Fig. 3.8; Anderson and Ohmart 1982). With no tillage, growth averaged much less than 2 m. The advantage of deep tillage is that it permits rapid root penetration to the water table. The effect of deep tillage was not pronounced during the first six months, but thereafter saplings with deep tillage outgrew those planted without the benefits of tillage (Fig. 3.9). Tillage also affected survival of trees (Table 3.3). Among 112 trees of

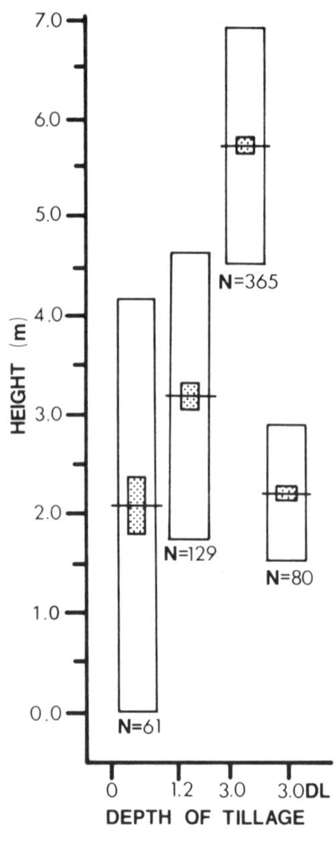

Figure 3.8 Mean height after two growing seasons of cottonwood trees with tillage to various depths planted on dredge spoil along the lower Colorado River. The horizontal line represents the mean; the large rectangle, one standard deviation of the mean; the small rectangle, two standard errors of the mean. *DL* = tillage to 3 m provided at beginning of second year.

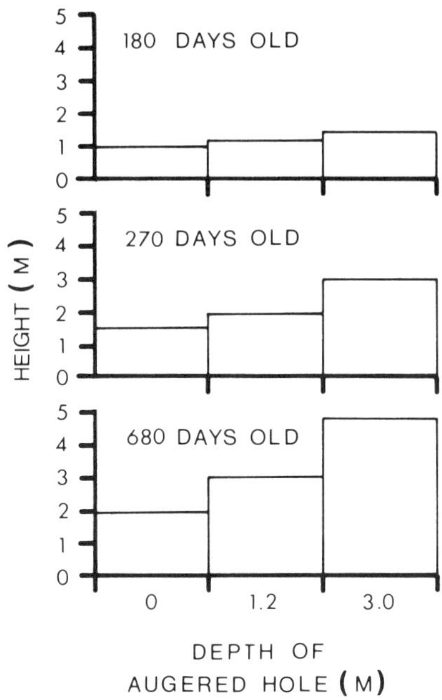

Figure 3.9 Effect of tillage on growth of cottonwood trees at three time intervals after planting in sandy soil along the lower Colorado River.

three species planted with no tillage, 43 (38%) died by the end of the second growing season; among 772 trees provided with tillage to 3 m and which suffered little or no competition, 20 (3%) died.

Impact of Weeds

Weeds growing in the immediate vicinity of the planted trees, which presumably provide competition for the saplings, were associated with tree mortality ranging from 5 to 71% (Table 3.3). When weeds were present and tillage was to 3 m, mortality of planted trees was high. When weeds were absent and tillage was to 3 m, mortality of planted trees ranged from 0 to 7%. The presence of grass also retarded growth in willows (Fig. 3.10; Anderson and Ohmart 1982). With tillage to 3 m, growth of willows was significantly greater than when there was no tillage, but growth was significantly less when weeds were present.

Eliminating Salt Cedar

We cleared a 20-ha area that supported a mean of 345 salt cedar trees per ha. This clearing involved pushing the debris into piles and burning it. The area was then root-ripped about 30 cm below the surface with the cutting edge of the ripper. The area was then leveled.

Table 3.3 Mortality of trees on dredge spoil under various combinations of environmental variables.

Species	N	*Depth of tillage (m)*	*Percent with competition*	*Days of irrigation*	*Percent mortality*
Cottonwood	61	0	16	207	42.6
Honey mesquite	32	0	25	212	18.8
Blue palo verde	19	0	42	201	57.9
Cottonwood	31	1.2	0	207	12.9
Blue palo verde	80	1.2	91	200	62.9
Willow	102	1.2	100	192	70.6
Cottonwood	27	1.2	100	159	29.6
Cottonwood	67	1.2	100	240	4.5
Honey mesquite	41	1.2	100	278	26.8
Honey mesquite	21	2.1	0	225	4.8
Blue palo verde	19	2.1	36	198	47.4
Willow	38	3	0	188	0
Willow	118	3	0	240	0
Cottonwood	339	3	0	240	0
Honey mesquite	154	3	0	220	7.1
Blue palo verde	123	3	6	198	7.3
Willow	21	3	100	204	66.7
Honey mesquite	56	3	100	214	37.5
Total or Total $\bar{x}$	1349	2.5	32	220	19.0

Immediately after this treatment, in July 1978–March 1979, there was no sign of any living salt cedar trees. By October 1979, there were 59 salt cedar trees per ha, these having regenerated from rootstock that had not been killed. These trees were chopped off below ground level the following winter. At the same time the area was seeded with quail bush. This shrub was planted with the intent of enhancing wildlife and to provide competition for any salt cedar that might redevelop. Quail bush germinates in the winter when salt cedar is dormant. We hoped that by the time the salt cedar began to develop, the quail bush would have become so well established that it would outcompete much of the salt cedar for sunlight, water, and nutrients. Treatment in this manner for two consecutive winters resulted in an additional 9% reduction in the number of salt cedar trees per ha (Table 3.4).

Density of Soil

As soil density increased, mean growth of planted trees decreased, other factors being constant (Fig. 3.11). This varied somewhat from one tree species to another but was a good generalization for all tree species on our plots (Anderson and

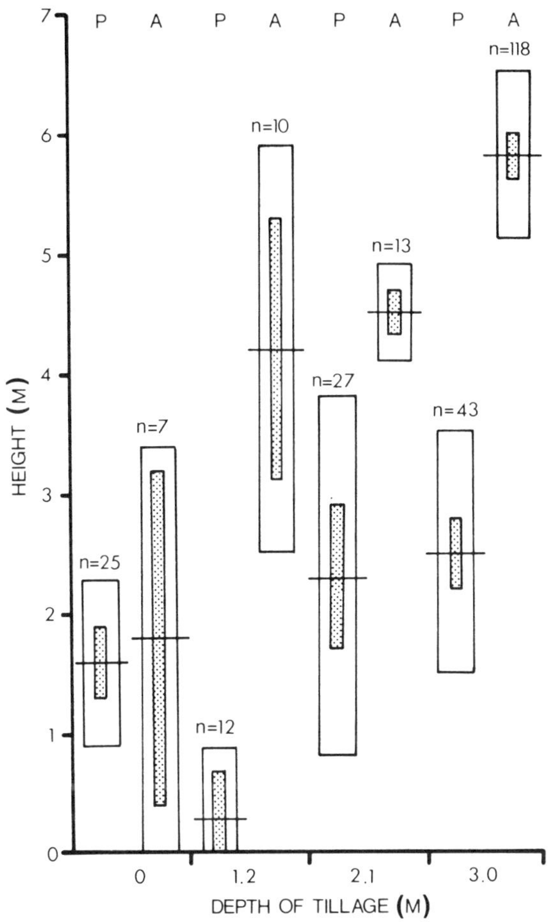

Figure 3.10 Mean height of willows when the extent of tillage and the presence of Bermuda grass varied but when the period of irrigation was similar. Symbols as in Figure 3.8. *P* and *A* refer to the presence or absence, respectively, of Bermuda grass.

Ohmart 1982). Mortality of trees also increased as soil density increased. Lack of growth of trees and their mortality were somewhat offset with deep tillage, and, in addition, longer periods of irrigation seemed to be beneficial.

Irrigation Requirements

Maximum survival of trees was attained with minimum irrigation when weeds were absent and tillage was to a depth of 3 m (Anderson and Ohmart 1982). Increased growth of trees was attained by extending the irrigation period when

Table 3.4 Number of salt cedar trees before clearing and percent reduction in salt cedar at various times after clearing and root-ripping on an experimental plot along the lower Colorado River.

	Number of salt cedar trees/ha	*Percent reduction*
Before clearing		
March 1978	345	0.0
After clearing		
July 1978	0	100.0
October 1979	59	82.9
October 1980	45	87.0
December 1981	29	91.6

weeds were present (Table 3.5). With 150 days of irrigation and 3 m tillage, mean height of trees was only 0.22 m, but with 480 days of irrigation mean height of trees was 1.75 m.

It is possible that 90 or even fewer days of irrigation are necessary when weeds are absent and tillage is to 3 m. A sample of six trees was irrigated for only 90 days and all of them survived and grew as much as trees planted under similar conditions but which were watered longer. Although this point needs clarification, if 90 days of irrigation was adequate, revegetation costs could be reduced by at least 40%.

DESCRIPTION AND EVALUATION OF IRRIGATION SYSTEMS USED

Description

We used two different drip irrigation systems on our revegetation sites. We provide a summary description; details are available elsewhere (Disano et al. in press). The system on the dredge-spoil site consisted of polyvinyl chloride (PVC) irrigation pipe. A well was drilled to 3 m and an electric pump was installed. Water was delivered from the pump through a vacuum-cleaning filter. This filter is an automatic backwash type with a hydraulic controller that detects pressure differential between the intake and outlet valves of the filter. From the filter, water entered a pressure-regulating valve that was set to maintain 35 psi in the irrigation system. Before entering the main line, water passed two pressure-sensitive electric cut-off switches. One switch was set at a high pressure of 40 psi, the other was set at a low pressure of 30 psi. When the pump was running and the irrigation control panel was set on automatic, an increase or decrease, respectively, in pressure at the settings would automatically break the electrical circuit and stop the pump. These automatic cut-off switches protected the integrity of the irrigation system in the event of failure in the pressure-regulating valve or a major break within the system.

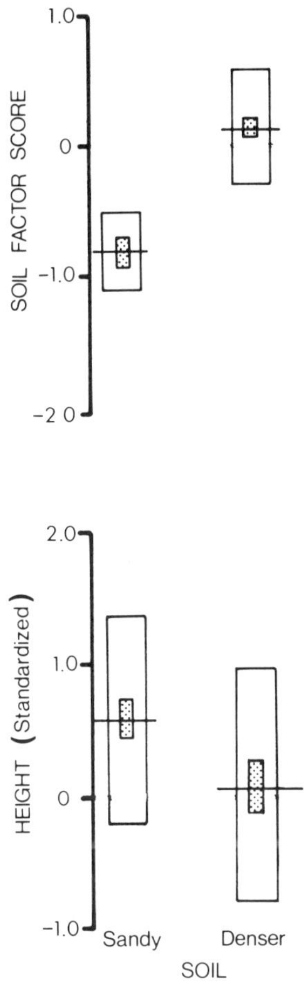

Figure 3.11 Mean soil factor score compared to mean height of trees in sandy and dense soil. Symbols as in Figure 3.8.

The main body of the system consisted of a primary line PVC pipe 15 cm in diameter running from the pump to the river. At the river a butterfly valve was installed along with a bypass pressure-relief valve. The entire capacity of pumped water could be discharged into the river with the butterfly valve open. This valve was a final safeguard; if all others failed, the valve would open at 50 psi and shunt water into the river, thereby relieving pressure on the system.

Lateral lines emanating from the main line delivered water to the trees. The entire system was buried 38–51 cm deep. Risers for irrigation were placed every 6 m along the 17 laterals.

The emitters were pressure compensating and delivered 15 l/hr with a system pressure between 15 and 50 psi. Tubing was attached at each end of an emitter to provide a point source of water to the base of a tree. Any number of laterals

Table 3.5 Two-way ANOVA of effects of length of irrigation and tillage depth on height (m) of willows planted in association with Bermuda grass. $\bar{x}$ = mean height (m), *SD* = standard deviation, *SS* = sum of squares.

Depth of tillage (m)	*Days of irrigation*									
	150	*165*	*180*	*195*	*210*	*225*	*240*	*480*	$\bar{x}$	*SD*
0	0	0	0	0	0	0	0	1.64	0.21	0.58
1.2	0.89	0.36	0.29	0.98	1.21	0.78	0.20	1.67	0.80	0.50
2.1	0	0	2.0	0	0.96	1.93	3.10	1.93	1.24	1.18
3	0	0	0.01	0	0	1.10	1.38	1.75	0.53	0.75
$\bar{x}$	0.22	0.09	0.58	0.25	0.54	0.95	1.17	1.75		

Total SS = 22.30
Column SS = 8.94 $R^2 = 0.401$, $F = 3.04$, df = 7/21, $0.01 > P > 0.05$
Row SS = 4.59 $R^2 = 0.206$, $F = 3.65$, df = 3/21, $0.05 > P > 0.025$

could be shut off at the main line by closing a shut-off valve. Any of the 2,500 emitters could be capped temporarily or permanently to prevent further irrigation of trees.

The irrigation system on the salt cedar cleared site received water from the Colorado River. Water was delivered to the site through a cement-lined canal. A jack gate was installed to divert water into a holding pond which held a 7-day water supply and served as a settling basin. Water was pumped from the pond with an 8-hp two-cycle engine with a centrifugal pump. This delivered 341 l/min at 20 psi to the main line.

A manually cleared filter was installed between the pump and main line. The system consisted of regular 7.6-cm aluminum sprinkler irrigation pipe for the main line. Laterals were flexible black polyethylene tubing with an inside diameter of 1.5 cm. The site was divided into four sections which could be irrigated simultaneously or separately. Each section had 17 laterals traversing it. Installation of a ball valve at the junction of each lateral along the main line made it possible to shut off any number of lines in any section. Each lateral valve contained a pressure regulator which prevented main line water from entering the lateral at more than 20 psi.

Each tree was watered by paired microtubes; each microtube delivered water at a rate of 4 l/hr. Irrigation to an individual tree could be stopped by removing the microtubing and installing "goof" plugs. In order to provide 4-l/hr delivery from each microtube, the length of the tube varied depending on where it was located. Using the flow characteristics of the system, appropriate microtube lengths were determined by computer analysis.

Evaluation of Irrigation Systems

If an electrical pump is to be used, electricity must be readily available. Locations for gasoline- or diesel-powered pumps are limited only by the availability of surface water or well water. The cost of pumping with electricity is comparable to that for gasoline-powered pumping, but the purchase cost of an electrical pump is about 20 times greater than that of a gasoline pump.

PVC pipe must be buried to prevent development of algae that clogs emitters and also to prevent deterioration of pipe that occurs when PVC pipe is exposed to sunlight. Burying pipe requires intensive labor and is therefore expensive. Black polyethylene tubing accumulates little algae and can be laid on the soil surface; in addition, it can be picked up and moved quickly to another area. Also, if polyethylene tubing is damaged it can be repaired more easily than buried PVC pipe. PVC pipe is about 10 times more costly than polyethylene tubing.

Choosing between the inexpensive microtubing and pressure-compensating emitters is more difficult. Although emitters initially cost six to eight times more than microtubing, they last longer and provide a more uniform distribution of water. Also, the cost of installation for emitters is about one-half that of microtubing.

For most desert riparian revegetation sites, an irrigation pump should be gasoline- or diesel-powered. Solar-powered pumps may also be feasible. For flexibility and durability, PVC pipe should be used in main irrigation lines; it must be buried in soil to about 30 cm. Lateral lines should be polyethylene tubing with pressure-compensating emitters.

EVALUATION OF REVEGETATION EFFORTS

Growth and Survival of the Vegetation

We compared growth of vegetation on the revegetation plot with the average for each vegetation measurement for Colorado River riparian vegetation. This was calculated by taking the mean measurement for each variable for the 23 vegetation types we recognized. On the basis of this mean the variables on the revegetation site were standardized; thereby values above and below the overall mean for riparian vegetation assume positive and negative values, respectively.

In July 1978 the revegetation site was cleared of salt cedar; native willows were left intact. Number of willows left intact was slightly above the average for willows normally present in riparian vegetation, which was only about 30 trees per ha. This emphasizes the paucity of willows in the average stand of riparian vegetation along the lower Colorado River. Although additional cottonwood, willow, and mesquite trees were planted, the standardized values on the site did not change because our survey included only the number of trees 3 m or more in height. None of the saplings had attained this height by fall 1981.

Irrigation was discontinued in summer 1980. After this time there was a decrease in foliage density. In addition, the precipitation level was substantially below average ($\bar{x}$ = 8.75 cm/year); thus foliage became much sparser on many plants and some died. This effect did not occur symmetrically over the site; some patches were affected more severely than others. This led to an increase in the horizontal patchiness of the area. The area remained in this condition until new vegetation was planted in winter 1980. By late summer 1980 many of the variables, especially foliage density and patchiness at 0.0–0.6 m, were above the riparian average (Table 3.6). From this information we feel confident that we know how to grow native vegetation under at least a moderate variety of environmental conditions. Furthermore, rapid growth can be expected under proper tillage and maintenance. Therefore, it is possible to drastically reduce the number of salt cedar trees in an area and suppress reinvasions by planting quail bush. Finally, revegetated areas can be planted according to a design that should be successful in quickly attracting a large density and diversity of birds.

Avian Use of Revegetated Areas

Most of the avian variables, also standardized with the mean for riparian vegetation, were above the average by the end of the first year (Anderson and Ohmart 1982). In 1981, even though the mean for vegetation variables was only slightly above the overall mean, the avian variables averaged much greater than the overall mean predicted just on the basis of the vegetation variables (Table 3.7). This is because several vegetational factors which were found to attract birds were present on the revegetation site in greater than average quantities. These included number of cottonwood and willow trees, foliage diversity in the horizontal plane (patchiness), total foliage density in some seasons, and number of shrubs. It is unusual to find such a wide array of vegetation variables to be above the riparian habitat average in a single area. Thus for three of the five seasons in 1981, avian densities and diversities on the revegetation plot tended to be far above the average for riparian vegetation along the lower Colorado River.

At all times of year the standardized avian variables increased as the vegetation on the revegetation plots developed. It seems clear that birds were responding to the changes in the vegetation. The following discussion considers, in some detail, the responses during each season of 1981.

Spring

In spring 1981 the mean for all avian variables was 1.30 (Table 3.7). The range of values among the eight avian variables was high. Densities of passerine granivores, number of Gambel Quail, and number of species of visiting insectivores were all >2.0. This means that fewer than 5% of the 23 existing natural vegetation types in the valley exceeded these values for visiting insectivores and virtually none exceeded the revegetation plot for the other two variables. Only doves were below average.

Summer

In summer 1981 the revegetation site was used less extensively than the average for riparian areas; five of the seven avian variables were below the riparian average (Table 3.7). Number of species of visiting insectivores was substantially above

Table 3.6 Vegetation variables on an experimental revegetation plot standardized to reflect relationship to Colorado River riparian vegetation ($\bar{x} = 0$). Clearing occurred in late summer 1978 and planting began in spring 1980. *FHD* = foliage height diversity, *CW* = Cottonwood, *HM* = honey mesquite, *SB* = screwbean mesquite, *W* = winter, *SP* = spring, *S* = summer, *LS* = late summer, and *F* = fall.

	Foliage density				*Horizontal patchiness*					*Density*					
	0–0.6 m	*1.5–3.0 m*	*≥4 m*	*Total*	*0–0.6 m*	*1.5–3.0 m*	*≥4 m*	*Total*	*FHD*	*CW*	*HM*	*SB*	*Shrubs*	$\bar{x}$	*SD*
LS,F,W SP 78–79	−0.95	−2.26	−1.06	−1.46	−0.54	−1.58	−0.94	−1.11	−0.53	1.01	−0.78	−0.55	−0.76	−0.91	0.78
S,LS,F,W,SP 79–80	−1.86	−2.99	−0.96	−1.73	−1.03	−1.30	−0.94	−1.27	0.67	1.01	−0.78	−0.55	−0.59	−0.97	1.06
S 80	−0.74	−1.84	−0.96	−0.59	0.93	−1.74	−0.72	−0.21	0.67	1.01	−0.78	−0.55	−0.59	−0.50	0.94
LS,F,W, 80–81	2.78	0.93	−0.99	0.14	0.96	−0.47	−0.94	0.03	−1.72	1.01	−0.78	−0.55	2.18	0.26	1.35
SP,S 81	2.27	0.48	−0.94	−0.06	0.23	−0.76	−0.94	−0.49	−1.13	1.01	−0.78	−0.55	2.53	0.11	1.25
LS,F,S,SP 81	1.31	0.49	−1.14	−0.09	1.20	−0.47	−0.27	0.24	−0.53	1.01	−0.78	−0.55	2.46	0.38	1.04

Table 3.7 Values of avian variables on an experimental plot standardized to reflect their relationship to average values ($\bar{x} = 0$) for riparian vegetation along the lower Colorado River. *VI* = visiting insectivores, *PRI* = permanent resident insectivores, *GR* = passerine granivores, *GQ* = Gambel Quail.

	Species			Densities						
	Total	*VI*	*PRI*	*VI*	*PRI*	*GR*	*GQ*	*Doves*	$\bar{x}$	*SD*
Spring										
1979	−1.11	−0.24	−2.58	−1.59	−3.04	−0.75	−1.68	−0.93	−1.49	0.94
1980	−1.11	−1.21	−1.48	−1.59	−2.57	0.13	−1.68	−1.24	−1.34	0.75
1981	1.08	2.02	−0.38	0.81	0.74	3.03	3.49	−0.37	1.30	1.44
Summer										
1979	0.34	0.30	0.23	−2.04	−1.16	—	0.79	−1.24	−0.40	1.07
1980	0.14	0.30	−0.86	−1.08	−1.53	—	−0.83	−0.61	−0.64	0.65
1981	0.14	1.06	−0.86	−0.17	−1.79	—	−0.06	−0.32	−0.29	0.88
Late summer										
1979	−0.33	0.23	−2.49	−0.83	−3.69	−1.30	−0.22	1.28	−0.92	1.57
1980	−1.08	−0.54	−0.41	0.30	−0.50	1.52	−0.15	2.79	0.24	1.19
1981	0.61	0.23	−0.93	0.21	−0.81	0.23	−0.11	0.00	−0.07	0.54
Fall										
1979	0.51	−1.24	−2.59	−0.97	−2.71	2.04	−1.92	0.38	−0.81	1.67
1980	3.16	1.43	0.82	2.27	2.10	4.00	1.38	0.16	1.92	1.25
1981	2.55	2.43	0.25	1.77	1.00	2.40	0.27	−1.18	1.19	1.34
Winter										
1978–79	−0.61	−0.10	−0.72	−0.55	−2.68	−1.01	−1.35	2.07	0.62	1.33
1979–80	−0.24	−0.42	−1.34	−1.40	−2.68	1.20	−0.48	1.26	−0.51	1.33
1980–81	2.54	2.18	0.50	0.94	2.68	3.92	2.88	−0.16	1.94	1.38

average, and density and diversity of permanent resident insectivores was substantially below average. This is partly explained by the very low number of dead trees (snags) to support nest holes for cavity-nesting species on the revegetation plot. Among visiting insectivores, Ash-throated Flycatchers (*Myiarchus cinerascens*), Brown-crested Flycatchers (*M. tyrannulus*), and Lucy Warblers (*Vermivora luciae*) are cavity nesters. Among permanent residents, the Ladder-backed Woodpecker (*Picoides scalaris*), Gila Woodpecker (*Melanerpes uropygialis*), and Northern Flicker (*Colaptes auratus*) are cavity nesters. Absence or low densities of these species contributed to reduced density and diversity values for insectivores during summer. As soon as the trunk diameter of the trees is of sufficient size some will be girdled to provide snags for cavity-nesting species.

Doves were below average because of the relatively small number of trees in which they could nest. Reasons for the reduced number of Gambel Quail are unclear. We predicted higher densities of quail because of the abundance of shrubs.

Late Summer

In late summer individuals of many avian species are found in habitats with which they are not associated in the main part of the breeding season. This may be due partly to the breakdown of territoriality and the wandering of birds hatched earlier in the year in at least some species. Other species may be dispersing as a result of social control mechanisms. Socially regulated species avoid large population decreases in the harsh season by, among other means, dispersing prior to the onset of winter, thereby leaving the most dominant individuals in prime habitats. Subordinates are pushed into suboptimal habitats where they survive as long as the food supply lasts or until they are eliminated by predators (Anderson et al. 1982). The arrival of passerine granivores in late summer contributed to the increased total species richness at that time, relative to summer.

Fall

In fall all of the avian variables, except number of doves, were above the average for all riparian vegetation (Table 3.7). Number of species of permanent resident insectivores and density of Gambel Quail were near average; densities of insectivores were 1–2 standard deviations (SD) above the mean, and total species richness, species of visiting insectivores, and density of granivores were 2–3 SD above the mean for the 23 riparian vegetation types.

Winter

In winter all avian species richness variables remained about the same as in fall relative to the average for riparian vegetation (Table 3.7). Densities of visiting insectivores decreased substantially. Possibly only individuals of some species present in fall remain in winter. Number of individuals for the other four avian density variables increased substantially relative to the mean for the 23 riparian vegetation types.

Rodent Use of Revegetated Areas

Data from the cleared site indicate that small mammal numbers increased after the area was cleared and native trees and shrubs were planted. In the four years before clearing, six species of small mammals were caught on the site. The average total capture rate was 11.3/270 trap nights. The cactus mouse was the most numerous, and deer mice were also common. After the site was cleared, total capture rates of rodents in 1978 and 1979 were lower than before clearing (Table 3.8). However, in 1980 when newly planted vegetation had been growing for nearly six months, capture rates of rodents were substantially higher than before the site was cleared. The deer mouse, as predicted by our model, was relatively abundant after disturbance when the vegetation was sparse. Also as predicted, the cactus mouse attained moderate densities. The Merriam's kangaroo rat, as predicted, was apparently attracted to the shrubs.

COST OF REVEGETATION

Comparative Costs on Two Sites

Our revegetation efforts were conducted on 30 ha of dredge spoil and on a 20-ha site from which salt cedar had been cleared. Our tree-growth experimental work was done primarily on the dredge-spoil area. The possibility of implementing a revegetation design which would attract a large density and diversity of wildlife

Table 3.8 Trap results for revegetation plots at three different times after salt cedar was cleared from the area.

Date	*Species*	*Captures/270 trap nights*
September 1978	Deer mouse	6.0
August 1979	Cactus mouse	1.0
	Deer mouse	3.0
	TOTAL	4.0
November 1980	Cactus mouse	14.4
	Deer mouse	15.9
	Hispid cotton rat	1.2
	Merriam's kangaroo rat	6.9
	TOTAL	38.4
October 1981	Cactus mouse	29.0
	Hispid cotton rat	4.0
	Merriam's kangaroo rat	15.0
	Desert pocket mouse	3.0
	White-throated woodrat	4.0
	TOTAL	55.0

was centered on the cleared site. However, the costs associated with each site varied, some of the most practical methods having been carried out on each site (Tables 3.9 and 3.10). For example, the cost per tree for installation of the irrigation system and getting it operational was much less on the cleared ($2.94/tree) than on the dredge-spoil site ($23.42/tree). Maintenance on the two irrigation systems was equal. Not only was the dredge-spoil irrigation system costlier to install, it was also costlier to purchase ($16.50/tree vs. $1.49/tree). Dealing with weed problems cost about $5.00/tree on the cleared area (Table 3.9), but in the sandy dredge-spoil site, where weeds had difficulty becoming established, this cost was only $1.23/tree (Table 3.10). On the dredge spoil the total cost per tree was about $138, and on the cleared site it was $99 (Anderson and Ohmart 1982).

Estimation of Costs when Sites and Irrigation Systems Are Carefully Selected

By selecting a site requiring minimal preparation and which would have minimal weed problems, site preparation and subsequent labor costs associated with weeding could be reduced by 50 to 75% (Table 3.11). An elaborate irrigation system (i.e., like the system on the dredge spoil) is unnecessary and impractical. Irrigation costs can be reduced by as much as 90% by choosing a simple irrigation system (e.g., as on the cleared site; Disano et al. in press). Use of such a system would reduce the cost of irrigation to $62/tree (Anderson and Ohmart 1982).

Sites where extensive clearing is required may also have more weeds. We have recommended selecting sandy areas with little pre-existing vegetation for tree planting. If revegetation is attempted in an area that must be cleared and that may have any combination of dense, multilayered, or saline soil, we recommend planting shrubs such as quail bush, which thrive under such conditions, rather than trees. Revegetation with quail bush after clearing salt cedar would enhance wildlife, especially in fall, winter, and spring, and should reduce evapotransporative water loss. Revegetation with quail bush would reduce costs to about $1,000/0.4 ha (Table 3.12). If tree planting in a certain area is deemed essential in spite of unfavorable environmental conditions (dense soil, multiple soil layers), financial considerations should be of secondary concern and funding agencies should be aware that the operation will be costly and that growth and survival of the trees may be substantially below maximum.

A carefully prepared list of required machinery should be included in revegetation plans and funds must be available for the purchase or rental of equipment. Local agricultural expertise can be useful. For example, we needed several pieces of equipment: a bulldozer capable of removing trees up to 7 m tall and for pulling a root ripper 30 cm below the soil surface; a mechanical auger capable of drilling 3 to 4 m deep; a hydraulically controlled blade for leveling; and a tractor large enough to pull the blade. Local experts provide specific information relative to the size of bulldozer and tractors required. We cannot overemphasize the impor-

Table 3.9 Costs of revegetating a refuge site (20 ha [50 a]) cleared of salt cedar.

	Days of operation/ man days/or units	*Basic cost*	*Total cost*	*Cost/ acre (0.4 ha)*	*Cost/ tree (2000 trees)*
I. Staff					
A. Labor					
1. Installation of irrigation system	108	$28.60	$ 3,089	$ 62	$ 1.54
2. Getting system operational	98	28.60	2,803	56	1.40
3. Planting trees, collection of seeds, planting seeds	100	28.60	2,860	57	1.43
4. Weeding	344	28.60	9,838	197	4.92
5. Daily operation and maintenance of system and care of trees, 8 months	800	28.60	22,880	458	11.44
6. Putting protective wire around trees	100	28.60	2,860	57	1.43
7. Measuring trees	108	28.60	3,089	62	1.54
B. Management and Supervision	—	—	12,700	254	6.35
C. Secretarial	—	—	6,000	120	3.00
D. Computer Work					
1. Key puncher	20	28.60	572	11	0.28
2. Programmer	—	—	2,000	40	1.00
Subtotal			68,691	1,374	34.33

II. Fringe Benefits	23%		15,799	316	7.90
III. Site Preparation					
A. Clearing	8	1,000	8,000	160	4.00
B. Root ripping	3	1,000	3,000	60	1.50
C. Leveling	7	500	3,500	70	1.75
D. Storage pond	7	1,000	7,000	140	3.50
E. Canal alteration	1	1,500	1,500	30	0.75
F. Slip plowing	400	2.50	1,000		0.50
G. Augering holes	1180	3.00	3,540	175	4.37
H. Backhoeing	420	10.00	4,200		2.10
Subtotal			31,740	635	18.47
IV. Irrigation Supplies					
A. Irrigation system	1	2,978	2,978	60	1.49
B. Spare parts	1	1,000	1,000	20	0.50
C. Pump	1	500	500	10	0.25
D. Gasoline for pump (gal)	1105	1.35	1,492	30	0.75
Subtotal			5,970	120	2.99
V. Truck					
A. Lease	1	4,800	4,800	96	2.40
B. Gasoline for truck (gal)	1105	1.35	1,492	30	0.75
Subtotal			6,292	126	3.15
VI. Trees	2000	2.00	4,000	80	2.00
VII. Travel	—	—	1,000	20	0.50
VIII. Miscellaneous (paper, copying, etc.)	—	—	1,000	20	0.50
Totals I–VIII			134,492	2,691	69.84
IX. Indirect Costs	41%	—	55,142	1,103	28.63
GRAND TOTAL			$189,634	$3,794	$98.47

Table 3.10 Costs of revegetating a dredge-spoil site (30 ha [75 a]).

	Days of operation/ man days/or units	*Basic cost*	*Total cost*	*Cost/ acre (0.4 ha)*	*Cost/ tree (2000 trees)*
I. Staff					
A. Labor					
1. Irrigation system installation	1208	$28.60	$ 34,549	$ 461	$ 17.27
2. Getting system operational	430	28.60	12,298	164	6.15
3. Tree planting	100	28.60	2,860	38	1.43
4. Weeding	86	28.60	2,460	33	1.23
5. Daily operation and maintenance of system and routine checks on trees	800	28.60	22,880	305	11.44
6. Putting fence around site	15	28.60	429	6	0.21
7. Measuring trees	108	28.60	3,089	41	1.54
B. Management and Supervision	—	—	19,050	254	9.53
C. Secretarial	—	—	6,000	80	3.00
D. Computer Work					
1. Key puncher and coder	20	28.60	572	8	0.29
2. Programmer	—	—	2,000	27	1.00
Subtotal			106,187	1,417	53.09

II. Fringe Benefits	23%		24,423	326	12.21
III. Site Preparation					
A. Burying irrigation pipe (Ditch-witch rental/mo.)	5	1,000	5,000	67	2.50
B. Augering holes	2000	3	6,000	80	3.00
Subtotal			11,000	147	5.50
IV. Irrigation Supplies					
A. Irrigation system	1	33,000	33,000	40	16.50
B. Spare parts, system alterations	1	4,500	4,500	60	2.25
C. Electricity for pump			1,445	19	0.72
Subtotal			38,945	519	19.47
V. Truck					
A. Lease	1	4,800	4,800	64	2.40
B. Gasoline for truck (gal)	800	1.35	1,080	14	0.54
Subtotal			5,880	78	2.94
VI. Trees	2000	2.00	4,000	53	2.00
VII. Fence (ft)	5000	0.64	3,200	43	1.60
VIII. Travel	—	—	1,000	13	0.50
IX. Miscellaneous	—	—	1,000	13	0.50
Totals I–IX			195,635	2,609	97.81
X. Indirect Costs	41%	—	80,210	1,070	40.10
	GRAND TOTAL		$275,845	$3,679	$137.91

Table 3.11 Costs of revegetating when site and irrigation system are carefully selected.

	Total cost	Cost/acre (0.4 ha)	Cost/tree (2000 trees)
I. Staff			
A. Labor			
1. Installation	$ 3,080	$ 62	$ 1.54
2. Getting system operational	2,800	56	1.40
3. Planting trees	2,860	57	1.43
4. Daily operation and maintenance (five months)	14,300	286	7.15
5. Measuring	1,500	30	0.75
B. Management and Supervision	12,700	254	6.35
C. Secretarial (half-time)	6,000	120	3.00
D. Computer			
1. Key puncher, coder	572	11	0.29
2. Programmer	2,000	40	1.00
Subtotal	45,812	916	22.91
II. Fringe Benefits (23%)	10,537	211	5.27
III. Site Preparation, Augering 2000 holes	6,000	120	3.00

IV. Irrigation Supplies			
A. System	2,978	60	1.49
B. Spare parts	1,000	20	0.50
C. Pump	500	10	0.25
D. Gasoline (63% of previous total, Table 3.10)	930	19	0.47
Subtotal	5,408	109	2.71
V. Truck			
A. Lease	4,800	96	2.40
B. Gasoline	1,080	22	0.54
Subtotal	5,880	118	2.94
VI. Trees	4,000	80	2.00
VII. Travel	1,000	20	0.50
VIII. Miscellaneous	1,000	20	0.50
Totals I–VIII	79,637	1,594	39.83
IX. Indirect Costs (41%)	32,774	654	16.33
GRAND TOTALS	$112,411	$2,248	$56.16

Table 3.12 Costs of revegetating with quail bush. It is assumed that a 20-ha (50-a) plot with moderately dense salt cedar is to be cleared.

	Cost/acre (0.4 ha)	*Total cost*
I. Staff		
A. Labor		
1. Installation	$ 62	$ 3,100
2. Getting system operational	56	2,800
3. Planting	35	1,750
B. Management and Supervision	5	250
C. Secretarial	5	250
Subtotal	163	8,150
II. Fringe Benefits	38	1,875
III. Site Preparation (clearing, leveling, etc.)	290	14,500
IV. Irrigation Supplies and Spare Parts	70	3,500
V. Truck and Gasoline	118	5,880
VI. Miscellaneous	20	1,000
Totals I–VI	699	34,905
VII. Indirect Costs (41%)	287	14,311
GRAND TOTAL	$986	$49,216

tance of including in any revegetation plan a detailed assessment of the equipment needed and a careful documentation of machinery specifications. Delays and higher costs are inevitable without careful planning.

We found that using ordinary farm machinery in relatively rough terrain led to many delays because of mechanical failure. We recommend scheduling about one day of down time for every full day of operation.

OTHER FACTORS ASSOCIATED WITH REVEGETATION PROJECTS

Contracting

Revegetation with high-value riparian wildlife habitats can be used to mitigate in kind and place for wildlife losses incurred in construction activities of action agencies. The agency responsible for mitigation may choose to have the work carried out under contract. Such work should be done by a reputable contractor of high integrity and proven abilities. Mitigation is expensive—often very expensive. Plans for mitigation for which adequate funds are not available should not be proposed, or if proposed, the shortage of funds should be explicit. The contracting agency should select a contractor whose mitigation proposal has a high chance of success rather than selecting the contractor with the lowest bid. The contracting agency should allow the contractor total freedom but should make intermittent field checks on progress. Persistent interference concerning elements of design, implementation, and maintenance can cause frustration and will probably result in curtailed progress.

Implementation of Mitigation Plans

The contractor should have sufficient funds to either buy or rent all equipment necessary to implement the design. Any other arrangements, including interagency cooperation, may be totally unsatisfactory relative to successful implementation, as we have discovered.

Timing of planting is important, but not necessarily critical, as we previously reported (Anderson and Ohmart 1979). In desert riparian areas, winter is a good time to initiate revegetation efforts. Evaporation is much lower, and thus thorough saturation of the soil from the surface to the water table is easier. By summer only enough water is needed to maintain a water-saturated soil column, plus the water used by the plants. Trees or shrubs planted in winter will have a well-developed root system by the onset of the hot summer and will suffer fewer side effects should the irrigation system fail.

In our revegetation efforts we have found that cuttings from wild stock started in a nursery have highest survival and growth rates. Using stock of unknown origins or from areas outside the general area to be revegetated is an open invitation to failure.

Monitoring

Initial revegetation efforts should be carefully monitored. We censused birds monthly and rodents seasonally; vegetation growth was measured quarterly on experimental plots. Each tree was marked with a numbered metal tag. Monitoring is critical; there should be adequate staff for data collection and for thorough analysis and sound interpretation of the data.

Monitoring methodologies should be kept constant throughout data gathering. After a period of time (i.e., after experience has been gained), it might be possible to predict that if the vegetation is developing according to design and wildlife is reacting in ways close to that predicted, all will go according to plan until the vegetation matures. However, pioneer efforts should be monitored until it is obvious that the desired objectives have been achieved.

ECONOMIC FEASIBILITY OF REVEGETATION

Knowing the cost of revegetation projects is not the same as knowing the economic feasibility of revegetation. Answers to questions concerning feasibility are largely value judgments. It seems clear that revegetation efforts are likely to be considered expensive. In our judgment a high degree of success in revegetating an area should be the major goal. To ensure success, the cost of essentials cannot be reduced. Staffing requirements should be viewed as a worthwhile investment; there will be a greater return on the dollar with too much staff than if there is not enough.

In the desert Southwest and probably elsewhere, root-ripping and augering (tillage) are essential for site preparation. Costs of irrigation can be reduced by using inexpensive but effective irrigation systems. Local farmers or agricultural extension service personnel have the best insight into the least expensive but most effective irrigation systems.

Although tilling the soil or augering holes for every tree is costly, perhaps \$200/ha, it is essential to ensure plant survival once the irrigation system is removed. Augering also reduces the time trees require water from three or more years to three to five months. This means use of much less water and reduced irrigation costs.

Earlier we estimated that revegetation of a 400-ha plot in the lower Colorado River valley, involving the clearing of salt cedar, would require 10 years (three years for clearing and planting, seven years for monitoring) and would cost \$3.5 to \$4.5 million (Anderson and Ohmart 1979). Careful planning could reduce this estimate by \$1 million. The first three years would require 90 to 95% of the expenditure. Those who place high values on wildlife and native vegetation might view this as an inexpensive price to pay. Others, whose values lie elsewhere, might view such an expense as exorbitant and not worthwhile.

In summary, we present a promising technique for mitigation for southwestern riparian habitats in kind and place. We have described the various steps preceding revegetation efforts in the hope that our generalized procedure can be of use in other areas. The basic procedure involved six steps (Table 3.13), ranging from preliminary data collection to monitoring of results.

Costs of revegetation may seem high to some individuals, but if a lesson is to be gleaned from our data it is this: Action agencies should explore all alternatives prior to destroying a reach of valuable riparian habitat. Should it be necessary to destroy it, they should be prepared to meet the high cost of replacing it in kind and in general proximity. More attention must be paid to losses of riparian vegetation because it consists of a distinctive and rapidly disappearing flora which supports a distinctive fauna. Proper watershed management practices, including the prevention of soil erosion, dictates that measures be taken now to preserve existing natural streamside vegetation and to replenish areas where it has already been destroyed.

ACKNOWLEDGMENTS

To conduct such a study as this has required the persistence, dedication, support, patience, and hard work of so many people that it would be almost impossible to list them all. We will attempt to list the most memorable among those who

Table 3.13 Outline of procedure involved in planning for mitigation.

1. Gathering solid base of data concerning the wildlife in the project area and in the area set aside for mitigation.
2. Making a thorough analysis of the data.
3. Creating predictive models with which to create, in theory, a design for the mitigation.
4. Designing required modifications:
   a. Site preparation (e.g., clearing, root-ripping, leveling, putting in irrigation system, etc.)
   b. Equipment needs
   c. Costs
   d. A careful analysis of probable delays and what these mean to the overall mitigation effort
5. Implementing design.
   a. Labor requirements
   b. Labor sources
6. Monitoring.
   a. Methods of gathering information
   b. Analytical and interpretive techniques
   c. Staff requirements

persistently toiled with us as this project moved from its genesis to completion. John and Louise Disano gave totally and unselfishly to the care of the revegetation sites and were always there to help when everything went wrong. Their contributions were of monumental importance to the accomplishment of our goals. In the U.S. Bureau of Reclamation, Ed Lundberg, Phil Sharpe, Manny Lopez, and Al Jonez provided moral and financial support throughout the study. Biologists such as Herb Guenther and Bill Butler were invaluable in giving constructive criticism and solace in times of crisis. In the Yuma Projects Office, Millie Elkins supported and helped throughout our endeavors. Subsequent regional directors, project managers, and those on the environmental staff are thanked. Personnel in the U.S. Fish and Wildlife Service such as W.O. Nelson, Jerry Stegman, Gordon Hansen, Joe Kathrein, Bob Hollock, Randy McNatt, and Wes Martin supported and guided us in times of confusion and chaos. The Fish and Wildlife Service YACC program provided much of the needed manpower to accomplish our goals. Bob Stegman, Kurt Hightower, Gordon Hansen, and Don Barry were extremely helpful and cooperative in providing us with YACC personnel. University of California personnel at Riverside in the Extension Service, especially Les Ede and Jule Meyer, helped in so many ways in resolving irrigation, salinity, and revegetation problems that we could not have succeeded without their help, and most important, they did not reject us when it might have appeared we were crazy.

Drafts of the final chapter were edited and typed by Cindy D. Zisner and edited and checked by Jane R. Durham and Susan M. Cook. Our computer help was provided by Linda C. Richardson. Illustrations were prepared by Melodie Carr, Rodney H. Ohmart, and Stephanie Lewis.

To all of these people and many others we owe our deepest gratitude. Finally, this work was funded by the U.S. Fish and Wildlife Service and U.S. Bureau of Reclamation Contract No. 1–07–34–X0176.

REFERENCES

Anderson, B.W., A. Higgins, and R.D. Ohmart. 1977. Avian use of saltcedar communities in the lower Colorado River Valley. In *Importance, Preservation, and Management of Riparian Habitat: A Symposium,* 128–36. Fort Collins, Co: USDA Forest Service Gen. Tech. Rept. RM-43.

Anderson, B.W., and R.D. Ohmart. 1976. *Vegetation Type Maps of the Lower Colorado River from Davis Dam to the Southerly International Boundary.* Boulder City, NV: USDI Bureau of Reclamation, Lower Colorado Region.

———. 1978. Phainopepla utilization of honey mesquite forests in the Colorado River Valley. *Condor* 80:334–38.

———. 1979. Riparian revegetation: An approach to mitigating for a disappearing habitat in the Southwest. In *The Mitigation Symposium: A National Workshop on Mitigating Losses of Fish and Wildlife Habitats,* 481–87. Fort Collins, CO: USDA Forest Service Gen. Tech. Rept. RM-65.

———. 1982. *Revegetation for Wildlife Enhancement Along the Lower Colorado River.* Boulder City, NV: USDI Bureau of Reclamation.

———. 1984a. *Vegetation Management Study for the Enhancement of Wildlife Along the Lower Colorado River.* Boulder City, NV: USDI Bureau of Reclamation.

———. In press. Avian use of revegetated areas. In *Proceedings of the California Riparian Systems Conference, September 17–19, 1981, University of California, Davis,* edited by R.E. Warner. California Water Resources Center Report No. 55. Davis, CA: University of California.

Anderson, B.W., R.D. Ohmart, and J. Disano. 1978. Revegetating the riparian floodplain for wildlife. In *Strategies for Protection and Management of Floodplain Wetlands and Other Riparian Ecosystems: Proceedings of the Symposium,* 318–31. U.S. Dept. of Agriculture, Washington, D.C.: USDA Forest Service Gen. Tech. Rept. WO-12.

Anderson, B.W., R.D. Ohmart, and S.D. Fretwell. 1982. Evidence for social regulation in some riparian bird populations. *Am Nat.* 120:340–52.

Anderson, B.W., R.D. Ohmart, and J. Rice. 1981. Seasonal changes in avian densities and diversities. In *Studies in Avian Biology No. 6, Estimating Numbers of Terrestrial Birds,* edited by C. J. Ralph and J. M. Scott, 262–64. Lawrence, Kans: Allen Press.

Brown, D.E., C.H. Lowe, and J.F. Hausler. 1977. Southwestern riparian communities: Their biotic importance and management in Arizona. In *Importance, Preservation, and Management of Riparian Habitat: A Symposium.* 201–11. *See* Anderson et al. 1977.

Carothers, S.W., R.R. Johnson, and S.W. Aitchison. 1974. Population structure and social organization of southwestern riparian birds. *Am. Zool.* 14:97–108.

Cohan, D.R., B.W. Anderson, and R.D. Ohmart. 1978. Avian population responses to salt cedar along the lower Colorado River. In *Strategies for Protection and Management of Floodplain Wetlands and Other Riparian Ecosystems: Proceedings of the Symposium,* 371–82. U.S. Dept. of Agriculture, Washington, D.C.: USDA Forest Service Gen. Tech. Rept. WO-12.

Disano, J., B.W. Anderson, and R.D. Ohmart. In press. Irrigation systems for revegetation. In *Proceedings of the California Riparian Systems Conference, September 17–19, 1981, University of California, Davis,* edited by R.E. Warner. California Water Resources Center Report No. 55. Davis, Calif: University of California.

Hubbard, J.P. 1977. Importance of riparian ecosystems: Biotic considerations. In *Importance, Preservation, and Management of Riparian Habitat: A Symposium,* 14–18. *See* Anderson et al. 1977. Rept. RM-43.

Johnson, R.R., and J.M. Simpson. 1971. Important birds from Blue Point cottonwoods, Maricopa County, Arizona. *Condor* 73:379–80.

Lowe, C. H. 1964. *Arizona's Natural Environment: Landscape and Habitats.* Tucson, Ariz: University of Arizona Press.

Ohmart, R.D., W.O. Deason, and C. Burke. 1977. A riparian case history: The Colorado River. In *Importance, Preservation, and Management of Riparian Habitat: A Symposium,* 35–47. *See* Anderson et al. 1977.

Phillips, A., J. Marshall, and G. Monson. 1964. *The Birds of Arizona.* Tucson, Ariz: University of Arizona Press.

Stevens, L.E., B.T. Brown, J.M. Simpson, and R.R. Johnson. 1977. The importance of riparian habitat to migrating birds. In *Importance, Preservation, and Management of Riparian Habitat: A Symposium,* 156–64. *See* Anderson et al. 1977.

Wauer, R.H. 1977. Significance of Rio Grande riparian systems upon the avifauna. In *Importance, Preservation, and Management of Riparian Habitat: A Symposium,* 165–74. *See* Anderson et al. 1977.

CHAPTER 4

Mechanisms of Colonization and Habitat Enhancement for Benthic Macroinvertebrates in Restored River Channels

James A. Gore
Faculty of Natural Sciences
University of Tulsa
Tulsa, Oklahoma 74104

Benthic macroinvertebrates comprise a large and diverse faunal community in most undisturbed running water ecosystems. As such, invertebrates represent a critical pathway for the transport and utilization of energy within that system (Cummins 1979). This flow of energy from headwaters to large-order rivers is a continuum of changes in species composition (representing alterations in feeding mechanisms and foraging habits) (Vannote et al. 1980). Alterations of the natural habitat, such as diversions, channel restructuring, or dredging of substrate material, have the potential of changing the energy dynamics of downstream faunal communities as well. Restoration of the benthic macroinvertebrate community to duplicate adjacent unstressed communities is essential to the maintenance of a stable restored ecosystem.

Aside from the value of restoration of benthic macroinvertebrates as energy processors in stream ecosystems, benthic invertebrates, particularly aquatic insects, are an important component of the ecosystem. For example, Allan (1981) estimated that 30 to 40% of the maximum daily ration of brook trout (*Salvelinus fontinalis*) was comprised of benthic macroinvertebrates. Even higher proportions have been reported by Elliott (1967, 1973, 1975). These food items were primarily taken by feeding fish from the general downstream drift of aquatic insects and other invertebrates. Drift distances vary between species of aquatic invertebrates (Waters 1972), but average drift distances appear to be about 10 meters (McLay 1970). Thus, if an impacted or denuded stream channel is longer than 10 meters in length, drift from an unstressed upstream source area would probably not be sufficient to sustain a viable fishery in the channel. Habitat restoration must be sufficient to attract invertebrate colonizers to establish a stable benthic community. The benthic community would, then, provide the energy source for the drift-feeding fish in the restored stream channel. Although drift feeding is important to salmonid species, Tippets and Moyle (1978) found that epibenthic feeding is also important as a food source. This is particularly true during winter months when ice cover may eliminate solar cues for drift activity and cold temperatures reduce metabolic activity of the invertebrates.

The importance of benthic invertebrates as food sources is not restricted to the cold water species, such as the salmonids. Hanson and Qadri (1979) have shown the importance of benthic invertebrates in the diets of the black crappie (*Pomoxis nigromaculatus*). Benthic macroinvertebrates can represent the major food supply to bottom feeding forage fish, especially the cyprinids and catastomids (DeSilva et al. 1980; Starnes and Starnes 1981). Forage fish are the important food source for many of the piscivorous warm-water fish species.

There have been few or no attempts to transplant whole communities of invertebrates to new sites. This process has been suggested to me by reclamation engineers at various times. Considering the difficulty in collecting "representative communities" for a transplant and physiological stress during transportation, transplantation does not seem to be a feasible reclamation alternative. However, considering the frequency of drift events and the large numbers of individuals in the drift (Hynes 1970; Waters 1972), colonization of a new channel is the easiest method for the establishment of a macroinvertebrate community in the restored stream. Other dispersal mechanisms (see the colonization section) also contribute colonizers to a new stream habitat. The goal of stream restoration managers becomes the enhancement of available habitat for colonizing invertebrates. If preferred habitat types are available to colonizers, a stable macroinvertebrate community can be established and maintained.

This chapter will provide the manager with information on the dispersal mechanisms of aquatic invertebrates, measurements of optimum habitat for invertebrates, and a synopsis of some typical reclamation efforts designed for macroinvertebrate habitat (further examples are cited by Starnes Chapter 7, and Burgess Chapter 8, this book).

BENTHIC MACROINVERTEBRATE DISPERSAL

Williams and Hynes (1976a,b; 1977) have performed extensive examinations of the mechanisms by which benthic organisms are able to repopulate or colonize stream habitat. In general, there are four primary sources of colonizers:

1. Drift of organisms from upstream source communities
2. Upstream migration of benthic invertebrates within the water
3. Movement from within the substrate or from adjacent bank storage areas
4. Colonization from aerial sources (that is, the oviposition by adult aquatic insect species)

Of the four mechanisms, drift of benthos has been shown to be the most important factor, contributing as much as 60% of the total fauna in a newly constructed habitat (Elliott 1967).

Drift is a continuous redistribution mechanism in stream ecosystems. Townsend and Hildrew (1976) found that within a stream community 82% of invertebrate movement was a result of drift within the water column of the stream. Drift is a diurnally periodic phenomenon of any stream which contains an established benthic invertebrate community (Waters 1972). Since invertebrates spend their foraging time at night (to avoid sight-oriented predators), initial investigations into the drift phenomenon were based on evidence of solar cues as signals for foraging and entry into the water column (Hughes 1966a,b). These initial observations have been supported by findings of decreased drift on moonlit nights (Anderson 1966) and increased drift during solar eclipse (Cadwallader and Eden 1977). However, recent studies indicate that dorsal light cues are but a small fraction of the mechanisms which induce drift. Suter and Williams (1977) observed no detectable differences in drift rate during the same solar eclipse reported by Cadwallader and Eden. In several different laboratory experiments, Ciborowski (1979) found little correlation between drift and photoperiod and proposed that internal physiological change rather than external cues were critical to drift rate. This seems to be supported by numerous observations of drift being highest in early spring and fall months (Stoneburner and Smock 1979; Cowell and Carew 1976). Ciborowski suggested that these were times of least active growth. Waters (1977) has proposed that drift, then, could also be used to estimate production of benthic invertebrates. Gore (1977) demonstrated that unusually large drift events were attributable to escape from changes in local habitat to less than optimum conditions. Bohle (1978) suggested that depletion of local food sources would result in an increase in drift rates. Peckarsky (1979, 1980) and Walton (1980) have indicated that predator-prey interactions result in addition of invertebrate individuals to downstream drift.

Regardless of the factors that determine the drift in a restored ecosystem, all of the research mentioned indicates that an undisturbed source area of colonizers must be available upstream of any stream section which will be restored and enhanced for colonizers. Indeed, if drift rates are variable and are highest in early spring and late summer months, it could be expected that restoration and habitat enhancement designed to be completed during these time periods would assure the most rapid recovery times to an equilibrium state.

Upstream migration of benthic invertebrates is probably underrated in its importance to the establishment of new communities. Bishop and Hynes (1969) found that upstream migration accounted for less than 10% of the new community while Brusven (1970) found it to be insignificant as a source to establish communities in denuded streams. However, Williams and Hynes (1976a) found that as much as 20% of new colonizers arrived from upstream migration. Within the Tongue River system of Wyoming and Montana, I have found that some species can invade an area 40 km upstream of their source in a period of six weeks or less (late instar nymphs of the dragonfly, *Ophiogomphus;* Gore 1977). These data also point to the need for a restoration specialist to consider maintenance of undisturbed downstream source areas for colonizers. This requirement may severely restrict the total length of a stream which can be restored at a given time. Some estimates of this restriction will be made in another section of this chapter.

In some stream systems, *movement within the substrate* to new habitat can provide as much as 20% of the new community members (Williams and Hynes 1976). This is particularly true in areas where the substratum and flood plain material is unconsolidated and porous. In river restoration projects, this source of potential colonizers will likely vary with the type of reconstruction efforts. In many cases where rerouting of channels, channelization, or placement of new substrate will occur during reclamation, there will be little contribution from substrate movement. However, when a preexisting channel is modified through habitat enhancement, both noninsect invertebrates and aquatic insect larvae, nymphs, and adults migrating from the hyporheos should accelerate the establishment of a benthic community. Protection and/or maintenance of a buffer strip along the shoreline of a channel to be reclaimed will probably increase the chances that hyporheic sources of colonizers will be maintained during stream alterations prior to restoration activity.

Colonization by ovipositing adult female aquatic insects is also an important factor in the maintenance and establishment of benthic communities. In general, the oviposition activity of adult insects is thought to compensate for the general downstream drift of the larval and nymphal stages of the life cycle (Hynes 1970). Otto and Svensson (1976) demonstrated drastic reductions in caddis fly populations when pupal populations were removed from benthic communities. The significance of aerial contribution is obviously limited to the primary periods of emergence of adult insects—the spring and autumn months (Williams and Hynes 1976b). Again, this temporal phenomenon suggests that optimum times for restoration of benthic communities will be during the spring and autumn months when contribution from drift and aerial activity of adult insects will be at their peaks. Because the majority of ovipositing adult aquatic insects show a general upstream flight pattern from point of emergence to point of egg deposition, the necessity of maintaining viable and stable communities downstream of the restored area is again indicated.

COLONIZATION BY MACROINVERTEBRATES

The substrate particles and associated physical and chemical characteristics of a restored stream channel, and particularly the newly constructed river channels of some reclamation projects act as bare "islands" to be colonized by benthic macroinvertebrates. The island biogeographic theories of MacArthur and Wilson (1967) have often been applied to stream ecosystems to model and predict establishment of stable communities. In general, colonizers will invade new habitats during a reasonably short period of time. Numbers of taxa level off at a maximum and, subsequent to a period of intense competition and the effects of extinction, decline to an equilibrium state (Simberloff 1978). MacArthur and Wilson argued that the period of time needed to attain equilibrium is a function of the distance between source island and intermediate "stepping stone" islands and the final recipient island. This basic theory assumes the linear relationships between source, stepping stone, and recipient islands. Many real world situations do not display this linearity, but rivers and streams are particularly well suited to examine colonization and the use of MacArthur-Wilson equations to predict the attainment of a stable community.

When colonization was examined using denuded or previously stressed stream ecosystems, maximum densities of benthic invertebrates were reported between 70 and 150 days (Cairns et al. 1971; Williams and Hynes 1977; Cherry et al. 1979; and Gore 1979). Minshall et al. (1983) found even longer periods of time for establishment of the final stable communities. I have also found similar times (300 to 500 days) for establishment of a stable benthic community in a completely restructured and habitat-enhanced stream (Gore 1982). When colonization was examined using implanted artificial substrates, shorter times of attainment of maximum density have been recorded—usually 21 to 30 days (Sheldon 1977; Khalaf and Tachet 1977; Meier et al. 1979; and Wise and Molles 1979). Ulfstrand (1968a,b) pointed out that many of these differences could be accounted for by differences in life history of individual species and the time of year in which the colonization occurred.

In addition to the variation in life histories, the distance of the recipient island (substrate of the reclaimed or restored channel) from the source of colonizers is a factor in attainment of maximum densities (Gore 1979, 1982; Minshall et al. 1983). An artificial substrate has an adjacent source and would be colonized much faster than the substrate at the downstream end of a lengthy river channel (i.e., it may be several hundred meters or more downstream of the source of drift contributed colonizers). This distance effect is an important factor to consider in the amount of restoration possible.

Of course, achievement of maximum densities does not necessarily imply reclamation success. The equilibrium community is attained after achievement of maximum densities and diversities and a period of community adjustment that includes arrival of more rare species, some amount of competition, and, possibly successful reproduction in the new environment (Simberloff 1978; Gore 1979, 1982; and Nichols 1980). An appropriate measure of this equilibrium state should include a measure of synchronous change in community structure between source and recipient areas.

MEASURING INVERTEBRATE HABITAT REQUIREMENTS

A critical element in restoration of macroinvertebrate habitat is the ability to reproduce key habitat characteristics of the environment occupied by invertebrate colonizers in the source areas.

It is generally acknowledged that the physical habitat characters which most directly control macroinvertebrate distributions in unstressed streams are velocity, depth, and substrate quality (Hynes 1970; Eriksen 1964; Williams and Mundie 1978; Gore and Judy 1981). Of course, water quality must also be maintained during restoration (Minshall and Minshall 1978; Ruggiero and Merchant 1979). Methods for maintenance of chemical water quality are discussed elsewhere in this book. The primary considerations here will be the physical restructuring and enhancement of habitat.

Substrate

Substrate composition is the most easily manipulated habitat characteristic in restoration of rivers and streams. To duplicate invertebrate source area substrate composition, it may be necessary to consider such factors as degree of imbeddedness of the particles, size of the particles, contour of the particles or substrate surface, and heterogeneity of the substrate types.

Aquatic macroinvertebrates are found in a wide variety of habitats. These range from rooted vegetation and periphyton (many gastropods, amphipods, isopods, coleoptera, and odonates) to sand and silt (many burrowing ephemeroptera and most chironomidae) to gravel and cobble substrates (most free-roaming pleroptera, ephemeroptera, trichoptera, and most filter-feeders of large particulates). Merritt and Cummins (1978) have listed the major habitat types for the families of aquatic insects. A similar habitat list for the major invertebrate foods of salmonids has been prepared by the Stream Enhancement Research Committee, Province of British Columbia (1980). In general, highest productivity and diversity of aquatic macroinvertebrates in lotic systems have been found in riffle habitats with medium cobble (256 mm diameter) and gravel substrates (Hynes 1970; Hart 1978; Gore and Judy 1981). These measurements have been taken in high gradient, undisturbed streams, for the most part. Ruggiero and Merchant (1979) have pointed out the importance of substrate characteristics when they noted that in some situations distribution of macroinvertebrates was more highly correlated with substrate than with water quality.

Figure 4.1 shows a device that I have proposed for use in measuring substrate profile (Gore 1978). This device (with some modifications) has been used successfully by other researchers as well (V. Resh 1983, University of California, personal communication). It has the advantage of accounting for particle size of the major substrate types, degree of imbeddedness, and overall profile of the substrate in microhabitats of high productivity and diversity. When placed within a suitable bottom sampler, the standard deviation of the distance of the rods from the substrate provides a single numerical index of substrate composition (as the index increases in value, larger average particle sizes and more heterogeneous substrates are indicated).

Regardless of the device or method used to measure substrate composition (from a Wentworth particle sorter to photographic or visual observation and classification), reproduction of source area substrate is essential for habitat enhancement in the restored river section. Obviously, not all streams and rivers will have cobbled

Figure 4.1 This device is used to assess variability of substrate profile. Standard deviation of the mean length of rods above the plexiglass surface indicates degree of roughness (see Gore 1978).

substrates as original or reclaimed substrates. In many prairie river systems, for example, common substrates are often sand and gravels. It has been suggested that an additional factor of substrate compaction and/or porosity should be measured in soft sand substrates since many invertebrates in these environments are adapted to a burrowing existence (Allan Covich 1982, University of Oklahoma, personal communication). This measure could be accomplished with use of particle size determination and the measure of penetrance of a meter stick under constant pressure for a given time period, for example.

Structures to Alter Depth and Velocity

In an unregulated stream (most often encountered in reclamation and restoration projects) it is not possible to control discharge patterns within the channel. In this respect, hydrologic restoration must account for normal discharge rates in maintenance of meander pattern and pool/riffle frequency.

In some respects, however, depth and current velocity can be manipulated in a restored channel by the placement of structures. Locally, various structure types can act to alter the shape and substrate composition of pools, riffles, and gravel beds by modifying depth and velocity patterns. Luedtke et al. (1976) used iso-velocity and depth maps to show habitat improvement by various structures. For example, log-drop structures improved fish habitat by trapping some sediment and decreased macroinvertebrate density and diversity only slightly. Gabion deflectors, which increased velocity upstream of potential riffle areas removed sediment accumulations from cobbled substrates. The net effect was to increase numbers and kinds of macroinvertebrates. In particular, mayfly (Ephemeroptera) density and diversity increased dramatically. In general, the use of structures can be implemented for control of higher than normal suspended loads in a reclaimed river section. Thompson (1980) found similar results in streams in agricultural areas of Illinois. With implementation of surface mine regulation required reclamation, such as maintaining riffles and pools, meanders, and gradient, stable fish and macroinvertebrate communities were achieved in two to three years. Without these remedial efforts, recovery in the same time period was slight and judged not successful.

The addition of habitat enhancement structures provides attractive colonization islands for benthic macroinvertebrates. The mapping of velocity and depth distributions have been used to display the enhancement abilities of various structures (Gore and Johnson 1981). During reclamation of the Tongue River, which had been routed through an abandoned surface mine pit, the original channel was recut and graded to represent the premining channel. With the addition of substrate material and embankment materials, the channel did offer some amount of suitable habitat for macroinvertebrates (see Fig. 4.2). However, the addition of structures to create cover for colonizing fish also modified flow patterns and, in some instances, modified substrate character so that macroinvertebrate production was also enhanced.

These composite depth and velocity predictions are shown in Figures 4.3 through 4.5 which are based on estimations of highest density and diversity of macroinvertebrates being found at mean velocities of 76 cm/s and depths of 28 cm (Gore 1978; Gore and Judy 1981). After pooling and cavitation had occurred behind large boulders placed in the stream (approximately 200 days), formation of gravel bars downstream of these boulders added optimum macroinvertebrate habitat to the restructured stream section. Rubble piles of large cobble placed in the reclamation section or formed after the freeze-fracturing and/or erosion of some of the boulders in the channel also provided habitable gravel bars for benthic communities. Trees anchored to the embankment to act as "snags" and cover structures for fish also altered velocity patterns in the streams. However, the advantage of the snags to macroinvertebrates were found to be the deposition of finer sediments into beds in the wakes of these cover structures. These sediment beds proved to be suitable for Chironomid larvae and the burrowing mayfly, *Ephemera simulans* (Gore and Johnson 1980). Deposition of finer sediments had the added

Figure 4.2 Reconstructed channel with habitat enhancement structures (rip-rap, snags, cobbled substrate, boulder deflectors) in place. Early stages of revegetation can be seen.

benefit of helping to cement the cobbled substrate and reducing bedload movement during high-flow periods in the river channel.

It is not wise nor practical to construct a continuous riffle area for the entire length of the channel simply to increase benthic invertebrate production. Hasfurther (Chapter 2, this book) and Bishop (1980) have discussed estimating frequency of pools and riffles for hydrologic balance in reclamation of streams. In essence, it is necessary to maintain similar proportions of pools and riffles as in upstream unstressed areas. Although pools do not provide large amounts of suitable habitat for macroinvertebrates, pools do provide an important control factor for the habitat suitability of further downstream riffle areas. Many reclamation projects, particularly in the arid and semiarid regions of the western United States, will have severe problems with control of suspended sediment in streams from erosion and runoff from unvegetated or newly revegetated embankment areas. Until vegetation and other erosion control structures effectively reduce suspended load, pools in the reclaimed river section can act as effective sediment traps.

On a reclaimed section of the Tongue River through an operating coal surface mine, we found that a large pool of about 2.5 m depth placed at 400 m intervals was sufficient to trap suspended sediment being carried at 5 to 8 times normal loads. Riffles downstream of this pool remained essentially comparable to unim-

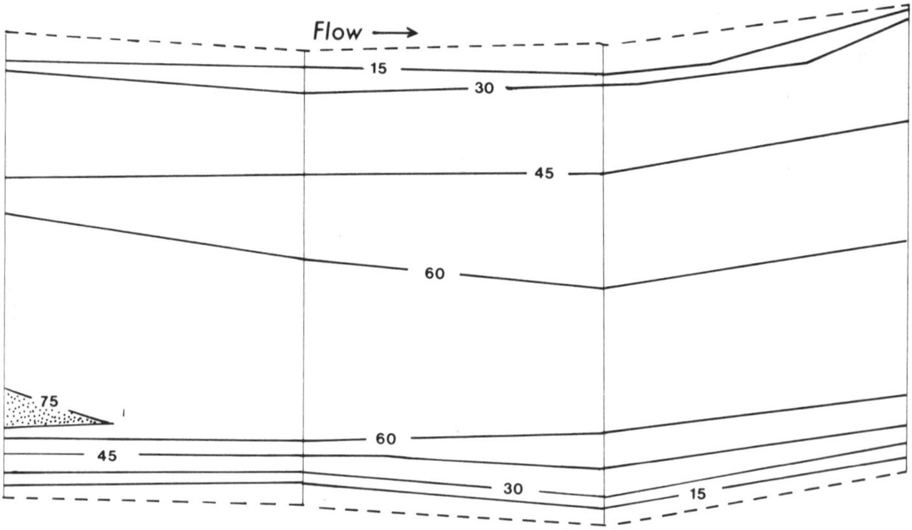

Figure 4.3 Computer derived iso-velocity map of an unrestored section of the Tongue River in Wyoming. This projection is based on channel configuration before placement of habitat enhancement structures and flow patterns at a discharge of 8.37 m³/s (295.65 cfs). Only a small section of this reach contains areas of optimum velocities for high community diversity (stippled area).

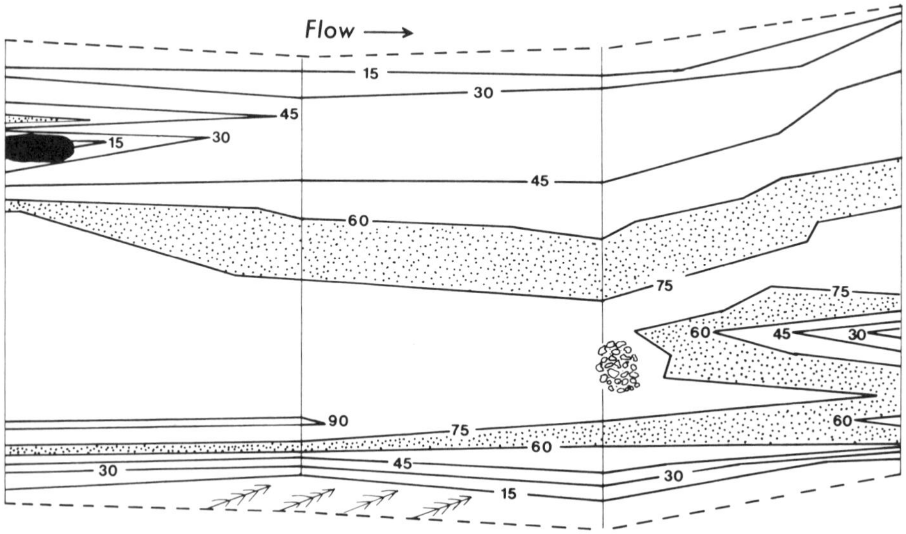

Figure 4.4 Iso-velocity map of the section (shown in Figure 3) of the Tongue River in Wyoming after placement of restoration structures and a flow duration of one year. This map projection is derived for a discharge of 8.37 m³/s (295.65 cfs). Note formation of suitable velocity areas in the wake of the rubble pile and along the flows deflected by the snagged trees.

pacted upstream sites (when comparing total suspended sediments) (Gore and Johnson 1980). Control of suspended sediment relieves potential loss of macroinvertebrates from scour of the substrate, reduction of light for periphyton production, and mechanical damage to gills and other structures of the invertebrates themselves. Luedtke et al. (1976) found that log-drop structures removed sediment in local areas of sediment deposition in high-gradient low-order streams. Gabion deflectors and/or channel diversion to a single unbraided channel increased average velocities in these small streams and increased the ability of the stream to carry suspended sediment. The net result was an increase in available macroinvertebrate habitat.

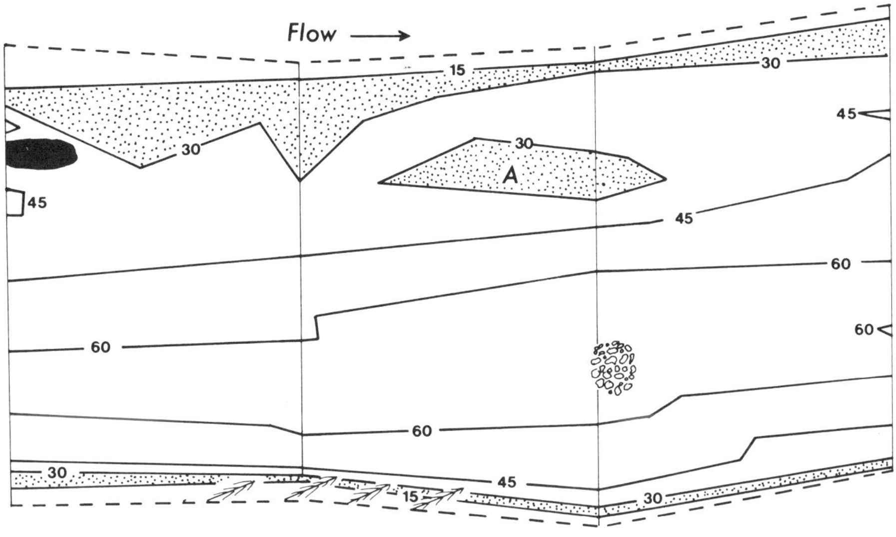

Figure 4.5 Iso-depth map of the section (shown in Figure 3) of the Tongue River in Wyoming after placement of restoration structures and a flow duration of one year. Map projection is derived for a discharge of 8.37 m³/s (295.65 cfs). Stippled areas represent optimum depth areas for macroinvertebrate community diversity. Section A is a gravel bar formed in the wake of the large boulder (2.5 m diameter) placed in the channel (upper left corner).

Devices which will provide for increases in average flow velocities should be considered a necessity in chronically sediment-polluted streams and rivers. In high-gradient low-order streams, these structures could include weirs, gabion deflectors, and wing deflectors. In lower gradient, larger order rivers, the same effects can be accomplished by proper placement of snags, effective embankment structures (rip-rap), and large wing deflectors. Gore and Johnson (1980) also found that construction of a narrow, slightly deeper channel through restructured riffle areas (an artificial *thalweg*) provided directed flow which acted to reduce sediment deposition and scour in high macroinvertebrate production areas.

PREDICTION OF RATES, TRENDS, AND STABLE COMMUNITIES

The goal of the reclamation biologist is to determine if stable macroinvertebrate communities have been established in restored river channels. Sheldon (1984) has proposed a set of models to describe larval aquatic insect colonization in stream ecosystems. These models are based on field experiences and the basic ideas of island biogeographic theory proposed by MacArthur and Wilson (1967). The models required some knowledge of immigration rates and intrinsic emigration probability in order to predict attainment of an equilibrium state. The equations are unique in that elements of competition between species are included as well as a resource tracking model to include the potential of increased periphyton production and detritus accumulation as colonization progresses. Sheldon's simulations predict that slow colonizers will exceed carrying capacities and decline rapidly while fast colonizers control their resource base and persist. Because colonization is controlled by many varied physical and biological events, Sheldon concluded that sweeping generalizations about the model were inappropriate without further testing and refinement of the models as they presently exist.

Despite the inability to produce general mathematical models for prediction of attainment of equilibrium communities, some basic trends have been reported that should aid reclamation specialists in observations and monitoring of restored benthic macroinvertebrate communities.

Just as different drift rates and distances have been reported for stream-dwelling invertebrates (McLay 1970; Waters 1972; Walton 1978), variable drift distances and rates have been reported for nonliving material that makes up the benthic environment (De La Cruz and Post 1977; Williams 1980). These noninvertebrate components add to habitat complexity and suitability for a great variety of benthic invertebrates. In some cases, it would be expected that certain invertebrate species would not establish populations on newly reclaimed streams until some of these noninvertebrate components were also established. Burrowing invertebrates were established in streams only after deposition of sediment on cobbled substrates (Gore and Johnson 1980).

Similarly, one might expect that grazing invertebrates would not establish permanent populations until sufficient periphyton growth had occurred. Indeed, Crisp and Gledhill (1970) and Gore (1979) have shown that colonization by larvae "rafting" with dislodged algal mats was a primary colonization mechanism in newly structured stream habitats. Williams (1980) found no correlations between colonization and the rates of deposition of course and fine particulate organic material. Gore (1979) predicted a sequential increase in habitat complexity with concurrent increases in benthic community diversity on substrate areas composed of "virgin" material (in this instance, the substrate material was quarried and delivered to the channel reclamation site). This observed increase in habitat complexity occurred most rapidly in the most upstream areas of the reclaimed reach. These observations verify previous reports of differential drift rates of noninvertebrate and nonliving benthic habitat material and indicate a potential time and distance limitation on the total length of stream that can be successfully reclaimed at a given time.

Yet, these same observations also suggest that restoration of benthic communities should be effectively accelerated with the placement of substrate material derived from adjacent flood plains, diversion, or previous channelization projects. In this manner, some amount of habitat complexity, particularly a food base from primary production, will be available at the commencement of colonization. Much of the cobbled substrate material from old river and stream beds will retain encysted algae, cyanophyte spores, and microbial colonies, as well as some accumulated organic detritus for use by the many grazer-scraper and collector-gatherer invertebrates that will colonize the new habitat. Sheldon's (1984) models predict an increase in colonization rates with an increased initial resource base. More rapid colonization rates have been reported when substrates derived from instream materials have been used as the colonization substrate (see for examples, Larimore et al. 1959; Fisher et al. 1982; Gray and Fisher 1981; Harrison 1966; and McAuliffe 1983).

As mentioned, MacArthur and Wilson (1967) have predicted impacts of distance between source and recipient islands in their models of island colonization. Without intermediate stepping stone islands, increased distances could be expected to result in reduced diversity and a delay in attaining community equilibrium on the recipient island. Since restructured stream and river channels represent quasi-directional colonization systems all substrate particles in the new channel may be considered to act as stepping stones to new substrate habitat further downstream. However, the effects of distance between source areas and the most downstream elements of the reclaimed section should be taken into consideration when attempting to assess reclamation success. As previously mentioned, benthic invertebrates exhibit great differences in drift frequency and drift distances per drift event as do the noninvertebrate components of drift in the lotic ecosystems.

One might expect to see some variability in community structure depending upon location of the sampling area and its distance from source areas of colonizers and the time since initiation of colonization. Gore (1979, 1982) has demonstrated this effect on a fairly short section of reclaimed channel of the Tongue River. For every increase in 200 m from the upstream source of drift colonizers, there was a lag of at least 75 days to attainment of maximum densities and diversities at the next further downstream sampling station. Indeed, the curves (Figures 4.6a,b) imply that this lag time, plotted logarithmically, is a doubling function with arithmetic increases in distance from the upstream source areas. Taken as sole predictors, these curves predict a recovery time of over 6.5 years for the most downstream substrate areas of reclaimed riffle of 1.2 km in length. Fortunately, these predictions of restoration success are unrealistic. Addition of tributary water, changes in surface runoff patterns, organic inputs, and other invertebrate dispersal mechanisms (particularly upstream migration and adult flights of aquatic insects for oviposition) will not allow this doubling factor to perpetuate *ad infinitum* with increased distance. Indeed, as stable communities are established on the stepping stone islands and areas in the upstream portions of the restored channel, these stable areas, in turn, become source areas of colonizers for the further downstream areas. Thus, distances

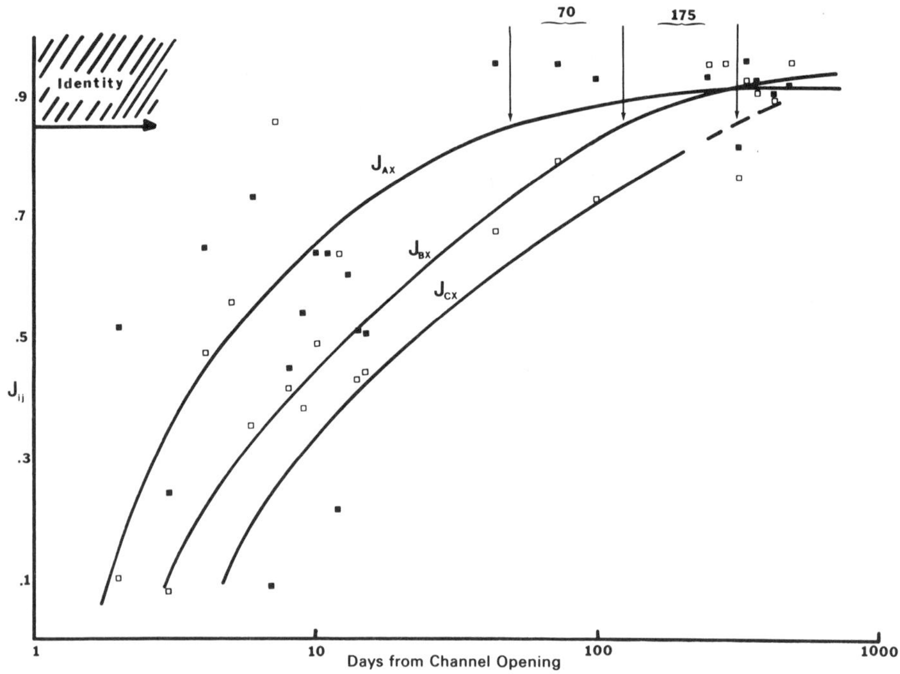

(a)

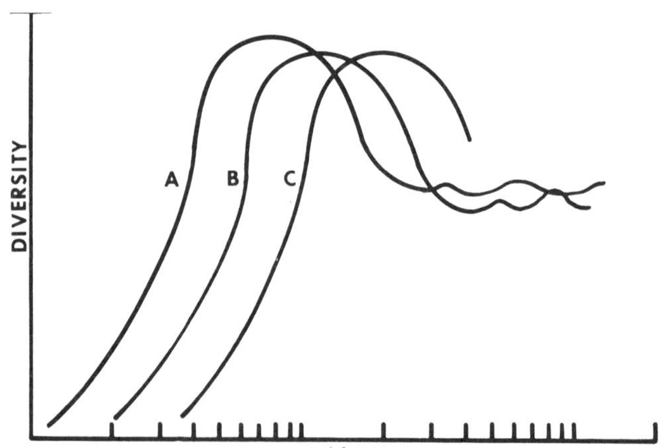

(b)

Figure 4.6(a) Trends in invertebrate colonization as a function of time after restoration has been initiated. Plotted as a function of Jaccard similarity (Jij) between successive points downstream in a restored river channel (A, B, C) and an unimpacted adjacent upstream community. Values greater than 0.85 indicate that the restored community is not significantly different from unimpacted areas. After Gore 1982. Reprinted with permission of *Hydrobiologia.* **(b)** The similarity plots of Figure 4.6a are translated into more standard predictive models similar to colonization curves.

between source area and recipient island substrates are continually being decreased during the colonization period.

Since dispersal mechanisms are a constant dynamic redistributional process (Waters 1972; Townsend and Hildrew 1976), the doubling effect I have described should be considerably dampened in field observations. Nevertheless, the curves in Figure 4.6 do demonstrate persistence in lag times, to some degree, until an equilibrium is reached. These lags can be reduced or altered depending upon the time of reclamation and recolonization. As previously mentioned, drift rates are higher during certain times of the year (highest from June to September; Williams 1980) because of changes in life cycle and activity. The demonstration of the lag times with increased distance serves to point out the necessity of proper planning of restoration projects to obtain the most rapid colonization by invertebrates.

Demonstration that a permanent and stable macroinvertebrate community has been attained is a desirable result of restoration activities. Gore (1979, 1982) and Minshall et al. (1983) have shown that this period of time to stability is fairly long and should not be confused with the amount of time observed to attainment of maximum densities. Even in fairly short sections of denuded or newly constructed stream channels, what might be considered stable communities were established in 600 to 800 days from initiation of colonization or restoration. As predicted by MacArthur and Wilson (1967) and Simberloff (1978) and demonstrated in the models of Sheldon (1984), a period of interspecific competition, stabilization of the resource base, and arrival of rare species occurs subsequent to attainment of maximum densities and prior to the stabilization of the community. Of course, in reclaimed areas of native substrate that has been previously inhabited by invertebrates and associated periphytic material, times may be reduced from the observations of Gore and Minshall et al., but general patterns should be expected to be maintained.

Figure 4.7 depicts the pattern of arrival of invertebrate colonizers by order and by functional group. As may be expected, collectors and grazers were early colonists in apparent synchrony with establishment of periphyton growth and detrital accumulations. The most recent colonizers to new benthic habitat were the "rare" species and the predators. Ecologically, this pattern makes sense and indicates that part of the period of dynamic adjustment after attainment of maximum densities is a period of predation and adjustment of population numbers (Simberloff 1978). This pattern also indicates that a rapid check for attainment of stable benthic communities in the restored channel would be the monitoring and comparison with unstressed areas of the densities and diversities of the major invertebrate predators.

The best monitoring regime for determination of stable communities and restoration success should involve comparisons between entire restored communities and communities in the unstressed source areas. Comparisons of communities by diversity indices (Shannon-Weaver or Simpson's for example) may not be sufficient to detect shifts in dominance or taxonomic structure. I have suggested that the use of an index of community similarity, like Jaccard's coefficient (see Sneath and Sokal 1973), made accurate predictions of establishment of communities similar

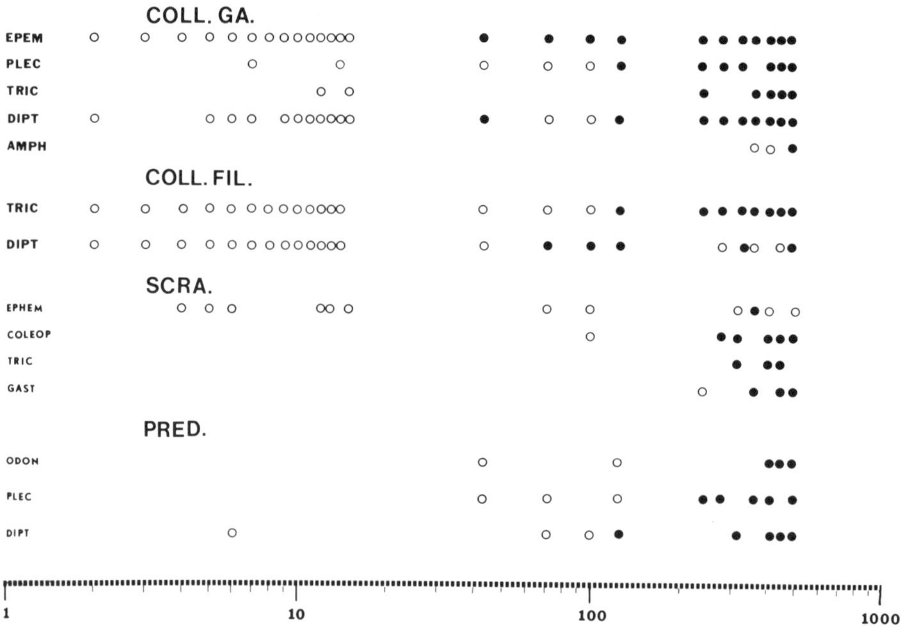

Figure 4.7 Trends in arrival of invertebrate functional groups as a function of time from restoration activity. Open circles indicate presence of the organism. Solid circles indicate presence in numbers comparable to densities in unimpacted communities. COLL. GA. = Collector-gatherers. COLL. FIL. = Collector-filterers. SCRA. = Scrapers. PRED. = Predators. Invertebrate groups examined were the Ephemeroptera (EPEM), Plecoptera (PLEC), Trichoptera (TRIC), Diptera (DIPT), Amphipods (AMPH), Coleoptera (COLEOP), Gastropods (GAST), and Odonata (ODON). After Gore 1982. Reprinted with permission of *Hydrobiologia.*

to undisturbed adjacent stream sections (Gore 1982). This type of analysis was sufficiently sensitive to detect short lag times between establishment of spring communities even after 200 days of colonization. Alternative methods for these comparisons which not only include measure of changes in taxonomic structure but assess attainment of comparable water quality are the various biotic indices, particularly those proposed by Hilsenhoff (1977) and Winget and Mangum (1979). Winget (Chapter 6 this book) has reviewed many of these methods in detail.

SUMMARY

The restoration of stable and viable benthic macroinvertebrate communities is an integral component of the effective restoration of a stream or river ecosystem. It is a particularly critical component if the benthos are known to supply a major portion of the food base for fish populations of the system. Of all of the requirements for reclamation and restoration of lotic ecosystems, benthic community restoration and recovery required the smallest amount of capital investment and least sophisticated of special structure development. The benefits to maintenance of higher trophic levels in the systems are well worth that investment. Only a few procedures need be followed:

1. Measurement and placement of substrate similar to the undisturbed source area of invertebrate colonizers
2. Placement of structures to control increased sediment loads
3. Placement of structures to ensure maintenance of stable and productive riffle areas
4. Monitoring of benthic recolonization
5. Assessment of restoration success by demonstration of stable communities similar to those in adjacent undisturbed source areas

REFERENCES

Allan, J.D. 1981. Determinants of diet of brook trout (*Salvelinus fontinalis*) in a mountain stream. *Can. J. Fish. Aquat. Sci.* 38:184–92.

Anderson, N.H. 1966. Depressant effect of moonlight on activity of aquatic insects. *Nature* 5020:319–20.

Bishop, J.E., and H.B.N. Hynes. 1969. Upstream movements of the benthic invertebrates in the Speed River, Ontario. *J. Fish. Res. Bd. Canada* 26:278–98.

Bishop, M.B. 1980. Geomorphic concepts and their application to ephemeral stream channel reclamation. In *Proc. 2nd Wyoming Mining Hydrology Symp.,* 249–61. University of Wyoming, Laramie.

Bohle, V.H.W. 1978. Beziehungen zwischen dem nahrunhsangebot, der drift und der raumlichen verteilung bei larven von *Baetis rhodani* (Pictet) (Ephemeroptera: Baetidae). *Arch. Hydrobiol.* 84:500–25.

Brusven, M.A. 1970. Drift periodicity and upstream dispersion of stream insects. *J. Ent. Soc. Brit. Col.* 67:48–59.

Burgess, S.A. 1985. Some effects of stream habitat improvement on the aquatic and riparian community of a small mountain stream. In *The Restoration of Rivers and Streams,* edited by J.A. Gore. Stoneham, MA: Butterworth Publishers pp. 223–46.

Cadwallader, P.L., and A.K. Eden. 1977. Effect of a total solar eclipse on invertebrate drift in Snobs Creek, Victoria. *Aust. J. Mar. Freshw. Res.* 28:799–805.

Cairns, J., Jr., J.S. Crossman, K.L. Dickson, and E.E. Herricks. 1971. The recovery of damaged streams. *ASB Bull.* 18:79–106.

Cherry, D.S., S.R. Larrick, R.K. Guthrie, E.M. Davis, and F.F. Sherberger. 1979. Recovery of invertebrate and vertebrate populations in a coal ash stressed drainage system. *J. Fish. Res. Bd. Canada* 36:1089–96.

Ciborowski, J.J.H. 1979. The effects of extended photoperiods on the drift of the mayfly *Ephemerella subvaria* McDunnough (Ephemeroptera: Ephemerellidae). *Hydrobiologia* 62:209–14.

Cowell, B.C., and W.C. Carew. 1976. Seasonal and diel periodicity in the drift of aquatic insects in a subtropical Florida stream. *Freshw. Biol.* 6:587–94.

Crisp, D.T., and T. Gledhill. 1970. A quantitative description of the recovery of the bottom fauna in a muddy reach of a mill stream in Southern England after draining and dredging. *Arch. Hydrobiol.* 67:502–41.

Cummins, K.W. 1979. The natural stream ecosystem. In *The Ecology of Regulated Streams,* edited by J.V. Ward and J.A. Stanford, 7–24. New York: Plenum Press.

De Silva, S.S., P.R.T. Cumaranatunga, and C.D. De Silva. 1980. Food, feeding ecology, and morphological features associated with feeding of four co-occurring cyprinids (Pisces: Cyprinidae). *Neth. J. Zoology* 30:54–73.

De La Cruz, A.A., and H.A. Post. 1977. Production and transport of organic matter in a woodland stream. *Arch. Hydrobiol.* 80:227–38.

Elliott, J.M. 1967. The food of trout (*Salmo trutta*) in a Dartmoor stream. *J. App. Ecol.* 4:59–71.

———. 1973. The food of brown and rainbow trout (*Salmo trutta* and *Salmo gairdneri*) in relation to the abundance of drifting invertebrates in a mountain stream. *Oecologia* 12:329–47.

———. 1975. Number of meals in a day, maximum weight of food consumed in a day, and maximum rate of feeding for brown trout (*Salmo trutta,* L.) *Freshw. Biol.* 5:287–303.

Eriksen, C. H. 1964. Benthic invertebrates and some substrate-current-oxygen interrelationships. *Pymatuning Symp. Ecol.* 4:98–115.

Fisher, S.G., L.J. Gray, N.B. Grimm, and D.E. Busch. 1982. Temporal succession in a desert stream ecosystem following flash flooding. *Ecol. Monogr.* 52:93–110.

Gore, J.A. 1977. Reservoir manipulations and benthic macroinvertebrates in a prairie river. *Hydrobiologia* 55:113–23.

Gore, J.A. 1978. A technique for predicting in-stream flow requirements of benthic macroinvertebrates. *Freshw. Biol.* 8:141–151.

———. 1979. Patterns of initial benthic recolonization of a reclaimed coal strip-mined river channel. *Can. J. Zool.* 57:2429–39.

———. 1982. Benthic invertebrate colonization: source distance effects on community composition. *Hydrobiologia* 94:183–93.

Gore, J.A., and L.S. Johnson. 1980. *Establishment of Biotic and Hydrologic Stability in a Reclaimed Coal Strip-mined River Channel.* Inst. of Energy and Env., University of Wyoming, Laramie.

———. 1981. Strip-mined river restoration. *Water Spectrum* 13:31–38.

Gore, J.A., and R.D. Judy, Jr. 1981. Predictive models of benthic macroinvertebrate density for use in in-stream flow studies and regulated flow management. *Can. J. Fish. Aquat. Sci.* 38:1363–70.

Gray, L.J., and S.G. Fisher. 1981. Postflood recolonization pathways of macroinvertebrates in a lowland Sonoran Desert stream. *Amer. Midl. Nat.* 106:249–57.

Hanson, J.M., and S.V. Qadri. 1979. Seasonal growth, food, and feeding habits of young-of-the-year black crappie in the Ottawa River. *Can. Field Nat.* 93:232–38.

Harrison, A.D. 1966. Recolonization of a Rhodesian stream after drought. *Arch. Hydrobiol.* 62:405–21.

Hart, D.D. 1978. Diversity in stream insects: regulation by rock size and microspatial complexity. *Verh. Internat. Verein. Limnol.* 20:1376–81.

Hasfurther, V.R. 1985. The use of meander parameters in restoring hydrologic balance to reclaimed stream beds. In *The Restoration of Rivers and Streams,* edited by J.A. Gore. Stoneham, MA: Butterworth Publishers pp. 21–40.

Hilsenhoff, W.L. 1977. *Use of Arthropods to Evaluate Water Quality of Streams.* Tech. Bull. No. 100. Madison: Wisconsin Dept. Nat. Res.

Hughes, D.A. 1966a. On the dorsal light response in a mayfly nymph. *Anim. Behav.* 14:13–16.

______. 1966b. The role of responses to light in the selection and maintenance of microhabitat by the nymphs of two species of mayfly. *Anim. Behav.* 14:17–33.

Hynes, H.B.N. 1970. *The Ecology of Running Waters.* Toronto, Ontario: Univ. of Toronto Press.

Khalaf, G. and H. Tachet. 1977. Les dynamiques de colonisation des substrats artificiels par les macroinvertebred d'un cours d'eau. *Annls. Limnol.* 31:169–90.

Larimore, R.W., W.F. Childers, and C. Heckrotte. 1959. Destruction and reestablishment of stream fish and invertebrates affected by drought. *Trans. Amer. Fish. Soc.* 88:261–85.

Luedtke, R.J., M.A. Brusven, and F.J. Watts. 1976. Benthic insect community changes in relation to in-stream alterations of a sediment-polluted stream. *Melanderia* 23:21–39.

MacArthur, R.H., and E.O. Wilson. 1967. *The Theory of Island Biogeography.* Princeton, N.J: Princeton Univ. Press.

McAuliffe, J.R. 1983. Colonization patterns, competition, and disturbance in stream communities, pp. 137–156. In *Stream Ecology: the Testing of General Ecological Theory in Stream Ecosystems,* edited by J.R. Barnes and G.W. Minshall. N.Y.: Plenum Press.

McLay, C. 1970. A theory concerning the distance traveled by animals entering the drift of a stream. *J. Fish. Res. Bd. Canada* 27:359–70.

Meier, P.G., D.L. Penrose, and L. Polak. 1979. The rate of colonization by macroinvertebrates on artificial substrate samplers. *Freshw. Biol.* 9:381–93.

Merritt, R.W., and K.W. Cummins. 1978. *An Introduction to the Aquatic Insects of North America.* Dubuque, Iowa: Kendall/Hunt Publ. Co.

Minshall, G.W., D.A. Andrews, and C.Y. Manuel-Faler. 1983. Macroinvertebrate recolonization of the Teton River, Idaho: further evidence for the application of island biogeographic theory to streams, pp. 279–297. In *Stream Ecology: The Testing of General Ecological Theory in Stream Ecosystems,* edited by J.R. Barnes and G.W. Minshall. N.Y.: Plenum Press.

Minshall, G.W., and J.N. Minshall. 1978. Further evidence on the role of chemical factors in determining the distribution of benthic invertebrates in the River Duddon. *Arch. Hydrobiol.* 83:324–355.

Nichols, F.H. 1980. Long-term stabilizing influence of "rare" species on benthic community structure in Puget Sound, Washington. 43rd Ann. Mtg. Amer. Soc. Limnol. Oceanogr., Knoxville, TN (presented paper).

Otto, C., and B.W. Svensson. 1976. Consequences of removal of pupae for a population of *Potamophylax cinbulartus* (Trichoptera) in a South Swedish stream. *Oikos* 27:40–43.

Peckarsky, B.L. 1979. Biological interactions as determinants of distributions of benthic invertebrates within the substrate of stony streams. *Limnol. Oceanogr.* 24:59–68.

______. 1980. Predator-prey interactions between stoneflies and mayflies: behavioral observations. *Ecology* 61:932–43.

Ruggiero, M.A., and H.C. Merchant. 1979. Water quality, substrate, and distribution of macroinvertebrates in the Patuxent River, Maryland. *Hydrobiologia* 64:183–89.

Sheldon, A.L. 1977. Colonization curves: application to stream insects on seminatural substrates. *Oikos* 28:256–61.

———. 1984. Colonization dynamics of aquatic insects. In *The Ecology of Aquatic Insects: a Life History and Habitat Approach,* edited by V.R. Resh and D.R. Rosenberg. N.Y.: Prager.

Simberloff, D. 1978. Using island biogeographic distributions to determine if colonization is stochastic. *Amer. Nat.* 112:713–26.

Sneath, P.H.A., and R.R. Sokal. 1973. *Numerical Taxonomy.* San Francisco: Freeman.

Starnes, L.B. 1983. Aquatic community response to techniques utilized to reclaim eastern U.S. coal surface mine-impacted streams. In *The Restoration of Rivers and Streams,* edited by J.A. Gore. Stoneham, MA: Butterworth Publishers pp. 193–222.

Starnes, L.B., and W.C. Starnes. 1981. Biology of the blackside dace *Phoxinus cumberlandensis. Amer. Midl. Nat.* 106:360–71.

Stoneburner, D.L., and L.A. Smock. 1979. Seasonal fluctuations of macroinvertebrate drift in a South Carolina piedmont stream. *Hydrobiologia* 63:49–56.

Stream Enhancement Research Committee. 1980. *Stream Enhancement Guide.* Vancouver, British Columbia: Govt. of Canada, Ministry of Environment, Fisheries and Oceans.

Suter, P.J., and W.D. Williams. 1977. Effect of a total solar eclipse on stream drift. *Aust. J. Mar. Freshw. Res.* 28:793–98.

Thompson, C.S. 1980. Effect of stream relocations on benthic macroinvertebrates and fishes in surface mine areas. Paper presented at the Spring 1980 Technical Session, Indiana Chapter, American Fisheries Society.

Tippets, W.E., and P.B. Moyle. 1978. Epibenthic feeding by rainbow trout (*Salmo gairdneri*) in the McCloud River, California. *J. Anim. Ecol.* 47:549–59.

Townsend, C.R., and A.G. Hildrew. 1976. Field experiment on the drifting, colonization, and continuous redistribution of stream benthos. *J. Anim. Ecol.* 45:759–72.

Ulfstrand, S. 1968a. Life cycles of benthic insects in Lapland streams (Ephemeroptera, Plecoptera, Trichoptera, Diptera Simuliidae). *Oikos* 19:167–90.

———. 1968b. Benthic animals in Lapland streams: A field study with particular reference to Ephemeroptera, Plecoptera, Trichoptera, and Diptera Simuliidae. *Oikos Suppl.* 10:1–120.

Vannote, R.L., G.W. Minshall, K.W. Cummins, J.R. Sedell, and C.E. Cushing. 1980. The river continuum concept. *Can. J. Fish. Aquat. Sci.* 37:130–37.

Walton, O.E., Jr. 1978. Substrate attachment by drifting aquatic insect larvae. *Ecology* 59(5):1023–30.

———. 1980. Invertebrate drift from predator-prey associations. *Ecology* 61:1486–97.

Waters, T.F. 1972. The drift of stream insects. *Ann. Rev. Ent.* 17:253–72.

———. 1977. Secondary production in inland waters. *Adv. Ecol. Res.* 10:91–164.

Williams, D.D. 1980. Temporal patterns in recolonization of stream benthos. *Arch. Hydrobiol.* 90:56–74.

Williams, D.D., and H.B.N. Hynes. 1976a. The recolonization mechanisms of stream benthos. *Oikos* 27:265–72.

———. 1976b. Stream habitat selection by aerially colonizing invertebrates. *Can. J. Zool.* 54:685–93.

———. 1977. Movements of benthos during the recolonization of temporary streams. *Oikos* 29:306–12.

Williams, D.D., and J.H. Mundie. 1978. Substrate size selection by stream invertebrates and the influence of sand. *Limnol. Oceanogr.* 23:1030–33.

Winget, R.N. 1985. Methods for determining successful reclamation of stream ecosystems. In *The Restoration of Rivers and Streams,* edited by J.A. Gore. Stoneham, MA: Butterworth Publishers pp. 165–92.

Winget, R.N., and F.A. Mangum. 1979. *Biotic Condition Index: Integrated Biological, Physical, and Chemical Stream Parameters for Management.* U.S. Dept. of Agriculture, Forest Service, Intermountain Region. Washington, D.C.: U.S. Government Printing Office.

Wise, D.H., and M.C. Molles, Jr. 1979. Colonization of artificial substrates by stream insects: influence of substrate size and diversity. *Hydrobiologia* 65:69–74.

CHAPTER 5

Stream Channel Modifications and Reclamation Structures to Enhance Fish Habitat

Thomas A. Wesche

Water Resources Research Center
University of Wyoming
P.O. Box 3067, University Station
Laramie, Wyoming 82071

The process of channel modification has played a major, although not always beneficial, role in the development of this country. Land drainage has been necessary to convert swampland into fertile, productive farmland. Dredging of our stream bottoms has led to the discovery of precious metals and also to the creation of navigable waterways to transport our people and products. As our cities developed, it was often found that the river that provided a ready source for water supply and waste disposal was also likely to periodically carry damaging flood waters. Hence, the need arose to redesign and, in some cases, relocate these streams. Channel realignment has also been necessary in numerous instances to provide suitable bridge crossings and right-of-ways for our highway system. Overgrazing of the riparian communities bordering our rivers and streams has led to possibly more subtle but nonetheless damaging impacts. In total, Arthur D. Little (1973) estimated that by 1972, over 200,000 miles of stream channel had been modified in the United States.

Given the sheer magnitude of such river manipulations and an increasing awareness by the public of the environmental ramifications of such acts, it is little wonder that engineers and biologists find themselves continually debating the pros and cons of channel modification. Whether it is called channelization, improvement, alteration, realignment, or stabilization, there will be definite impacts to the specific stream reach involved as well as possible upstream and downstream effects. Such impacts can be positive or negative depending upon the nature of the modification and the nature of one's interest in the river reach undergoing change. Potentially, the following characteristics of a reach could be altered:

- channel morphology
- channel hydraulics
- sediment erosion and deposition processes
- water quality
- habitat for aquatic biota
- the aquatic biota itself
- aesthetics
- recreation opportunities
- riparian communities
- the biota of the riparian communities

Historically, of the characteristics listed above, consideration was typically given only to the first two, and possibly the third. This has changed dramatically however, in recent years, as the concept of river restoration has become more widespread.

Nunnally (1978) stated, regarding river restoration:

> Many of the detrimental effects of channelization can be avoided, with little compromise in channel efficiency, by employing channel design guidelines that do not destroy the hydraulic and morphologic equilibria that natural streams possess. These guidelines include minimal straightening; promoting bank stability by leaving trees, minimizing channel reshaping, and employing bank stabilization techniques; and, emulating the morphology of natural stream channels.

The underlying tenent of the river restoration approach is that by thorough planning done before modification activity begins, a design simulating that of nature as closely as possible can be developed that not only alleviates the problem causing the needed modification, but also preserves many of the other valued reach characteristics. Too often in the past, the preservation of fish habitat, for example, was given little or no consideration until after the modification was completed. Later, when population levels were found to be declining due to the loss of habitat, attempts were made to artificially increase the carrying capacity of the reach by the addition of a variety of improvement structures. This is not to say that there is no place for structures such as wing deflectors or bank covers in habitat management. Rather, the point is that if proper planning had occurred during the design process, the need for these structures may not have been so great.

From a fisheries standpoint, a most simplistic view of the channelization process and associated impacts could be illustrated by the following flow diagram:

$$\Delta \begin{matrix}\text{Land and/or}\\ \text{Stream Use}\end{matrix} \rightarrow \Delta \begin{matrix}\text{Channel}\\ \text{Morphology}\end{matrix} \rightarrow \Delta\ \text{Hydraulics} \rightarrow \Delta\ \text{Habitat} \rightarrow \Delta\ \text{Population}$$

where Δ = change in
$\rightarrow$ = leads to (5.1)

The key to the river restoration approach is for the habitat biologist to have input into the process prior to a change in channel morphology brought about by modification, rather than after the habitat and population changes have already occurred. Generally, the organization of this chapter will follow the progression shown in the above diagram. After a brief review of the basic in-stream components of fish habitat (for brevity, this will focus on the salmonid family), the impacts of various channel modification activities on habitat diversity will be discussed. The concluding section of the chapter will then concentrate on channel restoration procedures and structures to enhance fish habitat, from a planning aspect as well as from a design and installation approach.

HABITAT COMPONENTS

The four fundamental components of salmonid habitat are acceptable water quality, food-producing areas, spawning–egg incubation areas, and cover. The extent to which each of these components is present in a given stream is dependent upon the stream's physical, chemical, and hydraulic characteristics. To provide a complete habitat, no matter how large or small the stream, requires the proper range of flows *through a suitable channel configuration,* preferably one the stream itself has formed. It is in this regard that channelization activities have the potential to devastate a stream habitat, unless adequate planning and reclamation are carried out.

Following are descriptions of these fundamental habitat components. As water quality considerations are discussed in other portions of the book, these will not be included. Rather, the emphasis is placed on providing general physical/hydraulic descriptions which may be used as design criteria for reclamation structures and practices.

Food-Producing Areas

There are generally two types of habitat exhibited in streams: riffles and pools (Odum 1959). Riffles are characterized by having a greater than average velocity, a less than average depth, and substrates composed of gravel-rubble. Pools are characterized by having a less than average velocity, a greater than average depth, and substrates composed of silt–fine gravel. Of the two, riffles are the primary food-producing areas. Careful examination of the parameters velocity, substrate, and depth will explain the reasons for this.

Velocity

According to Scott (1958) and Allen (1959), velocity is the most important parameter in determining distributional patterns of aquatic invertebrates. These invertebrates (benthos) live in a vertically constricted boundary layer (Pradtl's layer) between the water mass and the stream substrate (Giger 1973). At this level, water velocities would be at or near zero since velocity varies approximately as a parabola from zero at the bottom to a maximum at or near the surface (Linsley et al. 1975). Ambuhl (1959) stated that current velocity becomes important to the benthic invertebrate by governing the rate of oxygen renewal to the boundary layer. Logically, the faster the water current, the faster and more efficient the renewal rate. In fact, Ericksen (1966) felt that water velocity was perhaps of greater significance to respiration than the actual dissolved oxygen content of the water. Ambuhl (1959) showed the importance of velocity by showing that some species of invertebrate die quickly in oxygen-rich still water while living in oxygen-poor running water. Apparently, many of the swift-water invertebrates lack the morphological features and mechanisms which are present on many still-water forms for creating their own currents for respiration. Organisms in rapids communities do exhibit adaptation for maintaining position in swift water. Odum (1959) lists some of the more important adaptations: (1) permanent attachment to the substrate; (2) hooks and suckers; (3) sticky undersurfaces; (4) streamlined bodies; (5) flattened bodies; (6) positive rheotaxis (orient upstream); and (7) positive thigmotaxis (response to touch).

According to Ruttner (1953), the influence of water current is manifested in the quantity of organisms produced per unit area. Increased water velocities increase the exchange rate between the organism and its water supply, thereby promoting respiration and food acquisition (Giger 1973). Eriksen (1966) felt the importance of water current lies in its ability to renew the respiratory environment of forms that do not have the capability to do so for themselves.

Studies have been conducted that relate water velocity to numbers of organisms. Kennedy (1967) found the greatest numbers of organisms associated with velocities of 0.3–0.4 mps, with few invertebrates present in lower velocities. Needham and Usinger (1956) found the highest numbers of Ephemeroptera (mayflies) associated with velocities of 0.4–0.8 mps and Trichoptera (caddisflies) and Diptera (flies) with velocities of about 0.9 mps. Kimble and Wesche (1975), working on a small stream in southeastern Wyoming, determined that mayflies, caddisflies, and Plecoptera (stoneflies) exhibit preferences for mean water velocities greater than 0.15 mps. Based on limited studies in California, Hooper (1973) considered velocities of 0.5–1.1 mps as optimum for food production. Delisle and Eliason (1961) designated food-producing areas as those where current velocities near the bottom ranged from 0.15–0.9 mps. Giger (1973) felt that 0.15 mps was too marginal for food production and defined an ideal range of 0.3–0.6 mps. Regardless of the specific range of velocities, each of these studies points to the fact that food production is greater in riffles than in pools.

Substrate

Substrate size is directly related to water velocity, with larger materials (rubble, boulder) associated with faster currents, and smaller materials (silt, sand) associated with slower currents. The size of the material has been related by numerous investigators to the standing crops of benthic invertebrates. According to Pennak and Van Gerpen (1947) benthic invertebrates decrease in number in the series rubble, bedrock, gravel, sand. Kimble and Wesche (1975) reported a similar decrease in the series rubble, coarse gravel, sand and fine gravel, silt. Sprules (1947) noted that insect emergence decreased over substrates composed of rubble, gravel, muck, and sand. Sprules reported that in general, the diversity of available cover for bottom fauna appears to decrease as the size of inert substrate particles decreases. The majority of research conducted on substrate types has singled out rubble as the most productive. This larger substrate provides the insects with a firm surface to cling to and also provides some protection from the force of the current. These concepts of protection and attachment have been investigated by Ambuhl (1959), Sprules (1947), and Egglishaw (1964).

Depth

According to Kamler and Riedel (1960), water depth influences which habitat benthic organisms prefer. However, the exact influence of depth on food production remains largely unknown. Needham (1934) stated that depth influences the photosynthetic production of invertebrate food by regulating the light intensity. The deeper the water, the less the light penetration, the less the photosynthetic production of food, resulting in a decrease in invertebrate numbers. In their study of Prosser Creek in California, Needham and Usinger (1956) found the majority of organisms in relatively shallow, flat areas, with mayflies and caddisflies in depths of less than one foot. Kennedy (1967) reported the greatest numbers of organisms in depths of 0.08–0.15 m, with decreasing numbers at greater depths. The study by Kimble and Wesche (1975) indicated depth preferences of less than 0.3 m for mayflies, stoneflies, and caddisflies. In general, areas of highest productivity usually occur in trout streams at depths between 0.15 and 0.9 m, provided substrates and velocities are suitable (Hooper 1973).

Other factors thought to influence food production include stream size, light intensity, and stream gradient. However, the governing influence on a stream's invertebrate production is found in the parameters of velocity, substrate size, and depth. These parameters combine in the riffle sections of streams to provide optimal conditions for the majority of invertebrate species. Riffles are thus much more productive than pool areas.

Drift

Velocity and aquatic insects are also closely related in another way, that being in the delivery of food to the fish by the mechanism called *drift,* the movement of the organism downstream by the current. There has been much speculation as to the reasons for drift, some individuals feeling that it is a passive phenomenon while others feel it is an active, voluntary process.

Many investigators (Chapman 1966; Elliot 1967; Mundie 1969; Waters 1969; Everest and Chapman 1972; Good 1974) have shown a positive correlation between water velocity and the quantity of drift. Good (1974), in his studies of runoff on Nash Fork Creek in Wyoming, reported that drift rates increased with discharge through June 22–24, whereafter they decreased to lower levels. According to Waters (1969), water velocity is the major factor influencing the amount of drift, with increasing velocities increasing the drift up to the point of catastrophic conditions.

Other studies have shown increased drift at night (Chapman 1966; Elliott 1967; Chapman and Bjornn 1969; Dill 1969; Everest and Chapman 1972; Good 1974). This is largely explained by the fact that most benthic invertebrates are negatively phototactic, hiding during the daylight and becoming more active during the night. The invertebrates are thus more likely to enter the drift during this period of greater activity, for it is at this time that they become exposed to the current.

Reimers (1957) and Dimond (1967) both found evidence that drift was density related. Dimond found a definite relationship between the number of organisms in the drift and the increased bottom standing crop. This indicated that as the bottom became more crowded, the rate of drift increased. Bovee (1974) offered several interactions which may be involved in this phenomenon: (1) competition for food in dense populations could lead to more searching behavior by the animals, which makes them more susceptible to dislodgement; (2) territorial behavior might be practiced and individuals without territories comprise the drift; (3) high population densities might cause animals to spread into marginal habitats which are prematurely vacated. This last interaction suggests that invertebrates may voluntarily enter the drift to remove themselves from a crowded area. Elliott (1967) and Waters (1969) reported increased drift by certain species of mayfly during very low levels of water velocity. It is thought that this apparent voluntary entrance into the drift is made by the organism to create currents for respiration and is associated with drought conditions. Waters (1969) suggests that this type of drift may also redistribute the organisms to sites of more rapid current.

Regardless of whether the phenomenon of drift is voluntary or catastrophic in nature, the supply of drift available to salmonids has been found to be greater in areas of faster current velocity. Thus, if more food is available, it can be theorized that a fish will require less time and space to obtain his food, his territory size can be reduced, and population densities in a given area can be increased.

Spawning–Egg Incubation Areas

Spawning

Spawning habitat has been defined by numerous individuals (Thompson 1972; Hooper 1973; Hunter 1973; Smith 1973; Sams and Pearson 1963; Reiser and Wesche 1977) who have measured the hydraulic and physical parameters existing in the stream sections utilized by actively spawning fish. Parameters considered include water velocity, water depth, and substrate size. Generally, acceptable spawning areas exhibit water velocities between 0.15–0.9 mps, water depths of 0.15 m or greater, and substrate sizes between 0.6–7.6 cm. To a large degree, fish size will determine if an area is acceptable for spawning, with larger fish being able to dislodge larger substrate and endure swifter currents than smaller fish. According to Hunter (1973), interspecies spawning preferences for salmonids of the same size will be closer than intraspecies requirements for fish of varying size.

Incubation

The successful development of the incubating eggs is dependent upon certain chemical, hydraulic, and physical parameters.

Chemical. The most important chemical parameter relative to incubation is dissolved oxygen. The development of salmonid eggs is directly related to dissolved oxygen, with ever-increasing demands for it as the eggs develop, and a maximum requirement just prior to hatching (Hayes et al. 1951; McNeil, 1964). Recommended incubation flows for the Snake River in Hells Canyon were based on having at least 5.0 mg/l intragravel dissolved oxygen for the spawning salmonids (Thompson 1974). Hatchery operations are best carried out in water having a dissolved oxygen concentration of 7.0 mg/l (Bardach et al. 1972).

Hydraulic. Hydraulic parameters important in comprising a good incubation environment include the percolation rate of water through the spawning gravels, a pool-riffle sequence, and to a large degree, ground water seepage.

Because the percolation of water brings the necessary oxygen to the incubating eggs and removes the metabolic waste materials, the percolation rate influences the length of the incubation period and the relative sizes of new fry (Shumway et al. 1964). This, of course, is dependent on the dissolved oxygen concentration. In two redds with different percolation rates but with the same concentration of dissolved oxygen, conditions for embryonic development may be better in the area with the higher exchange rate of water (Coble, 1961).

A pool-riffle sequence in streams is important in providing cover, resting, and food-producing areas. The interchange area between a pool and riffle provides an excellent spawning environment, with velocities great enough to carry away silt and debris that may clog the redd substrate. In addition, the stream bottom at the lower end of a pool gradually assumes a convex appearance as the riffle area approaches, causing a downwelling of the current into the substrate. The convex nature of the tailspill also causes downwelling of water into the egg nest. Vaux (1962) has shown that both increased permeability and a convex streambed induce downwelling while a concave streambed causes upwelling. This movement of water into the gravel provides a constant supply of oxygen to the eggs and effectively removes metabolic waste materials. In addition, Stuart (1953) has suggested that this downward current may assist the female on the spawning grounds in maintaining her position against the force of the current.

Numerous investigators (Kendall 1929; White 1930; Greeley 1932; Hazzard 1932; Webster and Eiriksdotter 1976; and others) have shown that brown (*Salmo trutta*) and particularly brook trout (*Salvelinus fontinalis*) select spawning sites in areas with ground water seepage. This may also be true of other salmonids. Benson (1953) found a direct relationship between the amount of ground water, size of trout populations, and number of redds. It was thought that ground water would provide a constant flow over the eggs ensuring sufficient dissolved oxygen for development. Also, as ground water temperatures are often warmer than surface waters in the winter, the eggs are protected from freezing conditions and hatching time is reduced. Latta (1969) felt that in years of high ground water levels there is a greater survival of eggs and fry than in years with low levels. Hansen (1975) has shown that brown trout spawn in areas with and without ground water inflow in about equal numbers. Where ground water inflow was present, the brown trout preferred areas of intermediate surface-ground-water mix and avoided areas where strictly ground water flowed. The warmer ground water is able to hold less dissolved oxygen than the surface water and at the same time may cause an increased demand for oxygen by the developing embryo. Hansen (1975) suggested that the major benefit of ground water may be the wide range of hatching dates which ensures the survival and recruitment of new fish.

Physical. Primary physical parameters important in incubation include water temperature and permeability of the substrate.

To a large degree, the rate of egg development is dependent upon water temperatures, with the higher the temperature (to a point), the faster the egg development. For example, brown trout eggs will take 156 days to hatch at a temperature of 1.6°C but only 41 days at 10°C (Leitritz 1969).

The permeability of the substrate surrounding the eggs determines to an extent the percolation rate of water through the redd. This is important in two respects: (1) dissolved oxygen is brought to the developing eggs via the percolating water; and (2) metabolic wastes are removed from the developing eggs via the percolating water. According to McNeil and Ahnell (1964), the permeability of the material is dependent on: (a) density and viscosity of water; (b) porosity of stream bed; and (c) size, shape, and arrangement of solids. Permeability is considered high by McNeil and Ahnell (1964) when the bottom materials contain less than 5% by volume of sands and silts passing through a 0.833 mm sieve, and is low when the materials contain greater than 15%. The nest construction activity of the female, in effect, cleans the substrate of the very fine materials and results in gravel with a high permeability. However, certain land use practices such as timbering, road construction, and overgrazing can result in increased sediment loads being deposited in the interstices of the spawning gravels, reducing the permeability of the substrate and percolation of water to the developing eggs. This, of course, may result in an increase in egg mortality.

Thus, successful reproduction of salmonids depends on the presence in a stream of sufficient areas suitable for spawning that possess environments conducive to egg incubation.

Cover

Cover can be defined as those in-stream areas providing the fish protection from the effects of high current velocities and predation. Cover for fish in streams can be provided by overhanging vegetation, undercut banks, submerged vegetation, submerged objects (stumps, logs, roots, rocks), floating debris, and water turbulence (Giger 1973). The extent to which each of these forms is used is dependent upon species preference, and of course upon its availability in the stream. As was noted by Giger (1973), the cover requirements of mixed populations of salmonids are not easily determined. Shelter needs may vary diurnally (Kalleberg 1958; Allen 1969; Chapman and Bjornn 1969); by fish species (Hartman 1965; Ruggles 1966; Allen 1969; Chapman and Bjornn 1969; Lewis 1969; Pearson et al. 1970; Wesche 1973); and by fish size (Butler and Hawthorne 1968; Allen 1969; Chapman and Bjornn 1969; Wesche 1973, 1974, and 1980).

Overhead cover may consist of overhanging vegetation (trees, grasses, etc.), logs, or undercut banks. Many investigators (Newman 1956; Wickham 1967; Butler and Hawthorne 1968; Baldes and Vincent 1969; Chapman and Bjornn 1969; Lewis 1969) have shown that overhead cover is used by many species of salmonids, brown trout, brook trout, and rainbow trout (*Salmo gairdneri*) which exhibit photonegative behavior. Overhead cover is also utilized by species showing thigmotaxis (desire to be in close contact with an object). Giger (1973) cites another use of overhead cover, that of providing shadow areas along stream margins where water currents are frequently optimal for resting small fish.

Butler and Hawthorne (1968) and Lewis (1969) reported that brown trout utilize overhead cover to a greater degree than rainbow trout. Wesche (1973) determined that brown trout utilize overhead bank cover more so than in-stream rubble-boulder areas and devised a cover rating system whereby cover comparisons can be made on a stream section at different flows or on different stream sections at the same flow. Wesche (1974) began subsequent investigations into the relationship of cover to standing crop.

Submerged cover (e.g., stream substrate, aquatic vegetation, etc.) has been shown to be important in all stages of salmonids, from the newly hatched larva to the adult. Even salmonid eggs rely on submerged cover (substrate) for protection. Hoar et al. (1957) and Hartman (1965) reported that small salmonids recently emerged from the spawning gravel frequently hid under stones. Wesche (1973) noted that brown trout less than 15 cm used in-stream rubble-boulder areas to a greater degree than overhead cover. This suggests, as was noted earlier, that cover selection is in part based on fish size.

Generally, fish will establish a territory around the selected cover type. This tends to spread the fish population throughout the stream system, leading to a more efficient utilization of the food supply (Hunter 1973). It is within this microhabitat that the fish will spend the majority of its time, feeding and resting. According to Hunter, a fish that has succeeded in obtaining a station with favorable feeding conditions usually retains the territory partly by aggressive actions and partly because it grows faster than its neighbors which have less desirable stations. Hooper (1973) states that abundance of suitable cover determines the number of territories, and thus, the fish population.

That cover is indeed important is shown by the effects of its removal from different streams. Elser (1968) found 78% more trout in an unaltered stream section than in an altered section having 80% less cover. Boussu (1954) showed that losses in brush and undercuts from a stream section resulted in subsequent losses in the number and weight of resident trout. Peters and Alvord (1964) noted similar reductions in twelve Montana streams.

IMPACTS OF CHANNEL MODIFICATION ON HABITAT DIVERSITY

From the preceding discussion of habitat components, it becomes obvious that a key characteristic of any productive in-stream habitat is *diversity.* Considering the physical and hydraulic aspects, it is imperative that the proper blend of water depths, water velocities, and substrate types be present, together, to form the necessary food production, spawning-incubation, and cover areas that combine to form the complete habitat. In the case of cover, the composition and vigor of the riparian community is also critical. Having suitable depths and velocities available as spawning habitat for a given species is well and good, but if these depths and velocities are not present over the proper substrate size, the habitat value is diminished. Likewise, a stream reach consisting entirely of riffles and runs may have more than adequate food production and spawning capacity, but may not be capable of holding mature fish due to the absence of pools and the often associated cover. Thus, on a more microscopic level, the diversity of depths, velocities, and substrates must be present to adequately provide for each habitat component; while on a more macroscopic level, the diversity of habitat components must be available to form a complete habitat if the needs of all life stages of a given species or combination of species are to be met.

In recent years, the capability of biologists to obtain some measure of diversity has been greatly enhanced by the development of several methods for habitat assessment. Among these would be the Habitat Quality Index (HQI) of the Wyoming Game and Fish Department (Binns 1976, 1978, and 1979), the Habitat Suitability Index (HSI) models under development by the U.S. Fish and Wildlife Service's Western Energy and Land Use Team (Raleigh 1978 and 1982), the Physical Habitat Simulation (PHABSIM) System models developed by the USFWS Instream Flow Group (Bovee 1978; Bovee and Milhous 1978; and Milhous et al. 1981), and the Cover Rating Method developed by the Wyoming Water Resources Research Institute (Wesche 1980). Models or indices such as these allow for measurement and evaluation of the various habitat components before a river reach is modified and can provide valuable input into the planning process for the modification. Also, the application of such procedures can be used to identify critical habitat features that should be preserved and can facilitate habitat protection, mitigation, and/or postmodification improvement. The PHABSIM models are especially well suited for predicting impacts and also allowing cost-efficiency analysis for habitat improvement.

Following this brief discussion of the components of fish habitat, the need for preserving habitat diversity, and some methods for evaluation, a review of the general impacts of various types of channel modifications on in-stream habitat is appropriate before proceeding to habitat restoration planning, procedures, and structures.

There are numerous types and variations of development activities that can fall under the general heading of channel modification or channelization. For our purposes, mention of four major types of activities which can have major impacts on fish habitat should suffice. These are: channel enlargement, channel realignment, clearing and snagging, and bank stabilization. York (1978) presented the following definitions of these activities:

> *Channel enlargement.* The overall enlargement of channel cross section to increase its capacity to convey water and/or provide drainage.
>
> *Channel realignment.* Includes construction of a new channel and widening and/or deepening of the existing channel where the new alignment coincides with the existing channel. It usually includes straightening the channel to increase its capacity to convey water.
>
> *Clearing and snagging.* The removal of obstructions from the channel and river bank including the removal of vegetation and accumulations of bed material.
>
> *Bank stabilization.* A protective lining of concrete, timber, or large rock over all or part of the river bank to prevent erosion and/or increase the capacity of the channel to convey water.

Typically, the overall goal of most types of channel modification is to increase the discharge the channel can carry within its banks by altering the stage-discharge relationship, thereby facilitating the drainage of runoff water and protecting property located adjacent to the active channel. Inspection of the basic discharge equation reveals the hydraulic parameters that can be modified to realize this increase in discharge capacity:

$$Q = A \times V = (W \times D) \times V \tag{5.2}$$

where Q is discharge, A is cross-sectional area of the channel, W is width, D is depth, and V is velocity. Obviously, the channel cross section is increased by widening and deepening, while an increase in velocity is proportional to an increase in energy gradient (S) and hydraulic radius (R), and inversely proportional to channel roughness (n), as described by Manning's equation:

$$V = \frac{1.49\, S^{1/2}\, R^{2/3}}{n} \tag{5.3}$$

At first glance, this may not appear disastrous from the habitat aspect. Potentially, width, depth, and velocity, hydraulic parameters upon which habitat quality depends, are all increased. However, the reality of the situation can be quite different. While the capacity of the modified channel has been increased to handle peak flood flows, and if we assume for the sake of example that no flow augmentation is occurring, the net result may be that during low flow periods, which for western streams may be 75% or more of the time, water depths and velocities are reduced while only the water width or wetted perimeter is increased. Should depths and velocities be reduced below the minimum levels needed to meet the hydraulic requirements for fish spawning, incubation, resting, passage, and feeding, the habitat losses become obvious.

Numerous case studies have been conducted in recent years to specifically examine the impacts of channelization to aquatic habitats, the populations supported therein, and the bordering riparian zone. A sampling of these include the work of Wydoski and Helm (1980) on low gradient Utah streams; Chapman and Knudsen (1980) on small western Washington streams; Griswold et al. (1978) on warm-water streams in Ohio and Indiana; Schmal and Sanders (1978) on a Wisconsin marsh; Peters and Alvord (1964) and Elser (1968) on Montana trout streams; Barstow (1971) on wetland habitat in Tennessee; Marsh and Waters (1980) on the downstream effects of channelization in southwestern Minnesota; Stern and Stern (1980) on the effects of bank stabilization on smaller streams and rivers; Marzolf (1978) on potential effects of clearing and snagging on stream ecosystems; Maki et al. (1980) on effects on bottomland and swamp ecosystems in North Carolina; Barclay (1980) on impacts to riparian communities in Oklahoma; and Bulkley et al. (1976) on warm-water streams in Iowa.

Possibly even more critical to a fish population than reductions in water depths and velocities can be the reduction in channel roughness inherent in most channel modification projects. Numerous physical and biological factors contribute to roughness. Among these, those that have a direct relationship to habitat diversity include:

- channel configuration
- substrate type
- bank composition
- bank type
- bank configuration
- aquatic vegetation
- riparian vegetation
- snags

Typically, a straight channel reach offers less impedence to flow than does a meandering section. From a habitat perspective, however, the pool and associated undercut banks that may have been formed along the outside of a meander may be critical resting and cover areas for game fish. Likewise, a mixed substrate of cobble and boulders may contribute roughness to a channel, but in doing so provides additional rearing habitat as resting and feeding microhabitat for fish and key habitat for benthic macroinvertebrate production. As previously discussed, the value of undercut banks lined with snags and overhanging riparian vegetation is unquestioned as prime fish cover. However, such a bankside situation increases the overall channel roughness and is often replaced by smooth, clean banks in the modified channel. Aquatic vegetation can also have a good deal of influence on the stage-discharge relationship of a channel reach by increasing roughness while at the same time providing food and shelter for macroinvertebrates and fish life.

In summary then, it should be kept in mind that a major objective of many of the aquatic habitat reclamation treatments described in the next section of this chapter is to add some degree of roughness back into a modified channel. The key to doing this successfully is the combined skills of the habitat biologist, who will assess the habitat needs, and the project engineer, who will be able to assess the degree of reclamation that can be accommodated within the hydraulic constraints of the channel reach in question. Together, a feasible habitat improvement or rehabilitation plan can be designed that has as high a possibility of success as is economically possible.

HABITAT RECLAMATION TREATMENTS

The enhancement of in-stream aquatic habitat for fish life can be accomplished through a variety of approaches. Among these would be included:

- stream flow regulation (e.g., minimum flows, flushing flows, fluctuation control)
- watershed improvement and regulation of land use activities
- riparian management and enhancement
- overall channel design and alignment, including such treatments as installation of in-channel sediment basins, substrate manipulation, and development of artificial meanders
- stream bank stabilization and improvement
- obstruction removal
- biological enhancement, such as the planting of beaver in suitable areas
- water quality enhancement
- construction of spawning facilities
- installation of in-channel structures

The employment of any one or combination of these general approaches depends upon the particular problem at hand and the philosophy of the management agency involved.

As previous chapters in this book have dealt with such topics as riparian revegetation, water quality protection and restoration, applications of geomorphology to river reclamation, the use and design of meander patterns, and substrate manipulation to enhance benthic macroinvertebrate colonization, the primary focus here will be on in-channel reclamation structures and practices, the combination of which we will term *treatments,* to enhance fish habitat. The intent is not to downplay the importance of proper watershed management and avoid the debate which has been ongoing for several years regarding the merits of in-stream structures (Wydoski and Duff 1982; Haugen 1982; Platts and Rinne 1982). Rather, the objective is to provide the reader with information regarding the options available once the overall planning process has been completed and the decision has been made to utilize in-channel treatments. To this end, the discussion will be directed toward descriptions of the various types of structures which have been designed and tested, their intended purpose, an evaluation of their effectiveness in a variety of stream situations and over time, the selection of proper construction materials, and where possible, estimates of cost. To begin, it is first necessary to discuss the proper approach to planning a river reclamation project utilizing in-stream enhancement structures.

Planning

Patrick (1973) states: "Man must remember that the streams and the flood plains have evolved over long periods of time, and have developed a system that is best for the natural conditions at hand. He should study it carefully and make sure that modification follows the dictums of nature."

Such a philosophy should be adopted by anyone planning a channel modification project, whether it be the placement of a single-log dam in a small mountain stream or a lengthy channel relocation project on a major river. The question, of course, becomes, How does one go about determining what the "dictums of nature" are on the particular reach of stream in question, especially in light of the dynamic characteristics of unaltered streams, let alone one that has or is undergoing artificial changes in its watershed as a result of such activities as logging, grazing, road construction, urban development, mining and the like? Assuming an answer is found to the first question, the second question posed must be, How can we best reclaim or enhance this habitat in accordance with these "dictums of nature?" Once the answer is found to our second question, all that remains is the conduct of the appropriate treatments and hopefully, monitoring of these treatments over time to gauge their success or failure.

A good first step in the planning process would be to form an interdisciplinary team to help in addressing the two major questions posed above. Wydoski and Duff (1982) suggest that members of this team represent a variety of disciplines including not only fisheries biology, but also hydrology, hydraulic engineering, and soil science. Other specialists to possibly be included, depending upon the scope of the project, would be a vegetation specialist, someone trained in aesthetic evaluations and considerations, and a water quality specialist.

Once the team has been selected and coordination established, a sound second step would be for each specialist to develop a list of questions which need to be answered regarding their specific discipline and its relationship to the project at hand. Questions raised by the habitat biologists may follow along these lines, as modified from Claire (1980):

1. What fish species are now present?
2. What fish species are being managed for?
3. What are the habitat needs and preferences of these species?
4. What habitat conditions do we have at present?
5. What would be the preferred condition of the habitat?
6. What types of natural habitat do we have in presently unaltered stream sections?
7. How can the characteristics of the desired natural habitat best be restored or duplicated?
8. What habitat treatments are reasonable and practical given the constraints of the project and the limitations of the habitat?
9. What habitat treatments will enhance the habitat?

Generally, the answers to questions 1–6 can be found by the biologist, working independently of the rest of the team, through review of fish and wildlife agency records, field sampling of fish populations, and employment of one or several of the habitat evaluation techniques previously discussed. If the reclamation project is to be successful, questions 7–9 must be answered through close coordination with other team members.

The type of questions raised by other team members, primarily the hydrologist and the engineer, may include some of the following:

1. What are the shape and dimensions of the channel?
2. What are the shape and dimensions of the normal hydrograph through the stream reach to be reclaimed?
3. What are the extremes of flow over the period of record?
4. What are the probabilities of occurrence of various magnitudes of flood and low flows?
5. What is the natural pool-riffle ratio?
6. What is the spacing between successive pools and successive riffles?
7. What is the natural meander pattern and slope of the reach?
8. What is the composition of the stream bed sediments and the stream banks?
9. Is the stream reach, in which the reclaimed section is located, stable?
10. What natural or presently existing hazards to habitat quality (e.g., gully erosion, bank erosion, channel aggradation, channel degradation, pattern migration and change) need to be taken into consideration in the project design?
11. What is the mobility of the stream bed sediments?
12. Is the stream bed capable of enough scour to form pools or must structures be designed and installed to accomplish this?
13. What effect will various types of reclamation structures have on channel roughness and therefore the conveyance capacity of the channel?
14. Based upon the flood frequency analysis, what forces should the structures be designed to withstand for various anticipated flood frequencies?

15. Will the reclamation project affect the soil-water relationship between the stream and the riparian zone?
16. Will revegetation be required?
17. Given the type and condition of the stream bank soils and plant species, what will be the best revegetation plan?
18. To achieve revegetation, will artificial methods such as irrigation, fertilization, and fencing be required?
19. What effects will various treatment options have on the aesthetics of the reach?
20. What effects will various treatment options have on recreational use of the reach?

Certainly, the list of questions that need to be answered is potentially endless. The examples given above are offered as "food for thought" to anyone planning a habitat reclamation project. The specific questions that should be asked for given types of projects will, of course, vary, dependent upon the type of stream to be reclaimed, the extent of the reclamation, and the environmental, sociologic, and economic constraints of the project. The important point to be made, however, is that questions must be asked and answered as best as possible before arriving at the stream armed with shovels, chainsaws, tractors, and the like.

While it would be physically impossible, given the confines of this single chapter, to thoroughly discuss methods available to answer all of the questions raised above, there are several either new or relatively obscure references which should be mentioned that can aid the planning process. Skinner and Stone (1982) discussed the potential of using color infrared aerial photography for identifying ten instream hazards to trout habitat quality. Key indicators are given to assist in the identification process and possible techniques for mitigating each type of hazard are provided. Brooks (1974) and Lu (1975) had some degree of success in predicting the effects of various habitat reclamation treatments on a high mountain stream based upon the physical modeling of the stream reach in a laboratory flume. Cooper and Wesche (1976) designed and field-tested temporary reclamation structures to obtain some measure of the degree of habitat enhancement to be achieved before permanent structures were installed.

Heede (1979) demonstrated how theory and practical experience can be blended to assist hydrologists in predicting impacts of stream restoration projects when field time and on-site data are limited. Also, Heede (1980) discusses stream dynamics and their importance to the nonhydrologist. Such information can be critical to the biologist in the planning process for habitat reclamation projects.

Knowledge of the relative stability of a stream reach can be crucial to the success or failure of a reclamation project. Pfankuch (1975) has developed a quite simple procedure, entitled "Stream Reach Inventory and Channel Stability Evaluation," that allows the investigator to systematically evaluate the resistive capacity of low-order mountain streams to the detachment of bank and bed materials and gain knowledge of the capacity of streams to adjust and recover from potential flow or sediment production alterations. While not specifically designed for use in the siting of habitat reclamation structures, an understanding of the key features included in this index will provide valuable information needed throughout the planning process. Eifert and Wesche (1982) field-tested Pfankuch's procedure and found it to be a reliable indicator of trout habitat quality as well as channel stability. Also closely related to Pfankuch's work, Robinson (1982) has developed and tested the Wyoming Range Stream Inventory method for evaluating stream morphology in the context of user impacts to the riparian zone.

Goodwin (1979) presented a discussion of several empirical and theoretical approaches to determining the stability of artificial channels, while Lane and Foster (1980) developed procedures to predict channel morphology for small streams given changing land use and related this to sediment yield. Again, papers such as these can provide valuable input to the planning process.

An excellent example of the team approach to the successful conduct of a habitat reclamation and enhancement project is described by Brouha and Barnhart (1982). Working in the Upper Browns Creek watershed of northern California, the biologists involved identified the habitat deficiencies and needs, while the hydrologists analyzed the stream flow regimes to determine the needed flood and low-flow frequency information. Based upon these data, expertise in hydraulic engineering was utilized to determine the stream flow forces which reclamation structures would need to withstand. With this as input, the design engineer was able to determine the necessary specifications.

Based upon the results of the questions asked and the answers obtained under step two, the third step of the planning process would be finalization of the habitat reclamation plan. Participation is required here from all team members. Also, it is at this point in the process that care must be taken to ensure that all of the more practical aspects of the proposed project have been considered. Examples of the types of questions that must (or should) be answered include:

1. What is the cost-benefit ratio of what we are proposing to do?
2. What materials and equipment will be needed at the site?
3. What materials are naturally available at the site?
4. What are the personnel requirements for the installation?
5. Given local climatic and hydrologic conditions, when is the best time for installation?
6. Will private landowners need to be consulted regarding trespass rights and other possible matters of concern?

Once these three steps on the planning process have been completed, the next phase is the implementation of the plan. The following section describes the types of in-channel reclamation treatments which have been designed and tested over the years to reclaim and/or enhance fish habitat.

The final phase of the process, one which is often overlooked, is the careful monitoring and evaluation of the project following implementation. As stated by Wydoski and Duff (1982): "One of the most unfortunate and surprising features of stream habitat improvement has been the almost complete lack of quantitative evidence to base conclusions on the ultimate results."

Little guidance is given by the literature in regard to the proper timing of evaluations. Common sense dictates that the logical time for initial analysis of the hydraulic changes brought about by and the physical stability of the structures themselves would be immediately following the first ice-out and high water condition. Also critical could be an evaluation of conditions following a long period of uninterrupted low flow. Valuable information could be gathered regarding sediment deposition patterns and their possible short-term impacts on the habitat. Subsequent hydraulic and structural evaluations could be conducted following hydrologic events of unusual magnitude, both from the high-flow and low-flow aspects.

Regarding the evaluation of biologic response to channel reclamation, Gore and Johnson (1980) found that maximum densities of benthic macroinvertebrates were achieved in a completely reclaimed reach of the Tongue River in northern Wyoming within 70 to 100 days, while a stable macroinvertebrate community was attained within 300 days of water being turned into the new channel. By the end of the study (approximately one year after reclamation was completed), there was no indication that fish populations had reached maximum densities nor that a stable community had been attained. The authors suggested that if fish having a small home range, such as rock bass, are members of the local community, their presence in reclaimed habitat be used as an indicator of a stable community. Working with brook trout populations in Lawrence Creek, Wisconsin, Hunt (1976) found that the biomass of fish over 15 cm and production increased significantly during the first three years following habitat improvement. However, the greatest population responses were recorded during the second three-year period. Hence, we can tentatively conclude from these studies that while the evaluation of macroinvertebrate response can probably be done within one year of reclamation, to obtain a clear picture of the effect on fish populations, it may be best to wait at least four years.

While it is not always possible given the time, staffing, and budgetary restraints of resource management agencies, it is strongly recommended that whenever and wherever possible, the postmodification evaluation be conducted. Only through such research will we be able to learn from our past mistakes and make progress in the future.

Applications, Construction, Installation

A review of the published literature and in-house agency reports indicates that the most commonly used in-channel treatments are current deflectors, overpour structures such as dams and weirs, bank covers, and boulder placements. Other less commonly applied reclamation treatments that have been used successfully (and unsuccessfully) include digger logs, trash catchers, simple gabions, substrate manipulation, pool excavation, channel blocks/barriers, and beaver management. As one proceeds through this section, it should be kept in mind that history has clearly shown us that there is no need for a great variety of structures for stream habitat reclamation (Duff 1982). As there are really only several things that can be done to a stream or river to make it more accommodating for fish, an experienced person who pays heed to the "dictums of nature" and has an understanding of fish habitat can accomplish a great deal with only a relatively few simple types of deflectors, dams, and shelters.

Deflectors

Current deflectors have historically been one of the most commonly used in-channel treatments to improve fish habitat. In general, they are relatively easy to construct, inexpensive, easily modified to suit on-site conditions, built from a variety of materials, applicable to a wide range of stream sizes, adaptable for use with other treatments, multifunctional, and when properly planned, designed, and constructed, successful in improving habitat. Deflectors have and can be built with a variety of purposes in mind, including:

- directing current to key locations such as bank covers
- assisting in the development of meander patterns within the confining banks of channelized reaches
- deepening and narrowing channels
- scouring pools
- increasing water velocities
- removing silt from spawning gravels and critical areas for benthic invertebrate production
- protecting stream banks from erosion
- serving as barriers to keep flow out of side channels, thereby consolidating low flows
- encouraging development of riparian vegetation by means of silt bar formation
- helping to keep water temperatures cool
- enhancing pool-riffle ratios

As with any in-channel treatment we will discuss, there have been both successful and unsuccessful applications of current deflectors. White and Brynildson (1967) state that deflectors may be the best all-around devices for modifying stream channels, while Nelson et al. (1978) echo this view. Saunders and Smith (1962) reported that one year after the installation of deflectors and dams on Hayes Brook, the number of age I and older brook trout had doubled in the modified reach. Shetter et al. (1946) found that five years after twenty-four current deflectors were installed on Hunt Creek (Michigan), the number of good quality pools had increased from nine to twenty-nine, average pool depth had increased by 0.5 feet, and additional spawning gravel had been exposed. The authors concluded that the brook trout fishing had improved as a result of the in-channel devices. Hale (1969) altered over two miles of the West Branch of Split Rock Creek in Minnesota with a combination of deflectors, shelters, and dams. After three years of postinstallation evaluation, he concluded that: (1) the artificial alteration had improved the carrying and reproductive capacity of the reach for trout; (2) the alteration was cost effective when compared with stocking; and (3) the deflectors had their greatest effect on the substrate, increasing the percent of exposed gravels in the reach from 14 to 24%. Evaluation of rock jetties (deflectors) sixteen years after installation five channel widths apart on a channelized reach of Little Prickley Pear Creek (Montana) indicated that pool frequency, size, and depth were comparable to that found in unaltered sections, and that rainbow trout populations had been enhanced (Lere 1982). Tarzwell (1932 and 1936) found benthic invertebrate production in a Michigan stream to have increased 4- to 9-fold following deflector installation. Cause for the increase was attributed to silt bar formation below the structures and the subsequent development of beds of aquatic vegetation. Cooper and Wesche (1976) had success increasing brown trout habitat on a mountain stream impacted by dredging activity in southeast Wyoming through the installation of a 50-meter-long gabion deflector and flow consolidator. This structure functioned as an extension of a point bar which had developed in the widened channel and directed low flow to the opposite stable bank where artificial overhangs had been installed to provide cover. The massive structure functioned well during the first low-flow season and survived the subsequent ice-out and spring runoff. Unfortunately, beaver activity in the reach inundated the structure during its second summer and long-term evaluation could not be carried out.

While there have been numerous successful installations of current deflectors, there also have been failures. Rollefson and Erickson (1970) reported that following installation of a double-wing deflector on Cranney Creek (Wyoming), some silt was displaced and the channel was slightly deepened between the wings. However, the deposition which occurred immediately downstream from the structure negated any habitat gains. Lewis (1974) and Maughan et al. (1978) found that on West Virginia streams gabion deflectors were quite susceptible to damage and needed frequent repair. Knox (1982) found that of twenty-three log deflectors built in a channelized reach of Ten Mile Creek in Colorado, only five produced even marginal habitat while eighteen were of no value. Rock deflectors were also installed and one-half survived the flood flows of up to 31 m^3/sec, providing fair habitat. On portions of channelized reaches on the St. Regis River (Montana), Lere (1982) found the stream to be too steep and confined for rock jetties. Flood flows had reduced the ability of certain structures to concentrate the flow, while the pools formed provided little trout cover. Probable cause for the lack of success was attributed to the structures being spaced only two channel widths apart. Ehlers (1956) evaluated five log deflectors eighteen years after they were installed in a California stream. Three were still in excellent shape, while two had failed due to end- and undercutting.

A most critical step in the installation process is the selection of the proper in-channel location for placement of the structure. While the final siting decision will rely on the knowledge and good judgment of the project participants, much can be learned from the past experiences of others working in the field. The following list of current deflector siting criteria have been drawn from such examples:

Deflectors can be successful on streams of various size and are not limited to only smaller streams (Seehorn 1982).

Typical placement is in wider, shallow, lower gradient stream sections lacking pools and cover (Seehorn 1982).

Locations where the gradient exceeds 3% should be avoided (U.S. Forest Service 1969).

Either avoid reaches having vast flow fluctuations or design low-profile structures geared for the low-flow channel (U.S. Forest Service 1969; Cooper and Wesche 1976).

Don't build at the head of riffles as this may cause damming of the stream (White and Brynildson 1967; Nelson et al. 1978).

Avoid reaches that carry a great deal of debris as this may result in clogging or damming (Claire 1980).

The bank opposite the deflector must be stable (Claire 1980).

Avoid areas having a soft and/or unstable substrate (Barton and Winger 1973).

In straight reaches, alternating deflectors spaced 5–7 channel widths apart can produce a natural sinuous pattern of flow (Nelson et al. 1978; Lere 1982).

Avoid steep, high, eroded banks unless the plan calls for stabilizing the entire height of the bank (Seehorn 1982).

Be certain that bank conditions are such that it will be possible to anchor the ends of the structure 1.2–1.8 m into the bank (Seehorn 1982).

Point bars can provide an opportunity for constructing low-profile gabion or rock deflectors that consolidate and direct flow to the opposite stable bank (Cooper and Wesche 1976).

Greater cost efficiency can be achieved if natural materials (logs, rocks) are available at the site (Seehorn 1982).

If the outside bank is stable, a deflector placed on the inside of a bend can enhance a marginal pool (Wisconsin Dept. Natural Resources 1980; Seehorn 1982).

Avoid constricted channels having a high transport capability (Bailey 1982).

Current deflectors are generally quite easy to construct and are typically built of various combinations of logs, rocks, boulders, gabions, and wire mesh. Figure 5.1 illustrates the design of three simple types, the log-boulder deflector, the gabion deflector, and the *V*-type deflector. Construction details are provided. Figure 5.2 shows a simple rock-boulder deflector. The combination of the two deflectors at the upstream end, in Figure 5.1, forms a double-wing deflector, a modification which will be discussed later. Figure 5.3 provides a view of the habitat

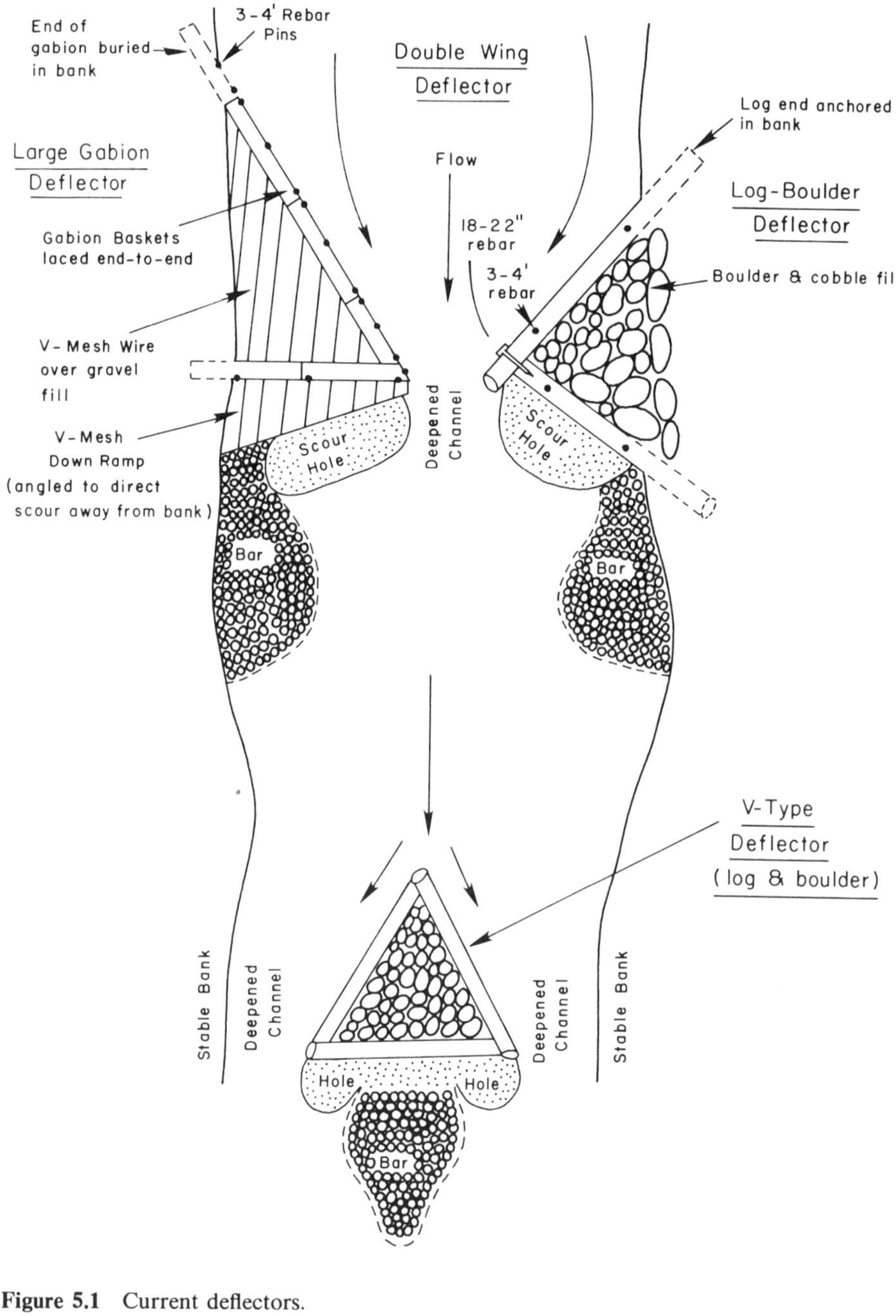

Figure 5.1 Current deflectors.

Figure 5.2 Log and boulder deflector with riprap stabilization on opposing bank.

that can be created by construction of a low profile gabion deflector on the streamward side of a long (50 m) point bar.

Several important characteristics of any current deflector that must be given consideration before the actual construction begins are the shape of the structure, its height, the angle of the deflector, the length it will extend into the channel, and the materials to be used. Regarding *shape,* several forms have been considered over the years, the most common being the peninsular wing (jetty) and the triangular wing (Figure 5.2). White and Brynildson (1967) recommend the use of this latter form because it reduces the tendency for erosion of the bank and bed behind the structure during high flow. Structure *height* is generally dictated by the elevation of the water surface at low flow. To avoid excessive damage to the structure itself and the opposite bank during high flow, the structure should not extend more than 0.15–0.3 m above the low-flow elevation (Seehorn 1982; Cooper and Wesche 1976; White and Brynildson 1967).

Typically, deflectors are *angled* downstream at approximately 45° from the current, while the back brace log is set at approximately 90° to the deflector log (Seehorn 1982; Swales 1982). Of course, these angles can vary depending upon the specific requirements of the project at hand. For setting the proper angle, Cooper and Wesche (1976) found temporary deflectors, made of hinged planks

(a)

(b)

Figure 5.3 **(a)** Low flow habitat before deflector-barrier construction. **(b)** Gabion cells in place, ready to be filled. **(c)** Completed gabion deflector-barrier.

(c)

Figure 5.3 *(concluded)*

and sand bags, to be quite helpful before the permanent structures were installed. Regarding the angle of the brace (downstream edge of deflector), the important consideration is that water overtopping the structure must be directed toward the stream, not the bank.

As with deflector angle, the *distance* that the deflector extends into the channel will vary from site to site, depending upon the specific results desired. In southeastern streams, Seehorn (1982) found that in most cases, the stream width had to be narrowed 70 to 80% to achieve good results. It should be noted, however, that the opposite banks were stabilized with a cover log that controlled the amount of erosion. Swales (1982) was successful on a small lowland river in England extending his deflectors one-third to one-half the distance across the channel. For general planning purposes then, a figure in the 50% range would probably be appropriate. On-site knowledge of relative bank stability, substrate size and compaction, and design flow and associated hydraulic characteristics would be necessary to determine more exact lengths.

Rock-Boulder Deflector. Probably the easiest type of current deflector to construct is the rock or boulder deflector. All that needs to be done is to shovel out a trench in the stream bed in the desired shape and set large interlocking angular boulders in place. The inside of the structure is then filled with smaller material (U.S. Bureau of Fisheries 1935). Claire (1980) recommends that dense, angular, nonerodible rock from 4- to 30-inch diameter be used, with at least 50% ranging from 6- to 24-inch diameter. For constructing rock jetties in Montana streams, Lere (1982) found that Type A riprap (at least 80% by weight has volume of at least 0.03 m^3) did poorly, while Type B riprap (at least 40% of total volume of rock of at least 0.11 m^3) was still performing successfully after sixteen years.

One of the most stable types of deflectors is the log-boulder structure illustrated in Figure 5.2. Installation instructions are as follows (modified from Wyoming Game and Fish Dept. 1982a; and Seehorn 1982):

1. Notch the ends of the posts and drill all holes on the bank for reasons of safety. If preferred instead of notching, a 0.4–0.6-m-long piece of rebar can be driven through the logs to connect them at the point.
2. Structures can be built one or two logs high, depending on specific needs. Seehorn recommends logs at least 0.3 m in diameter be used.
3. Dig the lower posts into stream bed about halfway.
4. Anchor logs 1.2–1.8 m into the bank, if possible.
5. Drive two 0.9–1.2-m-long pieces of rebar in each post. To provide added strength and prevent undercutting, the Wyoming Game and Fish Department recommends 2 or 3 additional lengths of rebar be driven through the point.
6. Nail with ring-shank nails.
7. Fill with rock and, if desired, cover with sod.

Gabions. The use of gabions for construction of in-channel structures has been debated for quite some time. The major objections to their use have been for reasons of aesthetics (U.S. Forest Service 1952) as well as their susceptibility to damage and frequent need for repair (Lewis 1974). Maughan et al. (1978) found that on Jennings Creek in the Southeast, repairs were needed every four to six years to replace broken wire sections and refill rock that had been lost. On the positive side, Johnson (1967) found that on Straight Creek in Montana the habitat provided by gabion deflectors was at least as good as natural habitat and that the undercuts formed under gabions as they twisted or settled provided good cover. Cooper and Wesche (1976) found that low-profile gabion structures were effective, easy to install, strong enough to withstand high discharge and fairly inexpensive. Also, when used on relatively wide, cobble-bottomed streams with well-developed point bars, the gabions tended to blend in quite well after they had been in place long enough to trap small debris particles.

Cooper and Wesche (1976) describe the step-by-step procedures for constructing the gabion deflector shown in Figure 5.3 as follows:

1. Using shovels, the stream bed was levelled where the deflectors would be.
2. Six 2.0 m x 0.8 m x 0.15 m gabion cells were cut from the larger factory gabions and placed end-to-end, in place, and laced to each other.
3. Where the ends of the structure abutted the banks of the high-flow channel,

a trench was dug back into the bank 0.3 m in order to anchor the structure and protect against erosion. Additional anchoring would be preferred, if possible.

4. The gabion cells were then filled with relatively clean cobble (0.1 m diameter). It is best to remove any sand or fine gravels before filling, as these smaller materials will wash out during high flow and cause sag.
5. Cells were filled level full, packed, then filled again until they were moderately rounded on top.
6. Lids were then laced on tightly. It was found that lacing lids to one edge of the basket before filling reduced overall lacing time because lacing through the packed cells was quite tedious.
7. After the cells were filled and closed, the area enclosed by the gabions was filled with cobble and covered with *V*-mesh fencing wire. This wire was secured on two sides to the gabion cells and on the bank side with rebar pins.
8. To resist unwanted scour, a downramp of *V*-mesh wire and cobble was then added to the downstream edge of the gabion.
9. To stabilize the streamside edge of the cells, rebar pins were driven through the filled gabions into the stream bed.

It is recommended that plastic-coated gabions be used, as galvanized material was found to oxidize readily. Expected life of the plastic-coated gabions was expected to be in excess of fifteen years.

Double-wing Deflector. While the three basic types of deflectors described above are the "standards of the industry," so to speak, several variations have been developed and tested which on certain occasions have proven successful. The most common of these would be the double-wing deflector, as illustrated in Figure 5.1. Very simply, this type consists of two current deflectors opposite each other in the stream channel. Construction of each is the same as that described above. Seehorn (1982) stated that the double-wing deflector is suitable for use in larger streams (greater than 30 ft wide), should narrow the stream by approximately 80% at the apex to be effective, is well suited to shallow sections where the gradient may be too steep for single deflectors and where the banks are too low to build a check dam, and costs less to build and maintain than a check dam. The U.S. Bureau of Land Management (1968) is somewhat more conservative and recommends the stream should not be narrowed more than 50%. The Wyoming Game and Fish Dept. (1982a) reports that results have been mixed with double-wing deflectors in scouring holes and that in smaller streams, formation of unwanted debris jams could be a problem.

Underpass Deflector. A second variation is the underpass deflector, or digger log, a very simple structure suitable for use on small streams. The purpose is to "blow out" silt and soft bottom materials, thereby scouring a pool (Everhart et al. 1975). Construction consists of setting the main log across the channel several centimeters up off the bottom. The ends are anchored well into the bank and braces on the downstream side can be added for additional stability. An extra benefit is that overhead cover is provided as well. As with double-wing deflectors, a major problem is formation of debris jams. Also, bank erosion and formation of unwanted silt bars downstream can reduce the effectiveness of underpass deflectors.

Current deflectors are often used in combination with other in-channel treatments. Cooper and Wesche (1976) and Seehorn (1982) report using deflectors to guide the flow to artificial covers placed on the opposite bank, while numerous authors report using them in combination with check dams on smaller streams to increase circulation in pools and enhance pool quality. Also, brush or artificial overhangs can be added to the downstream edge of deflectors to enhance their habitat value. Once the main structure is in place, a bit of imagination can go a long way toward improving even more the fisheries benefits of the project in a very cost-effective manner.

While it is difficult to estimate the costs and personnel requirements involved in constructing current deflectors, the literature does provide some guidance. Seehorn (1982) found that a four to six person crew could install 2 or 3 log and boulder deflectors and associated cover logs per day, while 1.5 to 2 double-wing deflectors could be built per day, depending on stream size. Knox (1982) reported that construction of twenty-two log and rock deflectors on the White River in Colorado averaged about $1500 per structure (1979 dollars), while Shaw (1982) indicated that installation of three *K*-dams and three double log deflectors cost $700 (1963 dollars). The total cost of construction materials for the modification work described by Cooper and Wesche (1976) was $1500 (1974 dollars). This included three gabion check dams, five deflectors of varying size (up to 50 m long), and numerous artificial overhangs. For a crew of three to four, actual construction time varied from less than one day for a small deflector up to five days for the 50 m structure.

Dams

Another often-used reclamation structure is the low-profile dam. Known by a variety of names, including check dam, weir, plunge, and overpour, this type of structure is generally used to create or enhance habitat on small, steep-gradient headwater streams. While possibly not as adaptable as the current deflector nor as easy to construct, dams can be multifunctional, relatively inexpensive, built from a variety of materials, and successful in improving habitat. Low-profile dams have and can be built with a variety of purposes in mind, including:

- deepening existing pools
- creating new pools above and/or below the structure
- collecting and holding spawning gravels upstream
- encouraging gravel bar formation for spawning below the structure
- raising water levels up to culverts to allow fish passage
- improving flow patterns and aiding flow recovery on intermittent streams

- trapping fine sediments on tributaries to prevent their movement into the mainstream
- aerating water
- slowing the current, thereby allowing organic debris to settle out and promote invertebrate production

Parallel to our discussion regarding current deflectors, the literature is dotted with examples of both the success and failure of dams for improving instream habitat. Duff (1982) reviewed the history of stream improvement in the United States and, based upon the trial and error efforts of such groups as the Civilian Conservation Corps, U.S. Bureau of Fisheries, U.S. Forest Service, and U.S. Bureau of Agricultural Engineering during the 1930s, concluded that probably no form of stream improvement has greater possibilities than the use of check dams. Rinne and Stefferud (1982) reported that the artificial pools created by dams in McKnight Creek (New Mexico) were 50 to 70% greater by volume than were natural pools, were 38 to 50% deeper, provided seven times more cover, and held half again as many Gila Trout by numbers and twice the biomass. Gard (1961), in a study comparing three types of dams on the headwaters of Sagehen Creek (California), considered the project successful as brook trout could be held after dam construction in previously barren waters. Rockett and Mueller (1968) reported that the construction of rock check dams was instrumental in creating additional habitat on Bear Trap Creek (Wyoming) which allowed the survival and maintenance of stocked rainbow trout. Spawning activity was found to increase from 1 or 2 spawning pairs per year to 35–40 pairs per year on Patrick Creek (California) following installation of rock and gabion weirs to collect and hold spawning gravels (Bailey 1982). Claire (1978) reported the successful use of log dams to raise water levels up to culverts to allow fish passage. The pools created were found to hold up to five times as many fish as adjacent natural pools. Otis (1974) described low dams as one of the most effective pool forming devices, while Boreman (1974), working on a New York stream, found that juvenile rainbow trout occupied pools formed by log overpours in equal numbers to natural pools. Maughan et al. (1978) evaluated log dams built in two Virginia streams thirty-five years prior and found them to be in excellent condition. The pool area in the improved sections was three to four times greater than in the natural channel, although no statistical difference was found between fish populations. A possible cause for this was higher fishing pressure in the improved reaches.

Certainly not all dam installations have met with the degree of success described above. Leusink (1965) found that the construction of low-elevation gabion dams on a Colorado stream did not impair or improve invertebrate production. Rockett (1979) reported that the installation of nine check dams on a small Wyoming stream did not increase the number of catchable-size fish. While the number of yearlings was increased, their growth rates and condition were reduced. It was felt that the food supply was not adequate for the increased number of yearlings being held. Richard (1963), evaluating seven log dams built on a California stream in 1955, concluded that the structures caused erosion, didn't help the fish population, and should be used only on streams having low to moderate flows where escape cover is needed. Log step-downs were also found to be unsuccessful on Benchmark Creek in Montana because they formed small pools and blocked fish passage (Johnson 1967), while on Ten Mile Creek (Colorado), only nine of thirty-nine log dams were still in fair to excellent condition four years after installation. The remainder were washed out, buried, or nonfunctioning. Of five rock dams built, only one survived (Knox 1982). Hutchinson (1978) reports poor results with gabions installed across the Siuslaw River in Oregon to create pools and collect gravel. The structures were installed over bedrock and were held down with steel pins and cable. A large percentage either washed out or rolled over, while those that remained tended to collect silt and debris rather than gravel.

There are two primary factors which govern the fate of any in-channel structure such as low dams: proper siting and proper construction. Drawing upon the experience gained from past successes and failures such as those described above, the following list of siting criteria has been developed:

- Generally, low dams are successful on smaller (1–9 m wide), high-gradient (0.5–20% slope) headwater streams not susceptible to excessive flood flows (peaks from approximately 2.8 to 5.7 m^3/sec) (Raleigh and Duff 1980; White and Brynildson 1967; U.S. Forest Service 1969; Wyoming Game and Fish Dept. 1982a; Seehorn 1982).
- A good location for placement is in a straight, narrow reach at the lower end of a steep break in the gradient (Seehorn 1982).
- The stream bed substrate should be stable.
- The banks should be stable and well defined.
- It should be possible to anchor both ends of the dam well into the banks (1–2 m).
- Successive structures should be placed no closer than 5–7 channel widths apart (White and Brynildson 1967).
- The reach selected should be pool-deficient.
- Water temperature regime should be such that if current is slowed, no harmful effects will result.
- A site should be selected which allows low dam height (0.3 m) to be maintained to allow fish passage but still enhance pools (Alvarado 1978).
- If passage is blocked, spawning gravels should be present between structures (Claire 1980).
- The availability of natural construction materials can make a project much more economically feasible.
- If heavy equipment is needed, access must be available.

There are three general types of low-profile dams that have been built and tested over the years, including rock-boulder dams, log dams, and gabion dams. The selection of the proper type for given situations will depend upon the specific objectives of the project at hand, the size and flow characteristics of the stream in question, available manpower and equipment, natural materials present at the site, economic constraints, and the desired lifespan of the structure.

Rock-boulder Dams. A typical design for a rock-boulder dam is presented in Figure 5.4. A structure such as this is ideal for very small streams, assuming that some large flat boulders are available as well as a good backhoe operator to set them securely in place (Wyoming Game and Fish Dept. 1982a). A seal is formed by packing finer gravels in front of and between the boulders. If a good seal can be obtained, habitat can be enhanced upstream from the structure in addition to the plunge pool created below. Aesthetically, the structure is natural looking and quite appealing.

Rock-boulder dams are generally quite easy to build and require less construction time than do log or gabion dams. Rockett and Mueller (1968) report building thirty-seven rock dams in only two days, while Gard (1961) averaged only 3.6 hours per structure on a stream ranging from 1.5 to 4.6 m wide.

Problems that have been encountered with rock-boulder dams include:

- difficulty in sealing (Warner and Porter 1960; Gard 1961)
- a lack of stability and durability in high runoff streams (Bender 1978; Gard 1961; Ehlers 1956)
- collapsing back into the plunge pool (Wyoming Game and Fish Dept. 1982a)

Log Dams. As shown in Figure 5.5, log dams can be designed in a variety of configurations, depending primarily upon the desired height and stability, as well as stream size. Over the past fifty years, numerous designs have been tested. The U.S. Bureau of Fisheries (1935) presents construction plans for ten different types of log dams. However, experience has shown that generally one of four types can be applied to most situations. These four are the single log dam, the *K*-dam, the wedge dam, and the plank or board dam (Duff 1982).

Regardless of the specific type of log dam to be built, adherence to the siting criteria given above, as well as to certain general construction criteria, will pay future dividends regarding structure utility and longevity. Following are some construction criteria which may prove beneficial:

- Secure anchoring of the structure is critical to its success. Log ends should be sunk 1 to 2 m into the banks if possible or at least ⅓ of channel width into the bank on both sides (Nelson et al. 1968; Alvarado 1978; Claire 1978).
- Undercutting is a main cause of failure. To reduce this risk, imbed the base log at least 0.15 m into the substrate. If the stream bed is soft and erodible, add a mudsill to the upstream face of the log for added stability, as shown in Figures 5.6 and 5.7 (Duff 1982). The placement of cobbles and boulders along the upper edge is also very beneficial.
- Endcutting is another major cause of failure. In addition to anchoring the

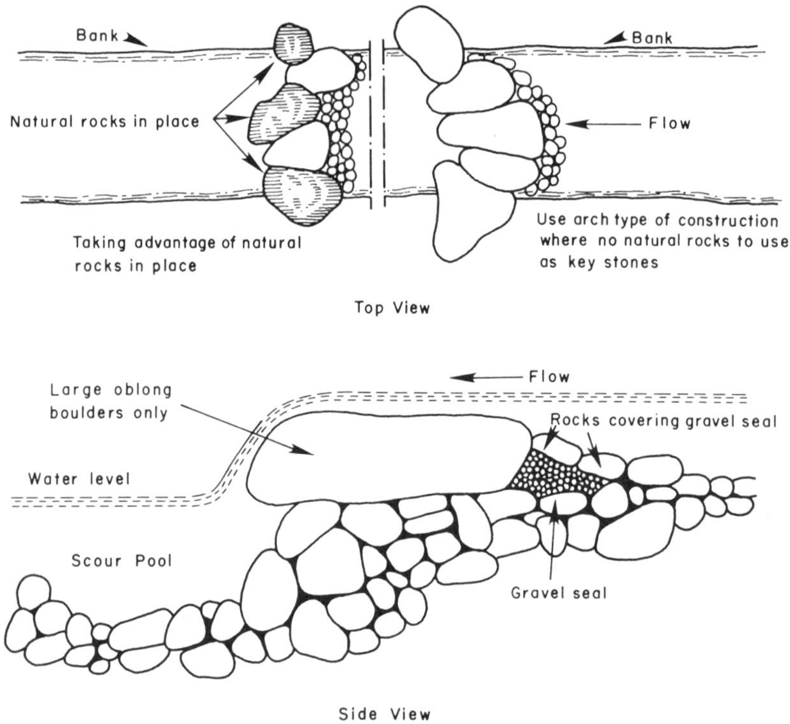

Figure 5.4 Boulder dams.

log ends into the banks, riprap should also be placed over each end. If additional stability is required, log and rock cribs (similar in design to current deflectors) can be constructed on each log end.

- The size of logs used will depend on stream size, availability at the site, desired height of structure, and availability of heavy equipment. The minimum size used should be no smaller than approximately 0.30 m in diameter. For a single log structure, this would allow 0.15 m to be imbedded and still leave a waterfall height of 0.15 m for scouring purposes.
- The type of log to be used will more than likely depend upon what is available at the site. Claire (1978) found western larch to be very durable, while Alvarado (1978) also recommends aspen or cottonwood. Ehlers (1956) reported success with white fir logs while the Wyoming Game and Fish Dept. (1982a) found railroad ties to be an excellent construction material.
- The key to longevity is to attempt to keep as much of the logs wet at all times as possible. Alvarado (1978) recommends keeping a small amount of overflow along the entire log to prevent rot and decay. This will of course require some experimentation in placing the log and also in designing the spillway, if one is desired to facilitate fish passage and centralize scour pool

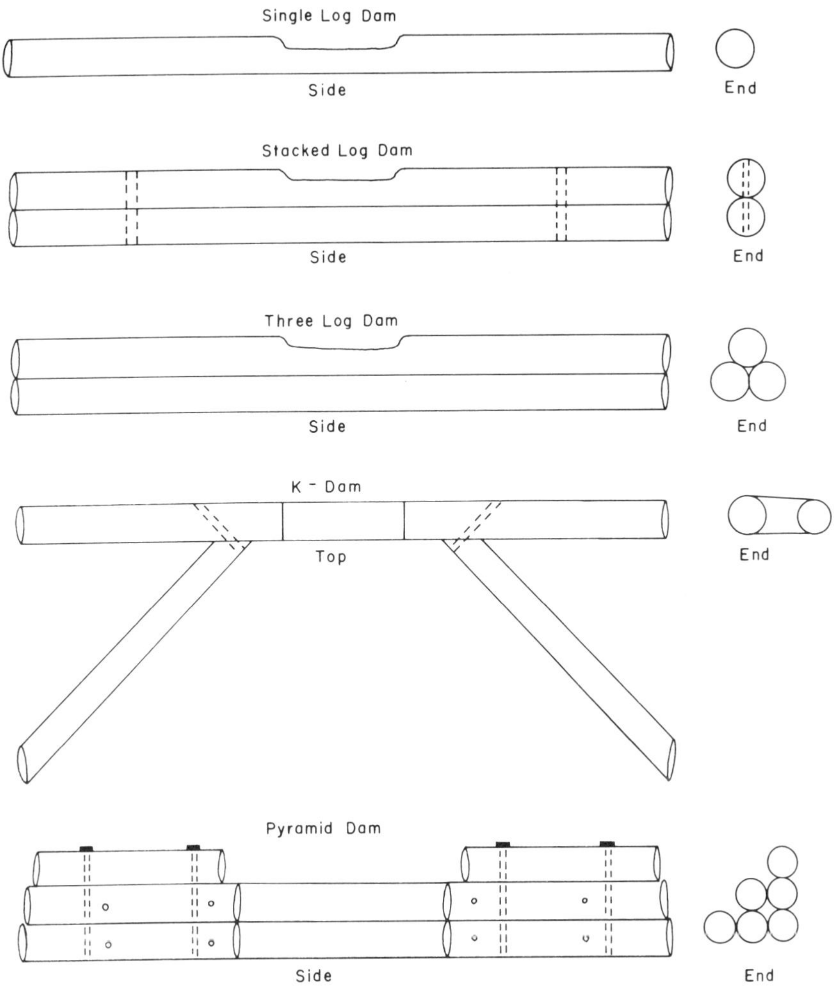

Figure 5.5 Various types of log dams.

development. Based upon the literature, life expectancies in the range of 20 to 40 years are not out of the question.

If the primary purpose of a dam is to raise the upstream water level, obtaining a good seal on the upstream face is the key to success. If the dam is several logs high, the logs should be hewn smooth so that they lie flat against each other, thus minimizing leakage. To lessen seepage under the structure, a wire and gravel mudsill should be added to the upstream face as shown in Figures 5.6 and 5.7. Bender (in Oregon Dept. of Fish and Wildlife, 1978) reports success by sloping gravel up the upper face of the logs, overlaying this with fence post-type stringers attached to the log parallel to the flow,

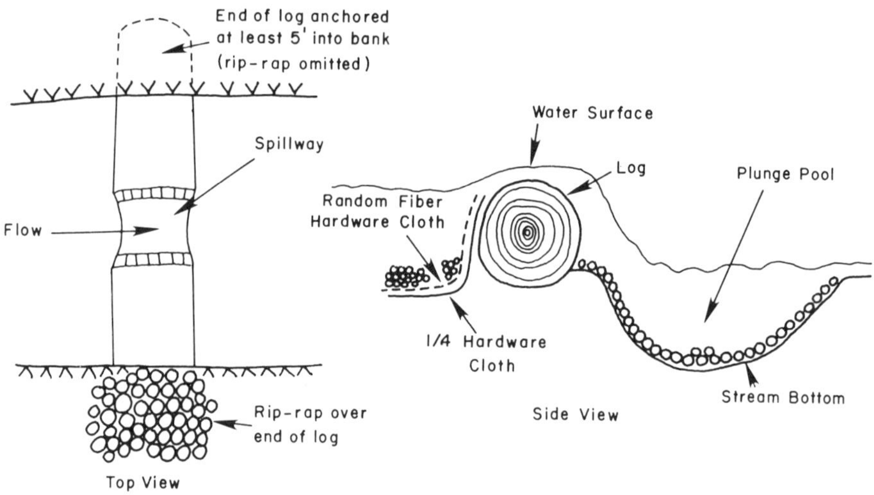

Figure 5.6 Single log dam.

covering the stringers with cyclone fencing, and then applying a final covering of gravel. Alvarado (1978) recommends that the height of the seal shouldn't exceed ⅔ the height of the logs. Maughan et al. (1978) found a plank fronting similar to that described by Bender to be more effective than using just wire mesh.

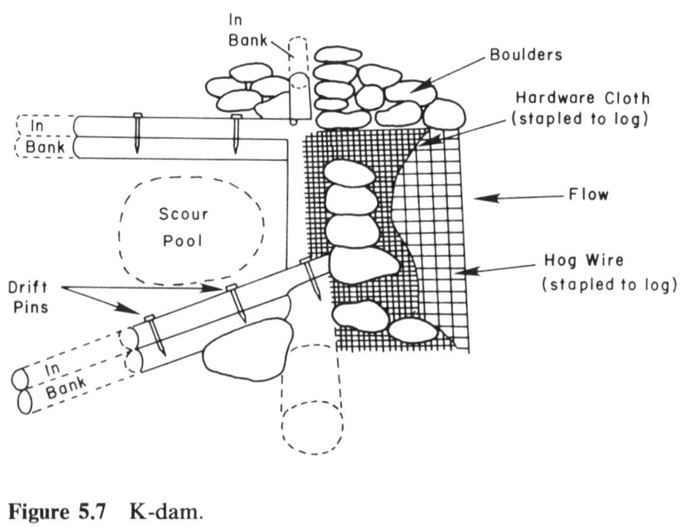

Figure 5.7 K-dam.

Probably the simplest, easiest to construct, most effective, and most commonly used log dam is the *single log dam.* Construction details are provided in Figure 5.6, while Figure 5.8 illustrates the installed structure. Generally, it is used in smaller streams up to 5 or 6 m wide when deepening of pools in the range of 0.15–0.3 m or formation of small scour pools is desired. Richard (1963) reported construction time as being 16 hours per structure when using logs approximately one meter in diameter, while Gard (1961) found that 4.5 hours per structure was required. Cost of the structures will of course vary with location, labor, stream width, and site characteristics, among other factors. Bailey (1982) reported that single log weirs averaged about $850 each, while Coffin (1982) found that single log dams cost $210 per structure, with $180 of this total being for labor, the remainder for materials.

A somewhat more stable variation of the single log dam is the *K-dam,* which derives its name from the configuration of the structure once the two downstream braces have been added (Figure 5.5). Figure 5.7 shows construction details with a wire and gravel mudsill in place. The Wyoming Game and Fish Department (1982a) reports the *K*-dam to be a "good solid structure with good results" when installed on a stream 7 to 9 m wide using a log 0.4 m in diameter. Richard (1963) also found the *K*-dam to be more sturdy than single log structures, although

Figure 5.8 Single log dam in place.

it required more construction time (41 hours compared to 16 hours). Seehorn (1982) reports that a crew of four to six people could construct 1 to 1.5 structures per day on streams 3 to 4 m wide.

The *plank or board dam* is another variation of the single log dam which has met with some success (Duff 1982). Also known as the Hewitt Ramp (White and Brynildson 1967), the principal advantage of the structure is that undercutting action is shifted from the base of the main log upstream to the point where the planks, which extend upstream from the top of the main log, intersect the streambed. This also protects the gravel seal on the upstream face of the main log. White and Brynildson caution that the structure should only be used on steep gradient reaches where water will not be impounded upstream for a distance greater than 5 channel widths. The important pool formed is the one scoured out below the structure. Disadvantages of the structure, as reported by the Wyoming Game & Fish Dept. (1982a), are that it is difficult and time consuming to install and seal, the result being an extremely costly structure.

The fourth general type of log dam is the *wedge or V-dam*. Generally, there are two advantages of this type over the *K*-dam. First, the wedge dam can be used on slightly larger streams because there are two shorter logs which form the wedge or *V*, and together span the entire width of the stream. With the *K*-dam, a single log must traverse the entire stream, making it more difficult to maneuver. Secondly, the wedge dam is less prone to undercutting (Seehorn 1982).

The design of the wedge dam is such that the wedge or *V* points upstream. The two main logs face upstream at 45° to the streamflow and are pinned together with rebar at midchannel. The ends of the main logs are, of course, well anchored into the banks. The two brace logs, very similar to the braces used on *K*-dams, are pinned to the main logs at approximately 90°, with the other ends well anchored into the banks. If necessary, a mudsill can be added to the upper faces of the main logs to prevent undercutting. All banks in the immediate vicinity of the structure must be well riprapped, especially on the upstream end where the wedge will tend to force the current toward the banks. Seehorn (1982) estimates that a crew of four to six can construct 1 to 1.5 wedge dams per day, depending upon stream size.

Gabion Check Dams. Gabion check dams have probably not been used as extensively as log dams primarily because they are quite expensive, time-consuming, and at times aesthetically unpleasing. Cooper and Wesche (1976) and the Wyoming Game & Fish Dept. (1982a), have, however, found that excellent habitat can be created. Possibly the best application of gabion dams would be in wide, shallow streams and rivers lacking pools and having an abundance of coarse gravels. As gabions can be laced together to form any width or configuration of structure desired, a low dam could be built without the aid of heavy equipment. By comparison, such equipment would definitely be needed to set a 10 m long by 0.4 m wide log in place.

A gabion check dam is illustrated at both low and high flow in Figure 5.9, while Figure 5.10 presents construction details. The following description of the

(a)

(b)

Figure 5.9 **(a)** Gabion check dam immediately after construction. **(b)** Gabion check dam at high flow.

step-by-step procedure for building an 8 m wide gabion dam is summarized from Cooper and Wesche (1976).

First the stream bed where the structure would lie was leveled. While this was being done by two crew members, others were preconstructing the gabion structure on the bank. This consisted of first cutting eight 2.0 m x 0.75 m x 0.15 m cells from the factory gabions. Four of these were laced end-to-end making the base of the dam 8 m long and 0.75 m wide. The other four baskets were also laced end-to-end and one of the lower edges was laced to the corresponding upper edge of the first row of baskets. This made a structure 0.3 m high, 0.75 m wide, and 8 m long.

The stream bed, where the upstream edge of the dam would be, was then dug down 0.3 m (the height of the structure) and sloped as shown in Figure 5.10. This allowed the two rows of baskets to be laid into the excavation in a sloping manner, thus creating the ramp and dam. The upper row of baskets was then folded back so that the lower row could be filled with rock from the dredge piles. When this was complete, the top row of baskets was once again folded over the bottom row and laced down so that it became the lid for the bottom row. The top row was then filled full with 0.1–0.15 m diameter rock. At this point, a layer of rubber mat was draped over the rock in the top row of baskets and tucked in around the edges of the gravel. This was done to completely seal the structures so no leakage could occur through the spaces between the fill rocks, and to facilitate the movement of high flows over the dam and resist scour. Another layer of rock was then placed over the mat to conceal it, and the lid was laced on to complete the structure. Wherever the ends of the structures abutted the stream banks, they were dug back into the banks to prevent erosion around the ends, covered with bank material, and then reinforced with boulders.

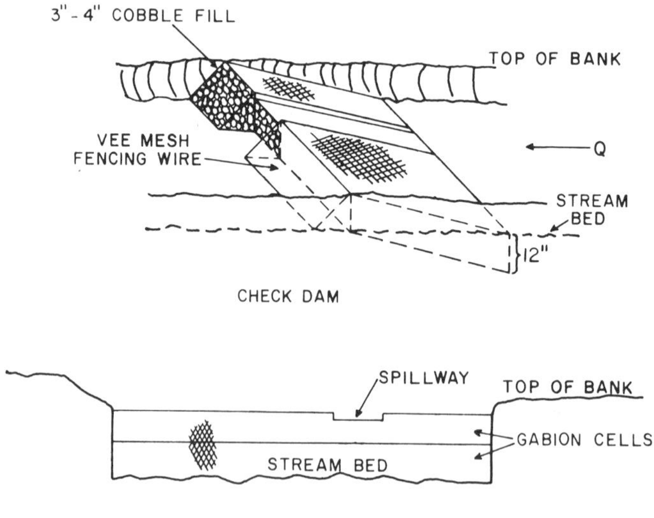

Figure 5.10 Design of gabion check dam.

Next, using the basic continuity equation $Q = VA$, the desired cross-sectional area of the spillway could be determined by knowing the minimum desired velocity through the spillway, and the discharge. Location of the spillway in the structure was determined by location of the thalweg through the structure area, while height of the spillway was determined by the height of the bank cover upstream from the dam. It was found that the spillway should be located as near the thalweg line as possible, but not right against the bank. This maintained the natural pattern of flow through the pool and still avoided scour at the channel banks, while maximizing the depths at the bank cover immediately upstream. Construction of the spillway was accomplished by first choosing the location based on the above criteria. The depths desired in the upstream pool, as well as the desired length of the pool, were then noted. When the width and depth of the spillway were thus decided, the upper basket of the dam was cut open at the desired place, and fill rocks were removed or hand placed until the correct spillway was formed. The rubber mat within the upper baskets was formed to fit the spillway, and the gabion was then molded to fit the depression and laced shut.

The final step in construction was the addition of the V-mesh downramp on the downstream edge to prevent undercutting of the check dam at high flow by not allowing an abrupt plunge to occur, except in the area of the spillway. The plunge pool formed below the spillway thus created a standing wave sufficient to aid trout in passage over the check dam even at low flow.

Beaver Introduction. A final type of treatment that can produce results similar to low-profile dams without any of the artificiality associated with such construction would be beaver introduction (reintroduction) and management. The influence of beaver on wildlife has long been recognized. In regard to salmonid habitat, beaver dams can play a key role in creating pools and escape cover as well as regulating flow, temperature, and sediment transport regimes. Especially in low-order, headwater streams, where low-profile dams are typically constructed, beaver activity can have a strong influence on the numbers and biomass of trout supported. Gard (1961) found that in some cases streams with beaver ponds had up to six times the total weight of salmonids than did adjacent habitat lacking ponds. In western mountain streams where the availability of riffles and spawning areas is typically not severely limiting, beaver ponds can provide critical rearing and overwintering habitat (Munther 1982). Also, by raising water tables, beaver dams can have a positive effect on the development or rehabilitation of riparian communities (Smith 1980), which in turn can provide additional bank cover and organic matter for the aquatic system.

Other In-Channel Treatments

Boulder Placement. The placement of individual boulders or boulder clusters is one of the simplest and most commonly applied in-channel treatments that can improve habitat on streams of any size. Generally, boulder placements are made with one or more of the following management objectives in mind (Claire 1980):

- provide additional rearing habitat;
- provide fish cover;
- improve pool-riffle ratios;
- restore meanders and pools in channelized reaches;
- protect eroded banks by deflecting flow.

While the literature detailing the results of boulder placements is not abundant, most applications appear to have been successful. Barton and Winger (1973) reported good results in forming holes on the Weber River in Utah, while on the St. Regis River in Montana, Lere (1982) found that after eight years a majority of the boulders were still functioning properly and that trout numbers were greatest in a river reach mitigated with random boulders. Knox (1982) found that random boulders placed in the Eagle River (Colorado) were successful in creating pool habitat in a channelized reach. In British Columbia, Haugen (1978) noted a twenty-fold increase in coho salmon numbers one year after rock clusters were installed on the Keough River, while in Wyoming, the Wyo. Game & Fish Dept. (1982b) reported boulder placements were successful in creating in-stream habitat in "rubbly glide" reaches of the Green River below Fontenelle Reservoir. Also, Kanaly (1971) found that the trout population in a channelized section of Rock Creek mitigated with large boulders quickly recovered to levels comparable with unaltered reaches.

Boulders can be placed either randomly or selectively, in clusters or individually. This will depend primarily upon the best judgment of the project biologist, the size of the stream, and the pattern of natural boulders in the river reach. Figure 5.11 illustrates several of the placement options available. While siting and construction considerations may not be as critical for boulder placement as for other in-channel treatments previously discussed, the following observations may be of help in the conduct of a successful project:

- Placement should be done during low flow to assure proper location and facilitate movement of heavy equipment in the channel.
- Boulder size will depend on stream size, flow characteristics, and bed stability, as well as the size of heavy equipment available for the project. Claire (1980) recommends rock in the 0.6 to 1.5 m diameter range, while Kanaly (1971) reported success using 1.5 m wide boulders. The U.S. Bureau of Land Management (1968) recommends that material in the 1 to 2 cubic meter size range be used.
- The harder the rock used, the better. Granite is much preferred over sandstone.
- Embedding the boulders a short distance into the streambed will result in a much more stable situation.
- Placement near streambanks should be done with caution to avoid erosion.
- The U.S. Forest Service (1969) has found that boulder placements will have their greatest effect on fish populations when employed in stream reaches having less than 20% of the area in pools.

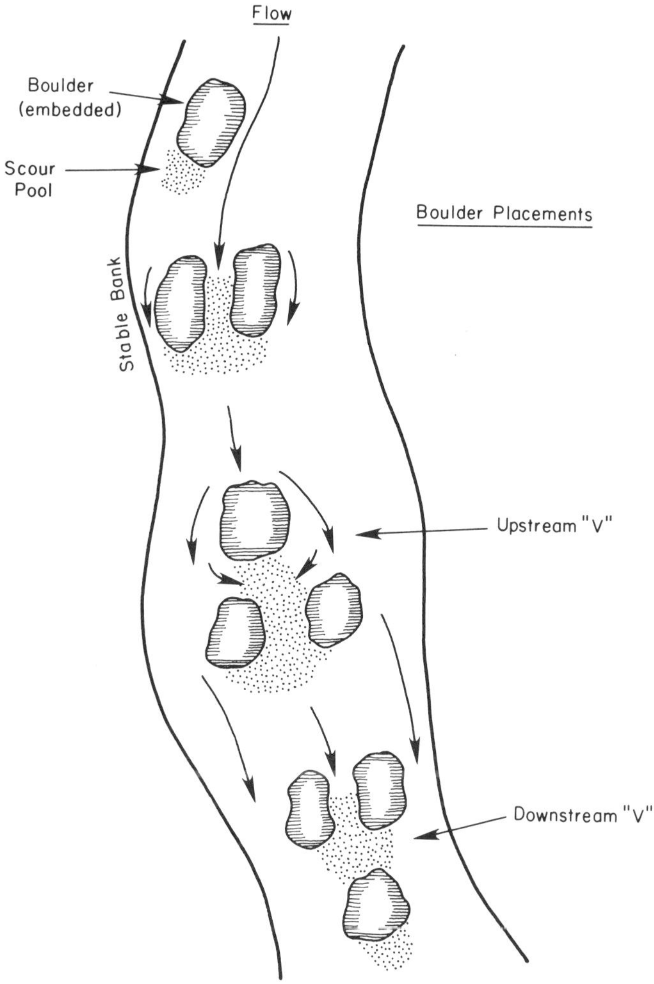

Figure 5.11 Boulder placements.

- The economics of any project will be greatly enhanced if natural material is available at the site.
- If the project stream is used for rafting and boating, this should be taken into consideration when locating the boulders.
- To avoid damage to the channel, the heavy equipment used should have rubber tires.

For locations on larger streams inaccessible to heavy equipment, Cooper and Wesche (1976) designed and tested a gabion structure, termed an *artificial boulder,* for creating additional holding water and cover for trout. Figures 5.12

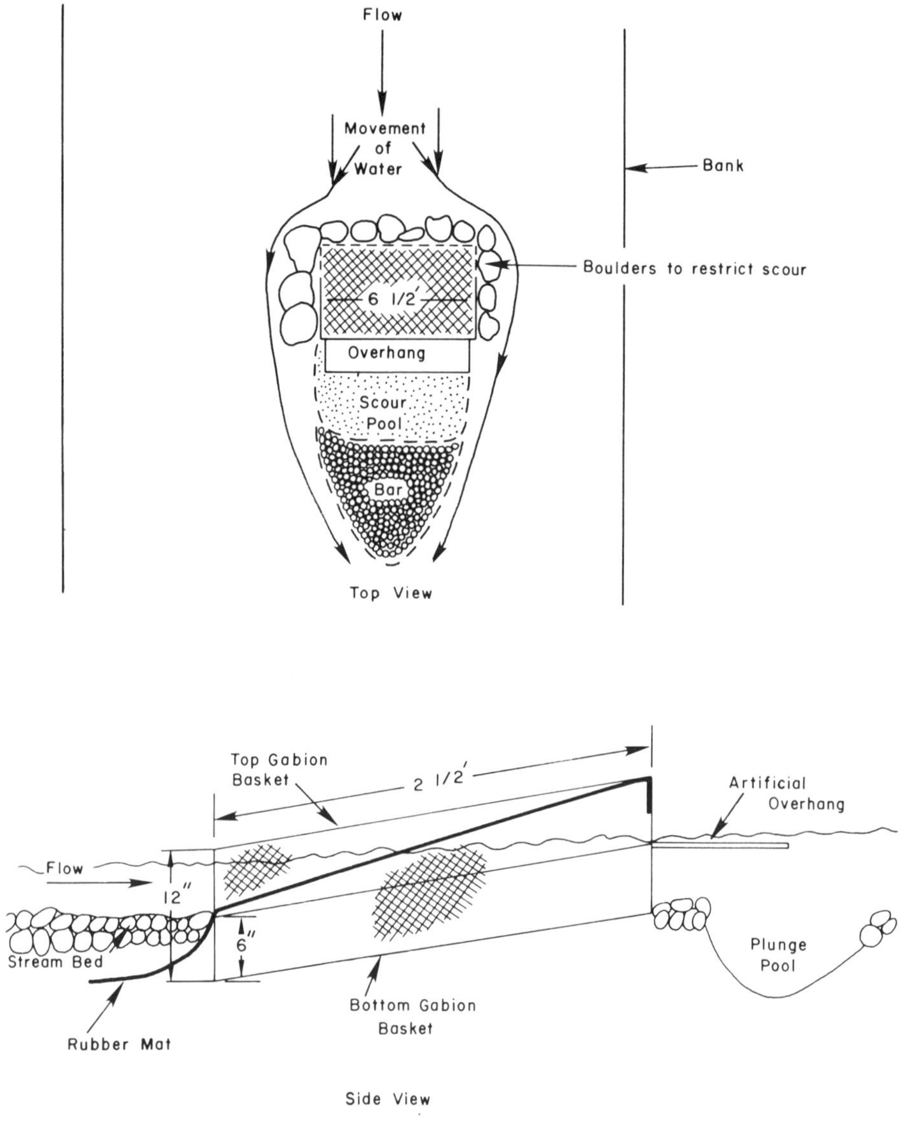

Figure 5.12 Design of artificial boulders.

and 5.13 present construction details of the structure, while Figure 5.14 shows a completed structure installed on the Blacks Fork River of southwest Wyoming. Testing of the structure on Douglas Creek and the Blacks Fork indicated that discharges in excess of 28 m^3/sec could be withstood (two out of eight did wash out). Also, the design of the structure does allow for attachment of an artificial overhang to provide overhead cover and formation of a gravel bar (spawning-sized gravel) on the downstream side (Wesche 1976). However, the structures

Figure 5.13 Lacing gabion cells together to form artificial boulder (rubber mat in place).

Figure 5.14 Completed artificial boulder.

are quite expensive to build, time consuming (at least eight hours per structure), and need to be extremely well anchored by 1–2-m-long rebar pins on both the upstream and downstream faces. For these reasons, application of artificial boulders is considered to be quite limited.

Trash Catchers. Trash catchers or barriers are easily constructed habitat reclamation devices that can be used to create pools, increase stream surface area, provide cover, slow velocities, and hold spawning gravel in place. While generally they are used in small, headwater streams to function much like low-profile dams, they can also be used in large streams to provide isolated pockets of cover. The primary use of trash catchers has been in the higher gradient mountain streams of the western United States (Wyoming Game and Fish Dept. 1982a; Navarre 1962; Knox 1982; Oregon Dept. of Fish and Wildlife 1978; Rinne and Stefferud 1982; Coffin 1982).

Navarre (1962) presents the following construction details for building a simple trash catcher across the entire width of a small stream:

1. The materials needed include steel fence posts (2 m long), 0.8 m hog wire (.08–.15 m mesh), and #12 galvanized tie wire.
2. The structure is sited using the criteria discussed above for low-profile dams.
3. The fence posts are cut in half and the 1 m sections are driven into the streambanks and bottom at 0.6 m intervals with approximately 0.2 meters left protruding above the low water level. The last post on each bank is placed just above the high water line.
4. The top of the hog wire is attached to the steel posts by a double strand of tie wire.
5. The hog wire is also attached to each post at two other locations.
6. The remaining hog wire is bent upstream and rocks are piled on the upstream edge to hold it in place.

Once constructed, the theory of operation is that the wire mesh will fill with silt, debris and gravel, thus forming a low dam. Navarre estimated that construction costs of trash catchers were only one-third to one-sixth of that required for log dams. While they are time and cost efficient as well as adequate producers of habitat, trash catchers can be aesthetically unpleasing if careful attention is not paid to detail (Figure 5.15). Also, their longevity can be hampered by the wire rusting out.

In larger streams (greater than 7 to 10 m wide), small, isolated trash catchers which do not traverse the entire channel width can be installed to create instream pockets of cover. For aesthetic purposes and to avoid collecting large pieces of drifting debris which could wash out the structure, the top should be located slightly below the low water line, as shown in Figure 5.16.

Half-logs. Another very simple but effective device for adding in-stream cover to almost any channel is the half-log. Johnson (1982) reports that electrofishing

Figure 5.15 Trash catcher on small stream.

in Wisconsin streams has shown that trout are found utilizing a high percentage of these structures. Construction details and overall channel layout are provided in Figure 5.16.

Substrate Development

As previously discussed, current deflectors and low-profile dams can be used in a variety of ways to improve the substrate component of a stream habitat. These include the removal of fine sediment deposits, the collection of spawning gravels upstream from structures, and the development of gravel bars downstream. Boulder placement to enhance cover has also been discussed. Another useful tool in habitat reclamation can be the introduction of properly sized substrate particles to enhance macroinvertebrate production (as discussed in the previous chapter) and spawning success.

There can be a variety of reasons why spawning gravels need to be added to a stream habitat, including flood scouring, dredging activity, channelization, and natural deficiencies. For spring-fed creeks not subject to exceptionally high runoff flows, the addition can be made by first determining the size of gravel required by the fish population; second, selecting favorable locations for gravel

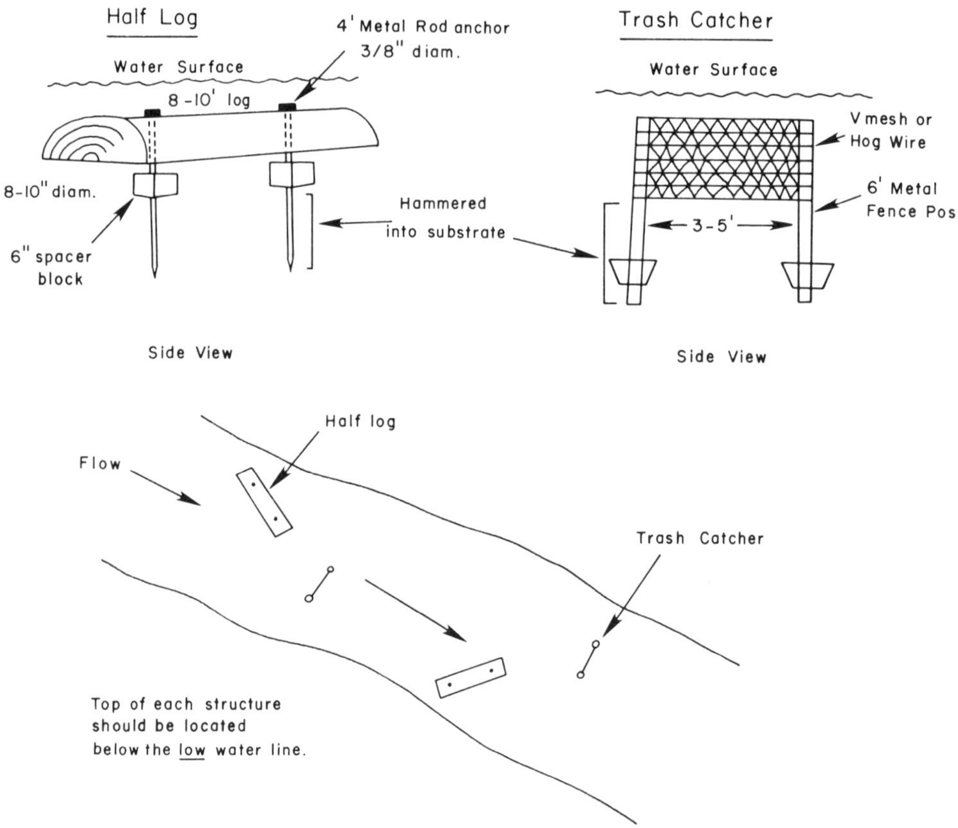

Figure 5.16 Design of half-logs and trash catchers on larger streams.

addition (pool-riffle interchanges are ideal); third, excavating the existing stream bed to a depth of 0.4 to 0.6 m to remove cobbles and other large particles that might interfere with redd construction; and fourth, filling the excavation with the proper sized gravels. Using a technique similar to this, the Wyoming Game & Fish Dept. (1982c) reported an increase in spawning activity of 4100% over ten years in Three Channel Spring Creek, a tributary of the Snake River.

On streams subject to high runoff, it will be necessary to install stabilizing structures such as gabions or logs firmly into the bed below the excavation to hold the gravels in place. The key to success will be the anchoring of the structures. Techniques to stabilize low-profile dams such as those described above should be employed. Both Fortune (1978) and Bender (1978) report successfully developing spawning beds in this manner.

Before proceeding with a spawning gravel introduction, Claire (1980) cautions that several factors should be given consideration. First, the reason why natural gravels are not available should be determined. If flow and gradient characteristics are such that the gravels are constantly being scoured out, introduction should not begin until suitable structures are in place to slacken the current. Second, if the watershed is in poor condition or is naturally prone to be a high producer of fine sediments, spawning success may be limited regardless of the temporary availability of gravels. Third, as the purchase and transport of gravels to the site can be expensive and time consuming, careful attention should be paid to selecting sites that are easily accessible and/or have suitable quantities of natural gravel available.

Bank Cover Treatments

The habitat value of streamside cover such as undercut banks, debris jams, and overhanging vegetation has been well documented in the literature and discussed earlier in this chapter. In habitat reclamation, there are four general types of treatments that can be applied to create or enhance available bank cover, including log/board overhangs (Figure 5.17), artificial overhangs of metal or fiberglass (Figure 5.18), tree/brush retards (Figure 5.19), and riprap (Figure 5.20). Each of these treatments provides excellent bank stabilization in addition to cover, with the possible exception of artificial overhangs. While each of these treatments can be applied

Figure 5.17 Log and board overhang.

Figure 5.18 Corrugated metal artificial overhang.

Figure 5.19 Tree revetment.

Figure 5.20 Riprap.

by themselves to enhance habitat, oftentimes they are installed in conjunction with other reclamation structures such as current deflectors or low-profile dams. If properly done, they can assume a natural appearance and blend well into the stream setting. Certainly, there is no place in river reclamation work for a bank treatment such as shown in Figure 5.21.

Log Overhang. The simplest type of log overhang is that described by Seehorn (1982). A cover log of suitable diameter (at least 0.3 m) and length is selected for the situation at hand. Several abutment logs are well anchored into the stream bank by digging and the cover log is pinned to these using rebar, thus forming the overhang. A variation of this would be to cut a long notch in the cover log (the notch will form the cover area for the fish), fit the log up against the bank to be treated (notched side down), and secure it to the stream bottom using 1.2 m lengths of rebar. Seehorn recommends placement of log covers on stream banks opposite current deflectors to prevent erosion as well as provide cover.

White and Brynildson (1967) describe a somewhat more elaborate log and board bank cover device, as pictured in Figure 5.17. General steps in construction are as follows:

1. Log pilings, spaced approximately 0.5 to 1 m out from the existing bank and approximately an equal distance from each other, are sunk (jetted) securely down into the stream bed.

2. Log or board stringers, securely anchored into the bank, are then run from the bank out to the pilings. Spikes or pins are then driven through the stringers into the tops of the pilings to secure them. This forms the support structure for the planks that serve as the overhang.

3. The planks are then laid over the stringers, parallel to the streamflow, and nailed in place.

4. Rock riprap is then applied over the planks as well as on the disturbed bank behind the device.

5. Soil and sod can then be applied on top of the riprap to complete the device.

While this type of overhang has been found to function well in midwestern streams having relatively stable flow patterns, results have been mixed in the West. The Wyoming Game & Fish Dept. (1982a) reports that while several have been successful, the tremendous stream flow fluctuation in Wyoming streams prevents them from functioning efficiently at all times.

Artificial Overhangs. Artificial overhangs constructed of corrugated steel and fiberglass were found to provide usable trout cover by Cooper and Wesche (1976). Figure 5.18 shows one such device installed while design detail is provided in

Figure 5.21 Car bodies for bank stabilization, ineffective in this case as well as aesthetically unpleasing.

Figure 5.22. Strap hinges were used in the construction to provide some degree of flexibility with changing flow and ice conditions. Investigations into the color of the overhangs concluded that a flat black or mottled brown rendered the structures quite inconspicuous, virtually invisible at times, depending upon light and background conditions. Anchoring of the structure was provided by driving the rebar pins securely into the stream bank. Overall, this type of artificial overhang was found to be effective, cost and time efficient, adaptable to a variety of bank conditions, and relatively stable. Installation must be done at low-flow to ensure that the structure will always remain under water. Several boulders placed under the overhang were found to enhance the cover provided and also served to protect against bank erosion. For locations where bank stabilization is the highest priority, overhangs of this type are easily used in conjunction with rock-filled gabions, as shown in Figure 5.23. As with any type of metal device used for river reclamation, particular attention must be paid to the aesthetics of the project.

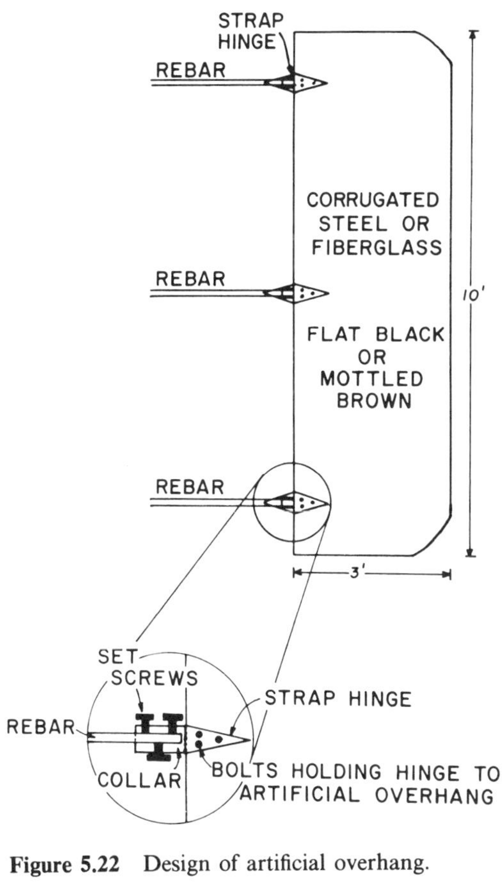

Figure 5.22 Design of artificial overhang.

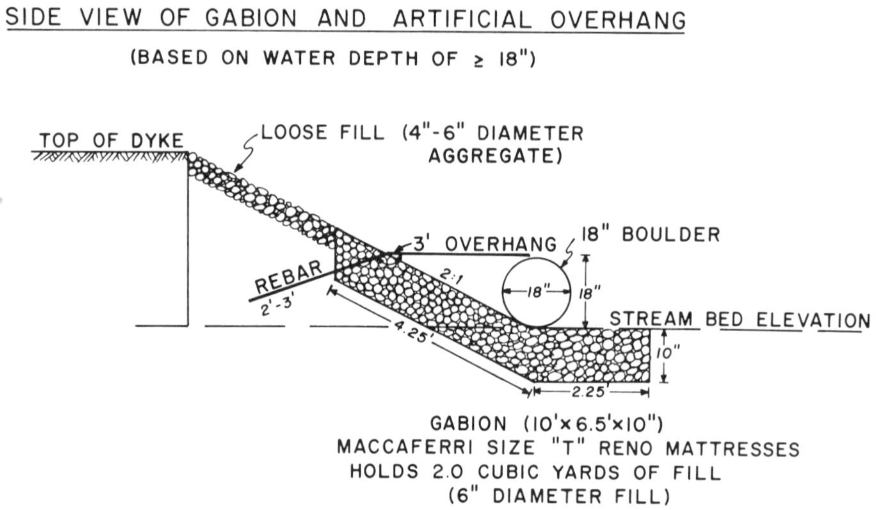

Figure 5.23 Combination of gabion and artificial overhang to provide bank stabilization and cover.

Tree Retards. A treatment that not only provides effective trout habitat but excellent bank stabilization as well is the use of tree retards (Figure 5.19). Sheeter and Claire (1981) found the use of juniper trees to be an effective method to stabilize high eroding banks on the South Fork of the John Day River in Oregon. The size of trees cut and used depended upon the personnel and equipment available. Generally, trees greater than 0.15 m in diameter required a tractor for placement. The junipers were placed, one per meter of bank, down over the edge, angling downstream, and their butts were tied to anchors located at least 1.5 m back from the bank's edge using #9 smooth wire. Two types of anchors were used, fence posts sunk deeply into the bank or a heavy cable line "dead manned" into the bank. Green trees with bushy crowns were preferred over slender ones. Using this procedure, the project was evaluated as a success. Mean water velocities near the bank were reduced by about two-thirds and silt deposits of up to 0.6 m deep were found in the tree tips the first year after placement, allowing native plant succession to occur. Only a 4% failure rate was noted, with improper anchoring and placement on outside curves the main reasons for failure.

The Wyoming Game & Fish Dept. (1982a) has also reported success using tree retards. Generally, conifer trees 9–18 m long and 0.25–0.4 m in diameter have been used, with the anchoring similar to that described above. If rock riprap is placed under the trees, the trees are overlapped about one-third. An overlap of one-half is recommended if riprap is not available.

Tree retards can be an excellent structure to install in newly channelized reaches. Generally, the materials needed will already be available at the site due to clearing activities; installation is relatively easy and not time consuming; excellent holding water can be immediately created to maintain a fish population while the new channel is developing and maturing; and, the result is aesthetically pleasing. Gore and Johnson (1980) reported that in a newly rechanneled portion of the Tongue River in Wyoming, such snags were the most frequently used areas by fish.

Riprap. A fourth method of lower bank treatment is by the placement of riprap (Figure 5.20). Not only does riprap provide fish cover and macroinvertebrate habitat, bank erosion can also be slowed or stopped, thereby allowing recovery of natural vegetation. Claire (1980) recommends that dense, angular, nonerodible pit run rock ranging in diameter from 0.1–0.8 m be used, with at least 50% of the material being in the 0.15–0.6 m size range. To improve the habitat and the aesthetic effects of riprap, the alignment should not be too straight or regular. For example, by varying the design only slightly, rock jetties, such as were previously discussed, can be formed. These can assist in pool deepening and habitat development, especially in new channels or channelized reaches.

To ensure the success of a riprap project, especially where long sections of bank are to be treated, it is strongly recommended that an engineer be directly involved. If this is not possible, advice should be sought from a source such as the State Highway Department. Their expertise can provide a more site-specific evaluation of such factors as bank slope; stream velocities; size, shape, gradation, and specific gravity of rock; and desired thickness of the covering. Such involvement will facilitate the development of the most time- and cost-efficient plan to meet the reclamation goals of the project.

Other bank treatments, such as revegetation and fencing, can also play a critical role in the success of any river reclamation project. Discussions of these treatments will be found in other chapters of this book.

REFERENCES

Allen, K.R. 1959. The distribution of stream bottom faunas. *Proc. New Zealand Ecol. Soc.* 6:5–8.

______. 1969. Limitations on production in salmonid populations in streams. In *Symposium on Salmon and Trout in Streams.* H.R. MacMillan Lectures in Fisheries. Vancouver: University of British Columbia.

Alvarado, R. 1978. *Minimum Design Standards for Log Weirs.* U.S. Forest Service Report, Pacific NW Region, Malheur National Forest.

Ambuhl, V.H. 1959. The significance of flow as an ecological factor. *Schweizerische Zeitschrift fuer Hydrologie XXI* (2):133–270. Translation by J. DeWitt, Humboldt State College, Arcata, Calif.

Arthur D. Little, Inc. 1973. "Statement by John M. Wilkinson, Arthur D. Little, Inc." *Hearings on Stream Channelization.* U.S. 92nd Congress. House of Representatives. Committee on Government Operations. Conservation and Natural Resources Subcommittee. March 20 and 21, 1973.

Bailey, R. 1982. Oregon. In *Proc. of Rocky Mt. Stream Habitat Management Workshop, Jackson, WY,* edited by R. Wiley, Wyoming Game and Fish Dept., Laramie. Sept. 7–10, 1982.

Baldes, R.J., and R.E. Vincent. 1969. Physical parameters of microhabitats occupied by brown trout in an experimental flume." *Trans Amer. Fish. Soc.* 98:230–38.

Barclay, J.S. 1980. *Impact of Stream Alteration on Riparian Communities in Southcentral Oklahoma.* FWS/OBS-80/17. Washington, D.C.: U.S. Government Printing Office.

Bardach, J.E., J.H. Ryther, and W.O. McLarney. 1972. *Aquaculture, the Farming and Husbandry of Freshwater and Marine Organisms.* New York: Wiley-Interscience.

Barstow, C.J. 1971. Impact of channelization on wetland habitat in the Obion-Forked Deer Basin, Tennessee. In *Amer. Fish. Soc. Spec. Publ. No. 2, North Central Division,* edited by E. Schneberger and J.L. Funk.

Barton, J.R., and P.V. Winger. 1973. *A Study of the Channelization of the Weber River, Summit County, Utah.* Utah Div. of Wildlife Resources and Utah State Dept. of Highways Report.

Bender, R. 1978. Log sills. p. 7. In: *Proc. Fish Habitat Improvement Workshop.* Ochoco Ranger Station, Oregon, Sept. 26–27, 1978. Oregon Dept. of Fish and Wildlife, Salem, Oregon.

Benson, N.G. 1953. The importance of groundwater to trout populations in the Pigeon River, Michigan. *Trans. 18th N. Amer. Wildl. Conf.,* 268–281.

Binns, N.A. 1976. Evaluation of habitat quality in Wyoming trout streams. Paper presented at Annual Meeting, Amer. Fish. Soc., Dearborn, Michigan, Sept. 1976.

———. 1979. *A Habitat Quality Index for Wyoming Trout Streams.* Fish. Res. Report No. 2. Cheyenne: Wyoming Game and Fish Dept.

———. 1978. Quantification of trout habitat in Wyoming. In *Trans. of the Bonneville Chapter, Amer. Fish. Soc., Salt Lake City, Utah,* edited by D. Duff.

Boreman, J. 1974. Effects of stream improvement on juvenile rainbow trout in Cayuga Inlet, New York. *Trans. Amer. Fish. Soc.* (3): 637–41.

Boussu, M.F. 1954. Relationship between trout populations and cover on a small stream. *J. Wildl. Mgmt.* 18:229–39.

Bovee, K.D. 1974. *The Determination, Assessment and Design of 'Instream Value' Studies for the Northern Great Plains Region.* Univ. of Montana Final Report on EPA Contract #68–01–2413, Univ. of Montana, Missoula.

———. 1978. The incremental method of assessing habitat potential for coolwater species, with management implications. *Amer. Fish. Soc. Special Publ.* 11:340–46.

Bovee, K.D., and R. Milhous. 1978. *Hydraulic Simulation in Instream Flow Studies: Theory and Technique.* Instream Flow Information Paper No. 5, FWS/OBS-78/33. Washington, D.C.: U.S. Government Printing Office.

Brooks, B.R. 1974. Stream modelling to determine the effects of channel modification for fish habitat improvement. Master's thesis, Univ. of Wyoming, Laramie.

Brouha, P. and R. Barnhart. 1982. Progress of the Browns Creek fish habitat development project. In *Proc. of Rocky Mt. Stream Habitat Management Workshop. See* Bailey 1982.

Bulkley, R.V., R.W. Bachmann, K.D. Carlander, H.L. Fierstine, L.R. King, and B.W. Menzel. 1976. *Warmwater Stream Alteration in Iowa: Extent, Effects on Habitat, Fish, and Fish Food, and Evaluation of Stream Improvement Structures.* Report on Project FWS/OBS-76/16. Ames: Iowa State University.

Butler, R.L., and V.M. Hawthorne. 1968. The reactions of dominant trout to changes in overhead artificial cover. *Trans. Amer. Fish. Soc.* 97:37–41.

Chapman, D.W. 1966. Food and space as regulators of salmonid populations in streams. *Amer. Natur.* 100:345–57.

Chapman, D.W., and T.C. Bjornn. 1969. Distribution of salmonids in streams, with special reference to food and feeding. In *Symposium on Salmon and Trout in Streams. See* Allen 1969.

Chapman, D.W., and E. Knudsen. 1980. Channelization and livestock impacts on salmonid habitat and biomass in western Washington. *Trans. Amer. Fish Soc.* 109(4):357–63.

Claire, E. 1978. Rock work, log sills, bank stabilization, flow recovery, fencing, pp. 2, 7, 8, and 9. *See* Bender 1978.

Claire, E.W. 1980. Stream habitat and riparian restoration techniques; guidelines to consider in their use. In *Proc. of Workshop for Design of Fish Habitat and Watershed Restoration Projects, County Squire, Oregon, March 10–14, 1980.*

Coble, D.W. 1961. Influence of water exchange and dissolved oxygen in redds on survival of steelhead trout embryos." *Trans. Amer. Fish. Soc.* 90:469–74.

Coffin, P.D. 1982. Northeastern Nevada stream and riparian habitat improvement projects. In *Proc. of Rocky Mt. Stream Habitat Management Workshop. See* Bailey 1982.

Cooper, C.O., and T.A. Wesche. 1976. *Stream Channel Modification to Enhance Trout Habitat Under Low Flow Conditions.* Water Resources Series No. 58. Laramie: University of Wyoming.

Delisle, G.E., and B.E. Eliason. 1961. *Stream Flows Required to Maintain Trout Populations in the Middle Fork Feather River Canyon.* Report No. 2. California Dept. of Fish and Game, Sacramento.

Dill, L.M. 1969. *Fulton River Fry Quality and Ecology Program.* Report of 1968 Studies. Vancouver: Dept. of Fish. of Canada.

Dimond, J.B. 1967. Evidence that drift of stream benthos is density related. *Ecology* 48:855–57.

Duff, D.A. 1982. Historical perspective of stream habitat improvement in the Rocky Mountain area. In *Proc. of Rocky Mt. Stream Habitat Management Workshop.* Wyoming Game and Fish Dept., Laramie. *See* Bailey 1982.

Egglishaw, H.J. 1964. The distributional relationship between the bottom fauna and plant detritus in streams. *Jour. Animal Ecol.* 33:463–76.

Ehlers, R. 1956. An evaluation of stream improvement devices constructed eighteen years ago. *Calif. Fish and Game* 42(3):203–17.

Eifert, W.H., and T.A. Wesche. 1982. *Evaluation of the Stream Reach Inventory and Channel Stability Index for Instream Habitat Analysis.* Water Resources Series No. 82. Laramie: Univ. of Wyoming.

Elliott, J.M. 1967. Invertebrate drift in a Dartmoor stream. *Arch Hydrobiol.* 63:202–37.

Elser, A.A. 1968. Fish populations of a trout stream in relation to major habitat zones and channel alterations. *Trans. Amer. Fish. Soc.* 97(4):389–97.

Eriksen, C.H. 1966. Benthic invertebrates and some substrate-current-oxygen interrelationships. In *Organism-Substrate Relationships in Streams.* Spec. Publ. No. 4, Pymatuning Lab. of Ecology, Univ. of Pittsburgh. Pittsburgh, Pa.: Univ. of Pittsburgh.

Everest, F.H., and D.W. Chapman. 1972. Habitat selection and spatial interaction by juvenile chinook salmon and steelhead trout in two Idaho streams. *J. Fish. Res. Bd. Canada* 29:91–100.

Everhart, W.H., A.W. Eipper, and W.D. Young. 1975. *Principles of Fishery Science.* Ithaca, N.Y.: Cornell University Press.

Fortune J. 1978. Gabions. *See* Bender 1978.

Gard, R. 1961. Creation of trout habitat by constructing small dams. *J. of Wildlife Mgmt.* 52(4):384–90.

Giger, R.D. 1973. *Streamflow Requirements of Salmonids.* Final Report on Project AFS 62-1. Oregon Wildlife Comm., Portland.

Good, W.R. 1974. *Some Effects of Spring Snowmelt Runoff on Aquatic Invertebrate Populations in a High Mountain Stream.* Water Resources Series No. 50. Laramie: Univ. of Wyoming.

Goodwin, C.N. 1979. *Effects of the Guernsey Silt Run Upon Canal Bank Stability and Seepage.* Water Resources Research Institute Report. Laramie: Univ. of Wyoming.

Gore, J.A., and L.S. Johnson. 1980. *Establishment of Biotic and Hydrologic Stability in a Reclaimed Coal Strip-Mined River Channel.* Water Resources Research Institute Report. Laramie: Univ. of Wyoming.

Greeley, J.R. 1932. Spawning habits of brook, brown, and rainbow trout and the problem of egg predators. *Trans. Amer. Fish. Soc.* 62:239–48.

Griswold, B.L., C. Edwards, L. Woods, and E. Weber. 1978. *Some Effects of Stream Channelization on Fish Populations, Macroinvertebrates, and Fishing in Ohio and Indiana.* FWS/OBS-77/46. Washington, D.C.: U.S. Government Printing Office.

Hale, J.G. 1969. An evaluation of trout stream habitat improvement in a North Shore tributary of Lake Superior. *Minn. Fish. Invest.* 5:37–50.

Hansen, E.A. 1975. Some effects of groundwater on brown trout redds. *Trans. Amer. Fish. Soc.* 104(1):100–10.

Hartman, G.F. 1965. The role of behavior in the ecology and interaction of underlying coho salmon (*Oncorhyncus kisutch*) and steelhead trout (*Salmo gairdneri*). *J. Fish. Res. Bd. Canada* 22:1035–81.

Haugen, G. 1978. Rock work. *See* Binder 1978.

Haugen, G.N. 1982. Views on stream habitat enhancement. In *Proc. of Rocky Mt. Stream Habitat Management Workshop. See* Bailey 1982.

Hayes, F.R., J.R. Wilmot, and D.A. Livingstone. 1951. The oxygen consumption of the salmon egg in relation to development and activity. *Jour. Exptl. Zool.* 116:377–95.

Hazzard, A.A. 1932. Some phases of the life history of the eastern brook trout (*Salvelinus fontinalis*). *Trans. Amer. Fish. Soc.* 62:344–50.

Heede, B.H. 1979. *Predicting Impact of a Restoration Project on River Dynamics—A Case History.* USDA Forest Service General Technical Report RM-62, Fort Collins, CO.

———. 1980. *Stream Dynamics: An Overview for Land Managers.* USDA Forest Service General Technical Report RM-72.

Hoar, W.S., M.H.A. Keenleyside, and R.G. Goodall. 1957. Reaction of juvenile Pacific salmon to light. *J. Fish. Res. Bd. Canada* 14:815–30.

Hooper, D.R. 1973. *Evaluation of the Effects of Flows on Trout Stream Ecology.* Emeryville, Calif.: Pacific Gas and Electric Co.

Hunt, R.L. 1976. A long-term evaluation of trout habitat development and its relation to improving management-related research. *Trans. Amer. Fish. Soc.* 105(3):361–64.

Hunter, J.W. 1973. *A Discussion of Game Fish in the State of Washington as Related to Water Requirements.* Fish Mgmt. Div. Report. Washington State Dept. of Game, Olympia.

Hutchinson, J. 1978. Gabions, spawning channels, blasting bedrock pools, pp. 4 and 6. *See* Bender 1978.

Johnson, M.O. 1982. Stream habitat management, Antigo area, Wisconsin. In *Proc. of Rocky Mt. Stream Habitat Management Workshop. See* Bailey 1982.

Johnson, R.L. 1967. *Evaluation of Stream Improvement Structures.* Job Completion Report on Project F-5-R-16. Helena: Montana Fish and Game Dept.

Kalleberg, H. 1958. Observations in a stream tank of territoriality and competition in juvenile salmon and trout (*Salmo salar* L. and *S. trutta* L.). *Inst. Freshw. Res. Drottningholm* 39: 55–98.

Kamler, E., and W. Riedel. 1960. A method for quantitative study of the bottom fauna of tartar streams. *Polskie Archiwum Hydrogiologic* TQM 7(21).

Kanaly, J. 1971. *Stream Improvement Evaluation in the Rock Creek Fishway, Carbon County (Preliminary)*. Admin. Report on Project 0571-08-6602. Cheyenne: Wyoming Game and Fish Dept.

Kendall, W.C. 1929. The fishes of the Cranberry Lake region. *Roosevelt Wildl. Bull.* 5(2).

Kennedy, H.D. 1967. *Seasonal Abundance of Aquatic Invertebrates and Their Utilization by Hatchery Reared Rainbow Trout.* Technical Paper No. 12. Washington, D.C.: U.S. Fish and Wildl. Serv.

Kimble, L.A., and T.A. Wesche. 1975. *Relationships Between Selected Physical Parameters and Benthic Community Structure in a Small Mountain Stream.* Water Resources Series No. 55. Laramie: University of Wyoming.

Knox, R.F. 1982. Stream habitat improvement in Colorado. In *Proc. of Rocky Mt. Stream Habitat Management Workshop. See* Bailey 1982.

Lane, L.J., and G.R. Foster. 1980. Modelling channel processes with changing land use. In *Proc. Symposium on Watershed Management, ASCE, Boise, Idaho, July 21–23, 1980.*

Latta, W.C. 1969. Some factors affecting survival of young-of-the-year trout (*S. fontinalis* Mitchell) in streams. In *Symposium on Salmon and Trout in Streams. See* Allen 1969.

Leitritz, E. 1969. *Trout and Salmon Culture.* Fish Bull. No. 107. California Dept. of Fish and Game, Sacramento.

Lere, M.E. 1982. The long term effectiveness of three types of stream improvement structures installed in Montana streams. Master's thesis, Montana State University, Bozeman.

Leusink, W.A. 1965. Alteration of stream habitat by constructing low elevation gabion dams. Master's thesis, Colorado State University, Ft. Collins.

Lewis, G.E. 1974. *Evaluation of Stream Improvement Structures Installed in District 11 of West Virginia Prior to 1973.* West Virginia Div. of Natural Resources Report.

Lewis, S.L. 1969. Physical factors influencing fish populations in pools of a trout stream. *Trans. Amer. Fish Soc.* 98:14–19.

Linsley, R.K., M.A. Kohler, and J.L.H. Paulhus. 1975. *Hydrology for Engineers.* New York: McGraw-Hill Book Co.

Lu, H.S. 1975. Criteria for placement of instream devices in high mountain streams. Master's thesis, University of Wyoming, Laramie.

McNeil, W.J. 1964. Effects of the spawning bed environment on reproduction of pink and chum salmon. *Fish Bull.* (U.S. Fish and Wildlife Service) 65(2):495–523.

McNeil, W.J., and W.H. Ahnell. 1964. *Success of Pink Salmon Spawning Relative to Size of Spawning Bed Materials.* Special Scientific Reprint. Fish. No. 469. U.S. Fish and Wildlife Service, Washington, D.C.

Maki, T.E., A.J. Weber, D.W. Hazel, S.C. Hunter, B.T. Hyberg, D.M. Flinehum, J.P. Lollis, J.B. Rognstad, and J.D. Gregory. 1980. *Effects of Stream Channelization on Bottomland and Swamp Forest Ecosystems.* Water Resources Research Institute Report No. 147. University of North Carolina, Raleigh.

Marsh, P.C., and T.F. Waters. 1980. Effects of agricultural drainage development on benthic invertebrates in undisturbed downstream reaches. *Trans. Amer. Fish. Soc.* 109:213–23.

Marzolf, G.R. 1978. *The Potential Effects of Clearing and Snagging on Stream Ecosystems.* FWS/OBS-78/14. Washington, D.C.: U.S. Government Printing Office.

Maughan, O.E., K.L. Nelson, and J.J. Ney. 1978. *Evaluation of Stream Improvement Prac-*

tices in Southeastern Streams. Bulletin 115. Blacksburg: Virginia Polytechnic Institute and State Univ.

Milhous, R.T., D.L. Wegner, and T. Waddle. 1981. *User's Guide to the Physical Habitat Simulation System.* Instream Flow Information Paper No. 11. FWS/OBS-81/43. Washington, D.C.: U.S. Government Printing Office.

Mundie, J.H. 1969. Ecological implications of the diet of juvenile coho in streams. In *Symposium on Salmon and Trout in Streams. See* Allen 1969.

Munther, G.L. 1982. Beaver management in grazed riparian ecosystems. In *Proc. of Rocky Mt. Stream Habitat Management Workshop. See* Bailey 1982.

Navarre, R.J. 1962. A new stream habitat improvement structure in New Mexico. *Trans. Amer. Fish. Soc.* 91(2): 228–29.

Needham, P.R. 1934. Quantitative studies of stream bottom foods. *Trans. Amer. Fish. Soc.* 64:238–47.

Needham, P.R., and R. Usinger. 1956. Variability in the macrofauna of a single riffle in Prosser Creek, California as indicated by a Surber sampler. *Hilgardia* 24:383–409.

Nelson, R.W., G.C. Horak, and J.E. Olson. 1978. *Western Reservoir and Stream Habitat Improvements Handbook.* USDI FWS/OBS-78/56. Western Energy and Land Use Team, Fort Collins, CO.

Newman, M.A. 1956. Social behavior and interspecific competition in two trout species. *Physiol. Zool.* 29:64–81.

Nunnally, N.R. 1978. Stream renovation: an alternative to channelization. *Environmental Management* 2(5):403–11.

Odum, E.P. 1959. *Fundamentals of Ecology.* 2d ed. Philadelphia and London: W.B. Saunders Co. 88–89.

Otis, M.B. 1974. Stream improvement. In *The Stream Conservation Handbook,* edited by J.M. Migel, 99–122. New York: Crown Publishers.

Patrick, R. 1973. Effects of channelization on the aquatic life of streams. In *Environmental Considerations in Planning, Design, and Construction.* Highway Research Board, Division of Engineering, National Research Council, National Academy of Sciences-National Academy of Engineering. Special Report 138. Washington, D.C.

Pearson, L.S., K.R. Conover, and R.E. Sams. 1970. Factors affecting the natural rearing of juvenile coho salmon during the summer low flow season. Unpublished report. Oregon Fish. Comm.

Pennak, R.W., and E.D. Van Gerpen. 1947. Bottom fauna production and physical nature of the substrate in a northern Colorado trout stream. *Ecology* 28(1):42–48.

Peters, J.C., and W. Alvord. 1964. Man-made channel alterations in thirteen Montana streams and rivers. *Trans. N.A. Wildl. Conf.* 29:93–102.

Pfankuch, D.J. 1975. *Stream Reach Inventory and Channel Stability Evaluation.* R1-75-002. U.S. Forest Service, Northern Region.

Platts, W.S., and J.N. Rinne. 1982. Riparian-stream protection and enhancement research in the Rocky Mountains. In *Proc. of Rocky Mt. Stream Habitat Management Workshop. See* Bailey 1982.

Raleigh, R.F. 1978. Habitat evaluation procedure for aquatic assessments, pp. 131–137. In *Proc. of Workshop on Methods for the Assessment and Prediction of Mineral Mining Impacts on Aquatic Communities: A Review and Analysis,* edited by W.T. Mason, Jr. USDI Fish and Wildlife Service. Washington, D.C.: U.S. Government Printing Office.

———. 1982. *Habitat Suitability Index Models: Brook Trout.* FWS/OBS-82/10.24. Washington, D.C.: U.S. Government Printing Office.

Raleigh, R.F., and D.A. Duff. 1980. Trout stream habitat improvement: ecology and management. In *Proc. of Wild Trout Symposium II,* edited by W. King.

Reimers, N. 1957. Some aspects of the relation between foods and trout survival. *Calif. Fish and Game* 43:43–69.

Reiser, D.W., and T.A. Wesche. 1977. *Determination of Physical and Hydraulic Preferences of Brown and Brook Trout in the Selection of Spawning Locations.* Water Resources Series No. 64. Laramie: Univ. of Wyoming.

Richard, J.B. 1963. *Long Stream Improvement Devices and Their Effects Upon the Fish Population, South Fork Mokelumne River, Calaveras County.* Inland Fisheries Admin. Report No. 63-7. California Dept. of Fish and Game, Sacramento.

Rinne, J.N., and J. Stefferud. 1982. Stream habitat improvement in the southeastern United States, Arizona, and New Mexico. In *Proc. of Rocky Mt. Stream Habitat Management Workshop. See* Bailey 1982.

Robinson, J.L. 1982. Development and testing of a stream morphology evaluation method for measuring user impact on riparian zones. Master's thesis, Univ. of Wyoming, Laramie.

Rockett, L.C. 1979. *The Influence of Habitat Improvement Structures on the Fish Population and Physical Characteristics of Blacktail Creek, Crook County, Wyoming.* Report on Project 3079-08-7601. Cheyenne: Wyoming Game and Fish Dept.

Rockett, L.C., and J.W. Mueller. 1968. *Stream Improvement Evaluations as Related to Fish Populations in Bear Trap Creek, Johnson County.* Admin. Report on Project 0367-08-6602. Cheyenne: Wyoming Game and Fish Dept.

Rollefson, M.D., and J.A. Erickson. 1970. *Seven-Year Study on Four Spring Creeks in Upper Star Valley with Special Reference to Brook Trout Stocking and Limited Stream Improvement Measures.* Admin. Report on Project 0169-08-6203. Cheyenne: Wyoming Game and Fish Dept.

Ruggles, C.P. 1966. Depth and velocity as a factor in stream rearing and production of juvenile coho salmon. *Canadian Fish Culturist* 38:37–53.

Ruttner, F. 1953. *Fundamentals of Ecology.* Translation by D.G. Frey and F.E.J. Frey. Toronto: Univ. of Toronto Press.

Sams, R.E. and L.S. Pearson. 1963. A study to develop methods for determining spawning flows for anadromous salmonids. Report of Oregon Fish Comm., Portland.

Saunders, J.W. and M.W. Smith. 1962. Physical alteration of stream habitat to improve brook trout production. *Trans. Amer. Fish. Soc.* 91:185–88.

Schmal, R.N., and D.F. Sanders. 1978. *Effects of Stream Channelization on Aquatic Macroinvertebrates, Buena Vista Marsh, Portage County, Wisconsin.* FWS/OBS-78/92. Washington, D.C.: U.S. Government Printing Office.

Scott, D. 1958. Ecological studies on the Trichoptera of the River Dean, Cheshire. *Arch. Hydrobiol* 54:340–92.

Seehorn, M.E. 1982. Trout stream improvements commonly used on southeastern National Forests. In *Proc. of Rocky Mt. Stream Habitat Management Workshop. See* Bailey 1982.

Shaw, M.A. 1982. Wasatch-Cache National Forest, Utah. In *Proc. of Rocky Mt. Stream Habitat Management Workshop. See* Bailey 1982.

Sheeter, G.R., and E.W. Claire. 1981. *Use of Juniper Trees to Stabilize Eroding Streambanks on the South Fork John Day River.* Technical Note OR-1, BLM 796-119 (2003). Washington, D.C.: U.S. Government Printing Office.

Shetter, D.S., O.H. Clark, and A.S. Hazard. 1946. The effects of deflectors in a section of a Michigan trout stream. *Trans. Amer. Fish. Soc.* 76:248–78.

Shumway, D.L., C.E. Warren, and P. Doudoroff. 1964. Influence of oxygen concentration and water movement on the growth of steelhead trout and coho salmon embryos. *Trans. Amer. Fish. Soc.* 93(4): 342–56.

Skinner, M.M., and M.D. Stone. 1982. *Identification of Instream Hazards to Trout Habitat Quality in Wyoming.* Completion Report on FWS/OBS 14-16-0009-78-80. Cheyenne: Wyoming Game and Fish Dept.

Smith, A.K. 1973. Development and application of spawning velocity and depth criteria for Oregon salmonids. *Trans. Amer. Fish. Soc.* 102:312–16.

Smith, B.H. 1980. Not all beaver are bad; or, an ecosystem approach to stream habitat management, with possible software. In *Proc. of 15th Annual Meeting, Colorado-Wyoming Chapter, Amer. Fish. Soc.*

Sprules, W.M. 1947. An ecological investigation of stream insects in Algonquin Park, Ontario. *Univ. of Toronto Studies, Biol. 56 Pub. Ont. Fish. Res. Lab.* 69:1–81.

Stern, D.H., and M.S. Stern. 1980. *Effects of Bank Stabilization on the Physical and Chemical Characteristics of Streams and Rivers: An Annotated Bibliography.* FWS/OBS-80/12. Washington, D.C.: U.S. Government Printing Office.

Stuart, T.A. 1953. Spawning Migration, Reproduction and Young Stages of Loch Trout (S. trutta). Report No. 5. Fresh. and Sal. Fish. Res., Home Dept., Scotland.

Swales, S. 1982. Notes on the construction, installation and environmental effects of habitat improvement in a small lowland river in Shropshire. *Fish. Mgmt.* 13(1):1–10.

Tarzwell, C.M. 1932. Trout stream improvement in Michigan. *Trans. Amer. Fish. Soc.* 61:48–57.

———. 1936. Experimental evidence on the value of trout stream improvements in Michigan. *Trans. Amer. Fish. Soc.* 66:177–87.

Thompson, K.E. 1972. Determining streamflows for fish life. In *Proc. Instream Flow Requirements Workshop, Pacific N.W. River Basins Comm., Portland, Oregon, 1972,* Vancouver, WA.

———. 1974. Salmonids. In *The Anatomy of a River,* edited by K. Bayha. Pacific N.W. River Basins Comm. Report. Vancouver, WA.

U.S. Bureau of Fisheries. 1935. *Methods for the Improvement of Streams.* Memorandum I-133. Washington, D.C.: U.S. Dept. of Commerce.

U.S. Bureau of Land Management. 1968. Stream preservation and improvement. Section 6760 of USDI, BLM Manual.

U.S. Forest Service. 1952. *Fish Stream Improvement Handbook.* 0-232379 Washington, D.C.: U.S. Government Printing Office.

———. 1969. *Wildlife Habitat Improvement Handbook.* FSH 2609.11. Washington, D.C.: U.S. Government Printing Office.

Vaux, W.B. 1962. *Interchange of Stream and Intergravel Water in a Salmon Spawning Riffle.* Spec. Sci. Rept. Fish. 405. Washington, D.C.: U.S. Fish and Wildlife Service.

Warner, K., and I.R. Porter. 1960. Experimental improvement of a bulldozed trout stream in northern Maine. *Trans. Amer. Fish. Soc.* 89:59–63.

Waters, T.F. 1969. Invertebrate drift-ecology and significance to stream fishes. In *Symposium on Salmon and Trout in Streams. See* Allen 1969.

Webster, D.A., and G. Eiriksdotter. 1976. Upwelling water as a factor influencing choice of spawning sites by brook trout (*Salvelinus fontinalis*). *Trans. Amer. Fish. Soc.* 105(3): 416–21.

Wesche, T.A. 1976. *Fall 1975 Evaluation of Blacks Fork River Improvement Structures.* Water Resources Research Institute Report. Laramie: Univ. of Wyoming.

______. 1973. *Parametric Determination of Minimum Stream Flow for Trout.* Water Resources Series No. 37. Laramie: Univ. of Wyoming.

______. 1974. *Relationship of Discharge Reductions to Available Trout Habitat for Recommending Suitable Streamflows.* Water Resources Series No. 53. Laramie: Univ. of Wyoming.

______. 1980. *The WRRI Trout Cover Rating Method, Development and Application.* Water Resources Series No. 78. Laramie: Univ. of Wyoming.

White, H.C. 1930. Some observations on eastern brook trout (*S. fontinalis*) of Prince Edward Island. *Trans. Amer. Fish. Soc.* 60:101–8.

White, R.J., and O.M. Brynildson. 1967. *Guidelines for Management of Trout Stream Habitat in Wisconsin.* Technical Bulletin No. 39. Madison: Wisconsin Dept. of Natural Resources.

Wickham, G.M. 1967. Physical microhabitat of trout. Master's thesis, Colorado State Univ., Ft. Collins.

Wisconsin Department of Natural Resources. 1980. *Wisconsin Trout Stream Habitat Management.* Madison, Wisconsin.

Wydoski, R.S., and D.A. Duff. 1982. A review of stream habitat improvement as a fishery management tool and its application to the Intermountain West. In *Proc. of Rocky Mt. Stream Habitat Management Workshop. See* Bailey 1982.

Wydoski, R.S., and W.T. Helm. 1980. *Effects of Alterations to Low Gradient Reaches of Utah Streams.* FWS/OBS-80/14. Washington, D.C.: U.S. Government Printing Office.

Wyoming Game and Fish Department. 1982a. Structures used in Wyoming. In *Proc. of Rocky Mt. Stream Habitat Management Workshop. See* Bailey 1982.

______. 1982b. Fontenelle reservoir tailwaters habitat development project. In *Proc. of Rocky Mt. Stream Habitat Management Workshop. See* Bailey 1982.

______. 1982c. Habitat improvement and gravel rejuvenation on a Snake River Tributary, Three Channel Spring Creek. In *Proc. of Rocky Mt. Stream Habitat Management Workshop. See* Bailey 1982.

York, T.H. 1978. *Impact Assessment of Water Resource Development Activities: A Dual Matrix Approach.* FWS/OBS-78/82. Washington, D.C.: U.S. Government Printing Office.

CHAPTER 6

Methods for Determining Successful Reclamation of Stream Ecosystems

Robert N. Winget, Ph.D.
Aquatic Ecology Laboratory
Department of Zoology
Brigham Young University
Provo, Utah 84602

Reclamation denotes the act of "rescuing from an undesirable or unhealthy state . . . making fit for cultivation or use" (Webster's Third New International Dictionary). Therefore, this chapter deals with evaluating the success of efforts to rescue streams from an undesirable state and transform them into healthy, productive, and usable resources.

Previous chapters of this book have addressed reclamation and management of the physical and chemical environs of streams—Hasfurther (meanders and hydrologic balance), Herricks (water quality), Anderson and Ohmart (riparian revegetation), Gore (physical habitat for macroinvertebrates), and Wesche (physical habitat for fish). This chapter describes several selected methods for the biological evaluation of the success of stream habitat and water quality reclamation and management measures.

It is assumed the "desirable state" of a stream ecosystem, the basis of evaluating success of reclamation, closely resembles that encountered under natural, unperturbed, unmodified stream ecosystems. Each stream, although possessing characteristics apparently similar to other streams, is unique. This uniqueness can be the result of present and historical differences (Curry 1977) in water quality (Hart and Fuller 1974; Vitousek 1977; Fortescue 1980), water temperature regimes (Hynes 1972; Macan 1974); upstream land and water use (Golterman 1975; Ringler and Hall 1975; Vitousek and Reiners 1975; Bakke 1977; Bormann and Likens 1979), stream gradient and elevation (Fisher and Likens 1973; Mueller-Dombois and Ellenberg 1974; Hawkes 1975; Reice 1977; Stoneburner 1977; Wiggins 1977), and annual hydrographs (Beaumont 1975).

It is also assumed that stream sections classified as similar, based upon physical and chemical stream descriptors, have similar biotic communities. This does not mean identical species in all similar communities, but rather corresponding species in identical functional groups (Pennak 1971). Aquatic species respond individually to environmental changes, each being found within definite conditions according to their evolutionary exposure or experience (Margalef 1969; Cairns et al. 1973; Hill 1975; Curry 1977). This environmental limitation to distribution or occurrence has been called *niche width* or *breadth* (Colwell and Futuyma 1971; Pielou 1972).

Changes in the physical or chemical environment will elicit adjustments in the biotic community of that environment (Hill 1975). Cairns et al. (1973) theorized that it would be difficult to modify or change community structure without also changing function. They further indicated that by managing or restoring structure, function would also be managed or restored. Severity and duration of environmental changes determine the magnitude of biotic changes (Margalef 1969). Changes may include shifts in relative numbers of individuals among species (Margalef 1958, 1969; Brillouin 1960; Shannon and Weaver 1963; Pielou 1966; Wilhm 1967, 1970; Wilhm and Dorris 1968; Cairns et al. 1973; Nuttall and Purves 1974; Hill 1975), alterations in rates and quantities of material cycles including energy flow (Odum 1969; Fisher and Likens 1973; Cummins 1974, 1977; Petersen and Cummins 1974), and changes in species composition (Gaufin 1973; Patrick 1969).

Following reclamation of a stream's physical and chemical environment, the biotic community should tend to assume the structure and function of a community in a similar unperturbed stream. This characteristic of communities to resume natural structure and function has been defined as the resilience of the community (Margalef 1969; Holling 1973; Hill 1975).

INDICES OF COMMUNITY STRUCTURE

Species Diversity, Evenness, and Redundancy

Many biotic indices have been formulated to describe effects of changing environmental conditions on biotic communities. *Dominance diversity indices,* collectively referred to as H (Margalef 1958; Brillouin 1960; Shannon and Weaver 1963; Cairns and Dickson 1971; Cairns et al. 1973), are probably the most widely used in impact assessment investigations. These indices are based upon information theory—maximum sample information exists when each individual in the sample belongs to a different species, and conversely, minimum information exists when all individuals belong to the same species. As pointed out by Wilhm and Dorris (1968) an unfavorable environmental factor, such as organic pollution, results in detectable changes in community structure. More information is supposedly contained in a natural than a polluted community. The polluted system is supposedly simplified and those species that survive encounter less competition and are usually able to increase in numbers. Redundancy increases in this case because the probability that an individual belongs to a previously recognized species increases and information per individual is reduced.

Masnik et al. (1976) reported no relationship between the proximity of an oil spill in Plum Creek and *H* values of either fish or macroinvertebrates. Hoehn et al. (1974) concluded that the effects of the oil spill were toxic effects since *H* did not change even though the total numbers of organisms decreased. This indicated each taxon within the community was affected equally rather than selectively. Hendricks et al. (1974) concluded that *H* was not adequate in defining impacts from a paper mill on the aquatic macroinvertebrates. Diversity as measured by *H* was more influenced by natural conditions of the river than from impacts of the paper mill. MacKay et al. (1973), after reviewing Shannon-Weaver index values from fifty unpolluted streams with a wide range of chemical, geological, and physical characteristics, concluded that *H* values are open to individual interpretations as to what levels are indicative of pollution.

Mathis (1968) concluded that high mountain streams and lowland moderately polluted streams have maximum *H* values. He theorized that moderate pollution may not lower *H* because moderately tolerant species probably replace fragile ones accompanied with an equalizing of number distribution over species present. Cook (1976) reported mild amounts of pollution may be associated with increases in *H* values. She also reported large amounts of organic enrichment lowered *H* by favoring only certain species. Harrel and Dorris (1968) reported that in areas of moderate pollution from oil field brines, *H* values increased above that expected in natural stream levels. But, in zones of heavy pollution, *H* decreased below natural levels. They also reported annual number of species and total community *H* (Patten 1962) increased in third, fourth, and fifth order streams but decreased in sixth order streams. Proceeding downstream, Otter Creek showed the same general pattern in *H* and redundancy as reported for streams that receive pollutional effluents. They emphasized the pattern of diversity in Otter Creek was due to physiographic stream succession, rather than the influence of effluents. Wilhm (1965) and Mathis (1965) reported similar findings.

To many, it is surprising to note the increase in *H* in moderately stressed or polluted situations compared with natural systems. Three probable reasons for this are: (1) when the perturbation causes fragile species to be replaced by more hardy species, and these account for a higher percentage of total numbers than the replaced fragile species, this would increase evenness and reduce redundancy; (2) when the perturbation is nonselective, such as a toxicant, it has more impact on the most numerous species causing a reduction in their numbers and increased community evenness; and (3) addition of a substance may provide a limiting chemical, allowing additional species to become established. High mountain streams are often devoid of essential nutrients and dissolved solids. As these streams proceed through various strata they often pick up essential elements and the community responds—number of species increases (Harrel and Dorris 1968) with an increase of total numbers more evenly distributed over available species. As streams proceed into the valleys, dissolved solids, salts, and nutrients often increase to nuisance levels, resulting in decreases in number of species and increased dominance of the more tolerant species.

The initial naming of the information content of a message as the diversity index might have been an oversimplification of a very complicated and intuitive diversity phenomenon (Zand 1976). It is apparent *H* values are open to criticism and any interpretation should be accompanied with substantial supportive data such as number of taxa (DeJong 1975), density, standing crop (Patrick 1973), rather complete species lists (Resh and Unzicker 1975), and physical and chemical descriptions of the systems (Mathis 1968; MacKay et al. 1973).

Most of the early work was concerned with problems resulting mainly from organic enrichment. Patrick (1973) reported that through analyzing ". . . shifts in species composition and structure of the community, the effect of a pollutant can be estimated." She supports the view that aquatic species [algae] differ in their resistance to various pollution types so that under one pollution a species would be called tolerant but under other conditions intolerant. Thus, species tolerance classifications based upon reaction to organic pollution are often invalid for other types of environmental changes.

Upon graphing number of species versus number of individuals per species, Partick (1973) determined that the structure of the resultant curve had great reliability in maintaining its shape. Under natural freshwater stream conditions a sample of a diatom community was represented by a truncated normal curve with an expected height and covering 10 to 12 density intervals. Organic pollution reduced the height of the curve but 13 to 15 density intervals were covered indicating excessive numbers for some species. Toxic pollution also reduced the height of the curve but only 11 to 13 density intervals were covered. She found that various pollutant types and intensities of pollution elicited different community responses and concluded, " . . . one should consider not only the structure of the . . . community but also the kinds of species and the total biomass." This system is an improvement over the dominance diversity indices in that it presents the number of species plus individuals, both largely ignored by *H*.

Hydrology, Geomorphology, and Community Structure Classifications

Workers have classified streams according to the geomorphology and hydrology of the systems (Horton 1945; Bruce and Clark 1966; Shreve 1966; Beaumont 1975; Hawkes 1975). Significant correlations between fish distributions and various stream categories have been reported. Other workers have expanded this concept to include structure and function of biotic communities (Fisher and Likens 1973; Cummins 1974, 1975; Minshall 1978). The stream continuum concept, the latest proposed system (Vannote et al. 1980), has done much to elucidate the general ecological progression of a stream as it proceeds from its headwaters to the valley floor. The continuum concept is a valuable tool in describing what to expect from a stream reach in relation to its place in a drainage.

AN INTEGRATED PHYSICAL, CHEMICAL, AND BIOLOGICAL METHODOLOGY

Several workers, realizing the importance of ecological information about individual species and their relative abundance in determining community diversity and condition have proposed various indices. These are based upon the tolerance-intolerance of species to environmental perturbations. Beck (1954, 1955) developed such an index, based largely upon the early work of Kolkwitz and Marsson (1908, 1909). Patrick (1968, 1973), Wurtz (1955), Chandler (1970), Chutter (1972), Cook (1976), Hilsenhoff (1977), and Winget and Mangum (1979) have also proposed biological classification systems.

In an unpublished review (Voshell 1980), it was concluded that only recently have biological classification systems been proposed that have the capabilities to synthesize a wide variety of ecological parameters and have predictive value for decision making and management. The overall integrative approach of two of these recent classification systems (Hilsenhoff 1977; Winget and Mangum 1979) offers much promise for the interpretation and application of data from benthic macroinvertebrate samples. The biotic index formulated by Hilsenhoff (1977) was summarized and evaluated by Hilsenhoff (1982) and Narf et al. (1984). It appears to be a valuable tool for evaluating water quality impacted by organic enrichment and low dissolved oxygen levels, but is not designed for use in monitoring other perturbations. The biotic condition index formulated by Winget and Mangum (1979) was selected for further discussion, as follows.

Purpose of the Biotic Condition Index

The purpose of the biotic condition index (BCI) is to provide a methodology for evaluating existing conditions of stream macroinvertebrate communities based upon their biological potential. Evaluations are based upon water quality, physical habitat, and aquatic biota data. The BCI is designed to be used in conjunction with water quality, physical habitat, and hydrological studies.

Physical and Chemical Independent Variables of the BCI

Data from 252 sets (four quantitative samples per set) of aquatic macroinvertebrate samples from ninety-eight stations on twenty-eight different streams in Utah and five other western states were compiled. Selected water quality and physical habitat parameter measurements for each were included. Correlation between aquatic community density, biomass, and diversity (dependent variables) and various physical and chemical stream characteristics (independent variables) were analyzed.

Four independent variables were selected for use in the *BCI* because: (1) there were strong relationships between them and the dependent variables; and (2) each represented a different type of influence on the community structure or function. The independent variables selected were stream channel percent gradient (stream maintenance and recovery component), substrate roughness (microhabitat heterogeneity component), total alkalinity (density and biomass component), and sulfate concentrations (water quality component).

Percent stream gradient was selected because of its positive correlation with macroinvertebrate community diversity and its relationship to the ability of a stream to maintain substrate quality. The stratum determines possible substrate materials available to the stream. Annual stream discharge, especially seasonal high and low flows, plus channel gradient, greatly influences the erosion and deposition patterns in a stream. Water velocity increases with gradient even if total discharge remains constant. Water temperature and dissolved oxygen are also related to gradient. Recovery from perturbations such as thermal pollution or oxygen depletion is quicker in streams with higher gradients and coarser substrates. Fisher and Likens (1973) reported gradient reflects energy flows or subsidy in oxygenation and food transport in streams. Huet (1959) considered gradient the primary feature characterizing different stream zones.

There was a strong relationship between macroinvertebrate community structure and *stream substrate roughness.* The preferred substrates for the greatest number of species were boulder and rubble with limited interstitial gravel, sand, and organic fines. Sand and silt dominant substrates were largely avoided. Gravel alone showed no correlation but mixed with rubble and boulder was highly selected. Due to increased microhabitat diversity there are more possibilities for community diversity in larger substrates than in finer substrates. Substrate roughness was calculated using the following information:

$$\text{Substrate Roughness} = \frac{(5 \times C_1) + (3 \times C_2) + C_3}{9} \tag{6.1}$$

where

Subscript:
1 = most dominant substrate type
2 = second most dominant substrate type
3 = third most dominant

Coarseness values (C):

Substrate size class	C
boulder >30.5 cm diam (>1 ft)	= 4
rubble 8–30 cm diam (3–12 in)	= 3
gravel 2.5–7.5 cm diam (1–3 in)	= 2
fines <2.5 cm diam (<1 in)	= 1

Total alkalinity was selected because of its relationship to community density and standing crop or biomass. Alkalinity is important in primary production, especially in the absence, or near absence of free carbon dioxide. Many species of algae can utilize bicarbonate ions as a CO_2 source for use in photosynthesis. Specific conductance showed similar correlations as total alkalinity but the relationships were not as strong. Water hardness could probably be substituted for total alkalinity.

Sulfate was selected because an increase in concentration is generally indicative of natural water quality deterioration and macroinvertebrate community diversity (number of taxa) correlated negatively with sulfate concentrations. Many geological formations contribute fine sediments and dissolved solids, including sulfates, to waters intersecting them. Agricultural irrigation return flows often have high sulfate levels, as do alkaline, saline, and thermal ground waters. It is not known whether the aquatic taxa are responding directly to sulfate or whether sulfate is an indicator of other conditions responsible for the noted community changes. Total dissolved solids could probably be used in place of sulfate.

For the ninety-eight stream stations used in this model, the mean gradient was 1.87% with a standard deviation (SD) of 1.30; mean substrate roughness was 2.46 (SD 0.58); mean total alkalinity was 167.2 mg/l $CaCO_3$ (SD 182.3 mg/l); and mean sulfate concentration was 95.2 mg/l SO_4 (SD 216.5 mg/l).

One measure of selectivity of a species for a particular range of an environmental parameter is the relationship between the mean level available to a taxon (the mean value of that parameter for all stations studied) and the mean level at the stations it occurred at. If the mean gradient for stations inhabited by a species is noticeably higher than the overall mean stream gradient (such as *Hesperoperla pacifica,* Figure 6.1), then that species was found more frequently at higher gradient

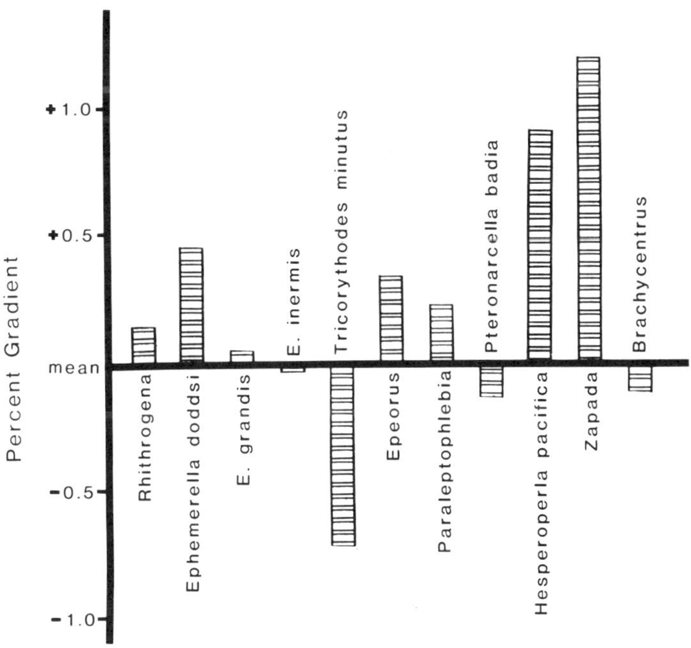

Figure 6.1. Relationship of mean percent stream gradient for stations at which each of 11 taxa occurred compared with mean for all stream stations studied.

stations than lower gradient stations. If the mean gradient for a species was nearly equal to the mean for all stations (such as *Ephemerella inermis*), care must be taken in interpreting that relationship. It may mean that species exhibited: (1) no selectivity for gradient; (2) strong selection for gradient equal to the overall mean; or (3) selectivity somewhere between. As in the case of *E. inermis*, selectivity was rather general except gradient less than 1.0% and greater than 4.0% were avoided.

Of eleven selected taxa, *Zapada* and *Hesperoperla pacifica* stoneflies showed the strongest positive variance in relation to mean gradient (Figure 6.1). *Tricorythodes minutus* exhibited a strong negative variance indicating a strong selection for lower gradient stream reaches. Of the eleven taxa shown, only *Tricorythodes minutus* showed a selection of substrates having less roughness than the overall average (Figure 6.2). *Ephemerella doddsi* showed the strongest selection for stations having higher than average substrate roughness. Nine of the eleven taxa illustrated were collected more frequently from stations with total alkalinity noticeably less than the total mean (Figure 6.3). This is not surprising since alkalinity concentrations correlated negatively with number of taxa but positively with community numbers and biomass. *Tricorythodes minutus* selected for higher alkalinity. Ten taxa exhibited strong selectivity for lower than average sulfate levels (Figure 6.4). *Tricorythodes minutus* was the only taxon selecting for stations with higher than average sulfate levels.

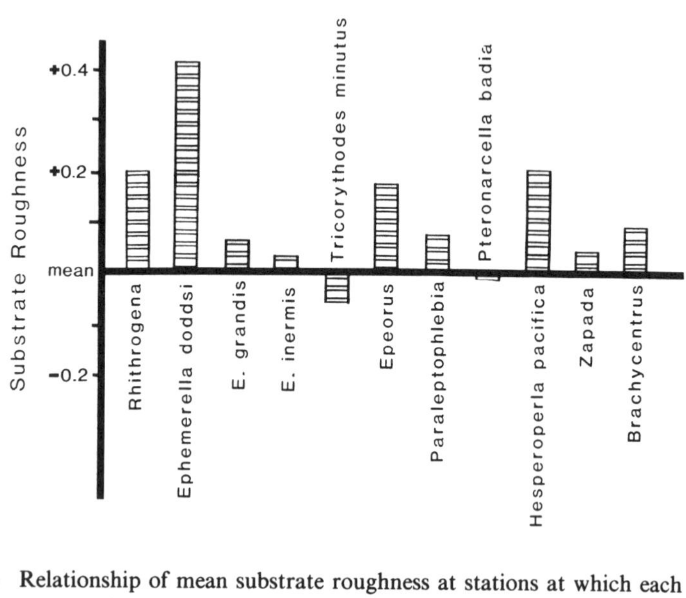

Figure 6.2 Relationship of mean substrate roughness at stations at which each of 11 taxa occurred compared with mean for all stream stations studied.

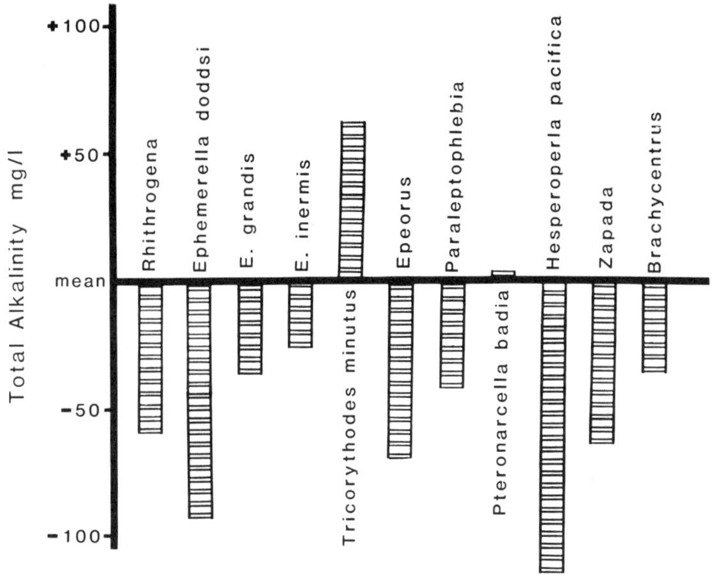

Figure 6.3 Relationship of mean total alkalinity concentrations at stations at which each of 11 taxa occurred compared with mean concentrations for all stations studied.

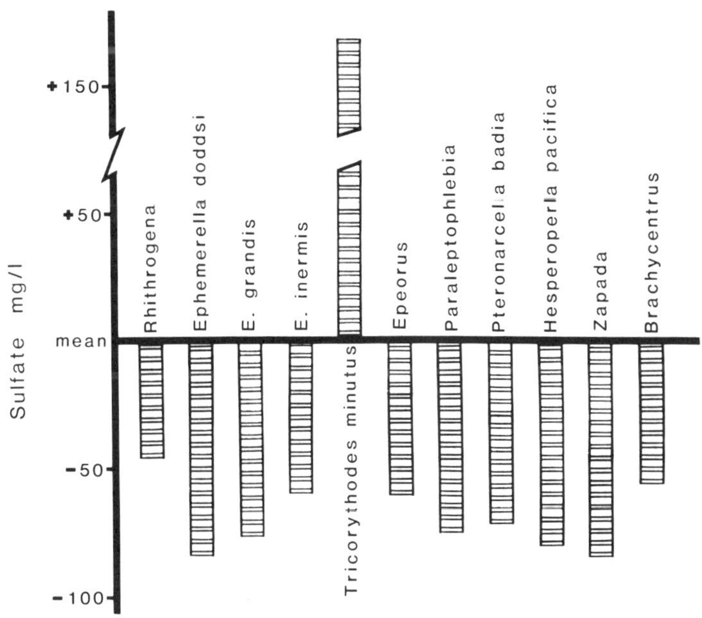

Figure 6.4 Relationship of mean sulfate concentrations at stations at which each of 11 taxa occurred compared with mean concentrations for all stations studied.

Tolerance Quotients—*TQ*

The tolerance quotient (*TQ*) is the product of four component values derived from a taxon's tolerance to low stream gradients, fine substrate materials, and high total alkalinity and sulfate levels. Tolerance quotients were calculated using the following information:

$$TQ_i = G_i \times SR_i \times TA_i \times S_i \tag{6.2}$$

where

G_i = gradient tolerance value for the ith species
SR_i = substrate roughness tolerance value for the ith species
TA_i = total alkalinity tolerance value for the ith species
S_i = sulfate tolerance value for the ith species

Gradient:	>3.0%	1.3–3.0%		<1.3%
Tolerance value:	1	2		3
Substrate roughness:	>3.0	2.5–3.0	1.9–2.4	<1.9
Tolerance value:	1	2	3	4
Total alkalinity:	<200 mg/l	200–300 mg/l		>300 mg/l
Tolerance value:	1	2		3
Sulfate:	<150 mg/l	150–300 mg/l		>300 mg/l
Tolerance value:	1	2		3

Determination of tolerance of a taxon to each independent variable is mostly objective being derived from actual sampling results. Of course there are instances when a taxon does not show a clear selection or avoidance according to the artificial categories selected (more categories for each independent variable would increase sensitivity of the TQs). In such cases values have to be assigned subjectively, hopefully based upon considerable experience of the worker or an outside consultant.

Tables 6.1, 6.2, and 6.3 show the basis for assigning tolerance quotients (*TQ*) to three species of *Ephemerella* mayflies, *E. doddsi, E. grandis,* and *E. inermis,* respectively. The values in the tables are ranked according to increasing density of each species per sample. When determining preference or avoidance for various independent variable values, more weight is given to those values in the higher density categories. *Ephemerella doddsi,* a species with narrow environmental tolerances, preferred gradients of 2.0 percent and greater ($G = 2$), substrate roughness

Table 6.1 Environmental selection by *Ephemerella doddsi* as defined by stream percent gradient, substrate roughness, water total alkalinity, and sulfate levels, ranked as to mean density of *E. doddsi* per sample set. Values given below are: *m* = mean number per meter square; *SD* = standard deviation of the mean; and *r* = range of mean number per sample set.

	Density	*Gradient %*	*Substrate roughness*	*Alkalinity mg/l*	*Sulfate mg/l*
m	11.3	2.1	3.0	46.8	11.3
SD	4.2	1.1	0.5	47.6	8.2
r	3–16	1.0–5.0	1.8–3.4	11–130	3–22
m	22.3	2.0	3.0	46.7	7.2
SD	2.3	0.9	0.4	52.4	6.0
r	17–22	1.0–4.3	2.2–3.2	11–150	3–21
m	41.7	2.7	2.8	67.4	13.0
SD	8.1	1.8	0.5	50.6	10.2
r	29–54	1.3–6.9	2.2–3.2	12–130	3–25
m	72.4	2.5	2.9	61.1	10.5
SD	10.8	1.4	0.5	52.6	9.8
r	59–86	1.0–5.0	2.2–3.2	14–150	3–34
m	116.3	2.1	2.8	78.0	10.1
SD	21.9	0.3	0.5	53.5	6.1
r	87–151	1.9–2.7	2.2–3.4	11–130	5–22
m	272.0	2.5	2.8	78.0	8.4
SD	102.1	1.5	0.4	54.7	6.7
r	161–484	1.0–6.9	2.2–3.2	12–130	3–21

between 2.5 and 3.0 (*SR* = 2), total alkalinities less than 150 mg/l (*TA* = 1), and sulfates less than 30 mg/l (*S* = 1). The resultant *TQ* for *E. doddsi* was 4. *Ephemerella grandis* was commonly found at stations with stream gradients of 1.0 (*G* = 3), low substrate roughness (*SR* = 4) and total alkalinities over 200 mg/l (*TA* = 2), but strongly avoided high sulfate levels (*S* = 1). The resultant *TQ* for *E. grandis* was 24. *Ephemerella inermis* is more tolerant of broader environmental conditions than either *E. doddsi* or *E. grandis.* The *TQ* for *E. inermis* was 48 (*G* = 3, *SR* = 4, *TA* = 2, and *S* = 2).

Tolerance quotients were originally calculated for fifty-four taxa. A cluster dendrogram, based upon the Jaccard Similarity Coefficient, depicts the frequency of co-occurrence of these fifty-four taxa (Figure 6.5). The clustering of taxa with similar *TQ*s illustrates the credibility of the determinants selected, the weighted values given each, and the *TQ*s assigned. From clustering analysis of numerous sample data plus other similar analyses of community associations, numerous taxa have been assigned *TQ*s. Table 6.4 contains a partial list of these taxa with their preliminary *TQ* values.

Table 6.2 Environmental selection by *Ephemerella grandis* as defined by stream percent gradient, substrate roughness, water total alkalinity, and sulfate levels, ranked as to mean density of *E. grandis* per sample set. Values given below are: *m* = mean number per meter square; *SD* = standard deviation of the mean; and *r* = range of mean number per sample set.

	Density	*Gradient %*	*Substrate roughness*	*Alkalinity mg/l*	*Sulfate mg/l*
m	13.2	2.5	2.8	66.6	10.7
SD	3.8	1.4	0.5	70.2	8.3
r	2–22	0.8–5.0	1.8–3.4	11–250	3–42
m	33.2	2.4	2.6	85.9	22.4
SD	4.3	1.0	0.4	57.7	25.4
r	23–43	0.7–4.5	1.8–3.4	8–240	3–111
m	64.8	1.5	2.6	143.2	24.2
SD	13.8	0.4	0.4	84.1	30.6
r	44–86	0.8–2.0	2.2–3.2	14–245	3–108
m	104.4	1.6	2.6	124.7	27.1
SD	14.7	0.5	0.7	94.3	36.7
r	86–129	0.7–2.0	1.6–3.2	14–205	3–111
m	180.3	1.5	2.6	160.5	22.0
SD	37.8	0.6	0.6	84.4	18.6
r	131–258	0.5–2.3	1.6–3.4	25–308	7–74
m	372.2	1.6	2.4	194.0	18.4
SD	78.2	0.4	0.4	51.2	17.2
r	259–495	1.0–2.3	1.8–3.4	70–260	3–61
m	662.5	1.6	2.4	178.5	9.5
SD	100.3	0.4	0.7	86.0	5.7
r	524–818	1.0–2.5	1.8–3.4	24–241	6–22
m	1877.7	1.4	2.4	198.7	14.5
SD	721.7	0.3	0.5	66.2	11.1
r	987–2388	1.0–2.0	1.8–3.2	25–260	6–36

Predicted Community Tolerance Quotients—*CTQp*

The predicted community tolerance quotient (*CTQp*) is the mean of the *TQ*s for a predicted macroinvertebrate community. Due to its individual water quality and physical habitat characteristics, each stream section is expected to support a community reflective of those traits. Similarity index matrices, correlation coefficient analyses, and Jaccard similarity coefficient clustering analyses were used to select representative taxa likely to occur together under various combinations of habitat and water quality conditions. Tolerance quotients for the taxa in the predicted community were summed and then divided by the number of representative taxa giving an arithmetic mean tolerance quotient for that community (*CTQp*). Table 6.5 contains a key for obtaining *CTQp* values for stream sections.

Table 6.3 Environmental selection by *Ephemerella inermis* as defined by stream percent gradient, substrate roughness, water total alkalinity, and sulfate levels, ranked as to mean density of *E. inermis* per sample set. Values given below are: *m* = mean number per meter square; *SD* = standard deviation of the mean; and *r* = range of mean number per sample set.

	Density	*Gradient %*	*Substrate roughness*	*Alkalinity mg/l*	*Sulfate mg/l*
m	27.6	2.7	2.4	133.1	42.2
SD	10.4	1.4	0.5	106.8	84.5
r	7–67	0.8–5.0	1.6–3.2	12–397	3–615
m	125.5	1.7	2.6	144.2	55.5
SD	18.6	0.8	0.5	101.1	112.4
r	73–183	0.3–4.0	1.6–3.4	11–341	5–782
m	223.2	1.8	2.7	117.3	18.9
SD	27.9	1.0	0.6	86.7	35.9
r	183–258	1.2–5.0	1.6–3.4	12–241	3–150
m	304.2	1.5	2.9	127.1	18.9
SD	32.9	0.4	0.5	116.6	18.4
r	258–355	0.7–2.1	1.6–3.4	11–330	3–74
m	442.3	1.7	2.6	113.7	18.9
SD	59.7	0.5	0.5	73.5	27.7
r	360–516	0.6–2.7	1.8–3.4	14–245	3–120
m	672.7	1.6	2.6	116.9	48.6
SD	98.7	0.9	0.6	90.8	70.4
r	528–839	0.2–4.0	1.6–3.4	12–240	3–195
m	1262.8	1.6	2.3	190.3	36.8
SD	294.8	0.6	0.5	73.0	48.2
r	853–1886	0.7–2.7	1.8–3.4	120–375	3–138
m	3654.7	1.4	2.4	197.3	23.1
SD	1524.2	0.5	0.6	56.4	25.9
r	1922–7718	0.7–2.7	1.8–3.4	70–260	3–111

Actual Community Tolerance Quotients—*CTQa*

Actual community tolerance quotients (*CTQa*) are simply the arithmetic means of the *TQ*s of sampled macroinvertebrates from a given station on a given date. The *CTQa* is directly affected by any change in species composition. If a fragile species (low *TQ*) is replaced by a moderately tolerant species, the *CTQa* will increase even though the number of species may remain the same. This is especially helpful under moderate environmental stresses that often result in increased density and/or evenness (increased *H* values)—conditions that have been misinterpreted as no impact or increased community condition.

There are certain guidelines to acquiring samples to be used in obtaining *CTQa* values. Samples should be quantitative with enough replicates to provide

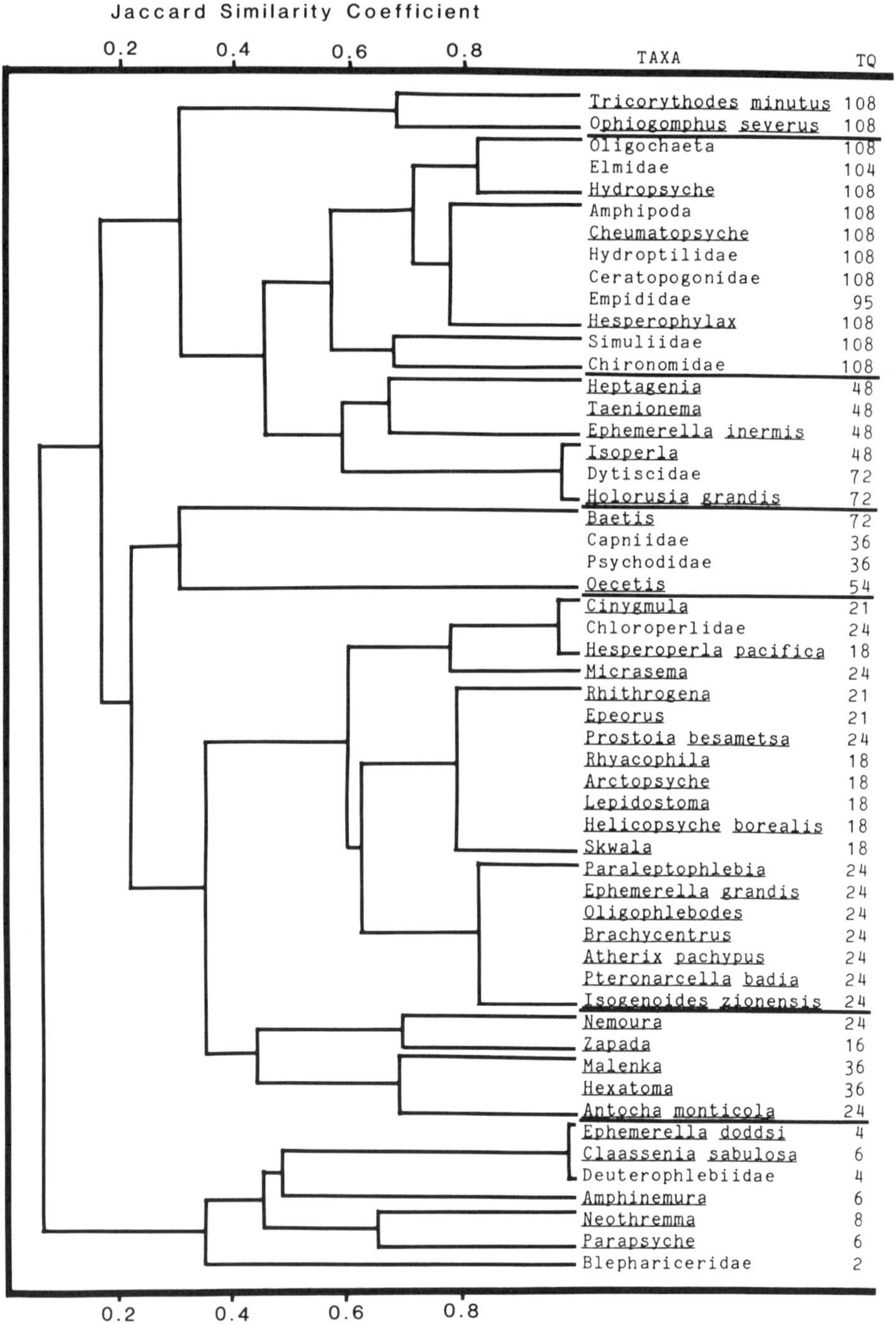

Figure 6.5 Dendrogram depicting the frequency of co-occurrence of 54 taxa, determined using the Jaccard Similarity Coefficient.

acceptable levels of variance among mean numbers per sample set. Any good quantitative sampler can be used, but it is advisable to use the same sampler for obtaining sample sets that are to be compared against each other. A Surber sampler modified to collect small instars and prevent loss of organisms due to backwash is shown in Figure 6.6. This sampler is good because of its simplicity and applicability in so many different habitat types.

It is advisable to select sample points in a stream so as to obtain maximum information while minimizing sample variance. One way to do this is to use the stratified random method described by Weber (1973) in which environmental variance is minimized by selecting for only one habitat type to take samples from. Rubble (cobble) riffles are the optimal stream habitat sites to sample. Rubble and gravel riffles are characterized ". . . by a unique physical stability which makes

Table 6.4 Tolerance quotients of selected aquatic macroinvertebrate taxa based upon tolerance to low stream gradients, fine substrate materials, total alkalinity, and sulfate.

Taxon	*TQ*	*Taxon*	*TQ*
Hirudinea	108	*Epeorus*	21
Oligochaeta	108	*Anepeorus*	48
Turbellaria	108	Leptophlebia	
Hydracarina	108	*Paraleptophlebia*	24
Asselidae	98	*Leptophlebia*	24
Amphipoda	98	*Choroterpes*	45
Megaloptera		*Traverella albertans*	45
Sialidae	72	Tricorythidae	
Corydalidae	72	*Tricorythodes minutus*	108
Lepidoptera		*Leptohyphes*	72
Pyralidae	72	Ephemerellidae	
Ephemeroptera		*Ephemerella*	48
Siphlonuridae		*E. grandis*	24
Ameletus	48	*E. doddsi*	4
Siphlonurus	72	*E. coloradensis*	18
Isonychia	48	*E. tibialis*	24
Baetidae		*E. inermis*	48
Baetis	72	*E. infrequens*	48
Callibaetis	72	*E. spinifera*	24
Pseudocloeon	72	*E. heterocaudata*	24
Centroptilum	36	*E. hecuba*	48
Dactylobaetis	36	Ephemeridae	
Paracloeodes	72	*Ephemera simulans*	45
Oligoneuriidae	36	*Hexagenia limbata*	45
Heptageniidae		Caenidae	
Heptagenia	54	*Caenis*	72
Stenonema	48	*Brachycerus*	72
Cinygmula	30	Polymitarcidae	
Rhithrogena	21	*Ephoron*	48

Table 6.4 (*Continued*)

Taxon	*TQ*	*Taxon*	*TQ*
Odonata		*Cultus aestivalis*	12
Cordulegastridae	72	*Isogenoides*	24
Gomphidae		*I. elongatus*	24
Gomphus	108	*I. zionensis*	24
Erpetogomphus compositus	72	*Kogotus modestus*	18
Ophiogomphus severus	108	*Pictetiella expansa*	18
Progomphus borealis	72	*Diura knowltoni*	24
Aeshnidae	72	*Isoperla*	48
Libellulidae	72	*I. ebria*	24
Agrionidae	108	*I. fulva*	48
Lestidae	108	*I. mormona*	48
Coenagrionidae		*I. quinquepunctata*	48
Argia	108	Chloroperlidae	24
Amphiagrion	72	Perlidae	
Enallagma	72	*Acroneuria abnormis*	6
Ischnura	72	*Claassenia sabulosa*	6
Coenagrion	72	*Hesperoperla pacifica*	18
Plecoptera		*Perlesta placida*	24
Nemouridae		*Doronuria theodora*	18
Amphinemura	6	Hemiptera	
Malenka	36	Belastomatidae	72
Prostoia besametsa	24	Corixidae	108
Podmosta	12	Gerridae	72
Zapada	16	Naucoridae	72
Nemoura	24	Notonectidae	108
Capniidae		Veliidae	72
Capnia	32	Mesoveliidae	72
Eucapnopsis	18	Macroveliidae	72
Isocapnia	24	Trichoptera	
Mesocapnia frisoni	32	Rhyacophilidae	
Utacapnia	18	*Rhyacophila*	18
Taeniopterygidae		*Atopsyche*	18
Taenionema	48	*Himalopsyche*	18
Doddsia	24	Glossosomatidae	
Oemopteryx	48	*Glossosoma*	24
Leuctridae		*Anagapetus*	24
Paraleuctra	18	*Protoptila*	32
Perlomymia	18	*Culoptila*	32
Pteronarcidae		Philopotamidae	
Pteronarcella badia	24	*Chimarra*	24
Pteronarcys californica	18	*Doliphilodes*	24
Pteronarcys princeps	24	*Wormaldia*	24
Perlodidae		Psychomyidae	
Megarcys signata	24	*Polycentropus*	108
Skwala parallela	18	*Nyctiophylax*	108

Table 6.4 (*Continued*)

Taxon	*TQ*	*Taxon*	*TQ*
Psychomyia	108	*Micrasema*	24
Tinodes	108	*Oligoplectrum*	24
Hydropsychidae		*Amiocentrus*	24
Hydropsyche	108	Helicopsychidae	
Cheumatopsyche	108	*Helicopsyche borealis*	18
Arctopsyche	18	Polycentropidae	
Smicridea	72	*Polycentropus*	72
Diplectrona	48	*Nictiophylax*	72
Macronema	48	Coleoptera	
Parapsyche	6	Haliplidae	54
Hydroptilidae	108	Dytiscidae	72
Limnephilidae		Hydrophilidae	72
Limnephilus	108	Elmidae	104
Dicosmoecus	24	Gyrinidae	108
Hesperophylax	108	Amphizoidae	24
Oligophlebodes	24	Hydraenidae	72
Apatania	18	Diptera	
Amphicosmoecus	18	Tipulidae	
Neothremma	8	*Antocha monticola*	24
Lenarchus	18	*Dicranota*	24
Chyranda	18	*Hexatoma*	36
Psychoglypha	24	*Eriocera*	36
Ecclisomyia	24	*Holorusia grandis*	72
Homophylax	18	*Helobia*	36
Allocosmoecus	18	*Tipula*	36
Asynarchus	108	Psychodidae	36
Clistorania	108	Blephariceridae	2
Grammotaulius	108	Deuterophlebia	4
Imania	48	Dixidae	108
Neophylax	24	Simuliidae	108
Onocosmoecus	18	Chironomidae	108
Pycnopsyche	72	Ceratopogonidae	108
Leptoceridae		Stratiomyidae	108
Oecetis	54	Tabanidae	108
Leptocella	54	Rhagionidae	
Triaenodes	54	*Atherix pachypus*	24
Mystacides	54	Dolichopodidae	108
Ceraclea	54	Empididae	95
Lepidostomatidae		Ephydridae	108
Lepidostoma	18	Muscidae	108
Brachycentridae		Syrphidae	108
Brachycentrus	24		

Table 6.5 A key giving predicted Community Tolerance Quotients (*CTQp*) for various combinations of gradient (%), substrates, alkalinity, and sulfate for any given stream reach.

	Key#	*CTQp*
1a. Stream gradient 0.1–1.2%	2	
2a. Substrate mostly boulder, rubble, & large gravel	3	
3a. Total alkalinity 0–199 mg/l	4	
4a. Sulfate 0–149 mg/l		53
4b. Sulfate 150–300 mg/l		67
4c. Sulfate >300 mg/l		88
3b. Total alkalinity 200–300 mg/l	5	
5a. Sulfate 0–149 mg/l		58
5b. Sulfate 150–300 mg/l		70
5c. Sulfate >300 mg/l		90
3c. Total alkalinity >300 mg/l	6	
6a. Sulfate 0–149 mg/l		65
6b. Sulfate 150–300 mg/l		80
6c. Sulfate >300 mg/l		95
2b. Substrate mostly fines (small gravel & sand)	7	
7a. Total alkalinity 0–199 mg/l	8	
8a. Sulfate 0–149 mg/l		59
8b. Sulfate 150–300 mg/l		80
8c. Sulfate >300 mg/l		92
7b. Total alkalinity 200–300 mg/l	9	
9a. Sulfate 0–149 mg/l		63
9b. Sulfate 150–300 mg/l		82
9c. Sulfate >300 mg/l		100
7c. Total alkalinity >300 mg/l	10	
10a. Sulfate 0–149 mg/l		80
10b. Sulfate 150–300 mg/l		95
10c. Sulfate >300 mg/l		105
1b. Stream gradient 1.3–3.0%	11	
11a. Substrate mostly boulder, rubble, & large gravel	12	
12a. Total alkalinity 0–199 mg/l	13	
13a. Sulfate 0–149 mg/l		50
13b. Sulfate 150–300 mg/l		65
13c. Sulfate >300 mg/l		88
12b. Total alkalinity 200–300 mg/l	14	
14a. Sulfate 0–149 mg/l		53
14b. Sulfate 150–300 mg/l		72
14c. Sulfate >300 mg/l		90
12c. Total alkalinity >300 mg/l	15	
15a. Sulfate 0–149 mg/l		70
15b. Sulfate 150–300 mg/l		85
15c. Sulfate >300 mg/l		100
11b. Substrate mostly fines (small gravel & sand)	16	
16a. Total alkalinity 0–199 mg/l	17	

Table 6.5 (*Continued*)

	Key#	*CTQp*
17a. Sulfate 0–149 mg/l		60
17b. Sulfate 150–300 mg/l		75
17c. Sulfate >300 mg/l		89
16b. Total alkalinity 200–300 mg/l	18	
18a. Sulfate 0–149 mg/l		63
18b. Sulfate 150–300 mg/l		80
18c. Sulfate >300 mg/l		95
16c. Total alkalinity >300 mg/l	19	
19a. Sulfate 0–149 mg/l		80
19b. Sulfate 150–300 mg/l		93
19c. Sulfate >300 mg/l		105
1c. Stream gradient >3.0%	20	
20a. Substrate mostly boulder, rubble, & large gravel	21	
21a. Total alkalinity 0–199 mg/l	22	
22a. Sulfate 0–149 mg/l		50
22b. Sulfate 150–300 mg/l		63
22c. Sulfate >300 mg/l		85
21b. Total alkalinity 200–300 mg/l	23	
23a. Sulfate 0–149 mg/l		50
23b. Sulfate 150–300 mg/l		65
23c. Sulfate >300 mg/l		90
21c. Total alkalinity >300 mg/l	24	
24a. Sulfate 0–149 mg/l		72
24b. Sulfate 150–300 mg/l		90
24c. Sulfate >300 mg/l		103
20b. Substrate mostly fines (small gravel & sand)	25	
25a. Total alkalinity 0–199 mg/l	26	
26a. Sulfate 0–149 mg/l		63
26b. Sulfate 150–300 mg/l		80
26c. Sulfate >300 mg/l		93
25b. Total alkalinity 200–300 mg/l	27	
27a. Sulfate 0–149 mg/l		75
27b. Sulfate 150–300 mg/l		88
27c. Sulfate >300 mg/l		100
25c. Total alkalinity >300 mg/l	28	
28a. Sulfate 0–149 mg/l		80
28b. Sulfate 150–300 mg/l		95
28c. Sulfate >300 mg/l		105

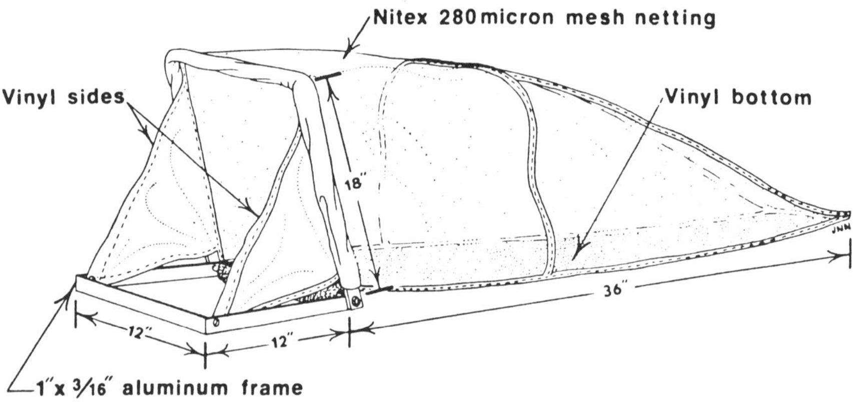

Figure 6.6 Modified Surber sampler. Enlarged bag prevents loss of sample materials due to backwash when sampling in deep or swift waters and fine mesh allows collection of small instar larvae.

them a desirable habitat for stream fauna (and) possesses a temporal stability" (DeMarch 1976).

When selecting times to sample, it should be noted that early spring and late fall samples appear to have less variability from year to year than do summer samples. DeMarch (1976) noted similar findings from a study relating seasonal substrate composition changes with changes in species distribution among substrates. Grant and MacKay (1969) concluded that species with similar habitat and trophic requirements avoided direct competition by staggering seasonal occurrence of their stages of development. Best results are obtained when comparing samples taken from the same season each year.

In streams of the western United States there are usually numerous species of mayflies, stoneflies, and caddisflies present. These species, when identified to the species or genus level, generally provide the manager with adequate information as to the condition of the aquatic community. It is then not imperative to identify all taxa to the generic or species level. With limited resources available, it is not always possible to spend the time required to identify representatives of some groups such as Chironomidae and Simuliidae. But enough community representatives must be identified to the species level that the *TQ*s obtained represent the actual tolerance level of the community. It is important that any samples compared with each other be identified to the same level. The taxonomic groups to be used in stream evaluations depend upon those species present in the streams plus the expertise of workers in the area. If enough information is available on the species of Chironomidae and experts and resources are available to identify collected specimens, then by all means use chironomid species.

Biotic Condition Index—BCI

In order to compare actual macroinvertebrate community condition against a predicted potential condition, divide the *CTQp* by the *CTQa* and multiply by 100. The result is the biotic condition index (*BCI*). The *BCI* is actually an index of percent of predicted. The *CTQa* is usually greater than the *CTQp* due to impacts from factors not considered in the four independent variables of this model. These other factors are usually related to perturbations, the extent of impact being reflected in the numerical *BCI*. If the *BCI* for a given set of samples was 90, that would indicate the actual macroinvertebrate community condition was 90% of the predicted. That stream section, with its gradient, substrate, total alkalinity, and sulfate was meeting only 90 percent of the predicted potential.

Dominance Community Tolerance Quotient—*CTQd*

A biotic community is often modified by drift from upstream or from small tributaries entering from above the sampling point. Thus species may be present in small stretches of a stream they generally wouldn't occupy. Their presence in samples could provide false information regarding community condition. These species are generally present in very low numbers in relation to resident species. Also, environmental changes often result in altered relative abundance of resident species as well as species composition changes. These changes in numbers per species are not reflected in the *CTQa* or *BCI*. Thus the *BCI* may not fully reflect community change. In order to test the validity of a *CTQa* and the resultant *BCI*, a dominance modified *CTQ* must be calculated using the following formula:

$$CTQd = \frac{\Sigma\,(TQ_i \times \log x_i)}{\Sigma \log x_i} \tag{6.3}$$

where

x_i = mean number per meter square of the ith species

If *CTQd* is within 3 of the *CTQa*, then *CTQa* is an adequate indication of community tolerance structure and the *BCI* is considered valid. If *CTQd* is higher than *CTQa* by 4 or more then the community is dominated by the more tolerant taxa indicating environmental stress not indicated by the *BCI*.

Management Strategy

Aquatic communities in selected rubble-riffle habitats reflect both total stream habitat condition and water quality, but changes in water quality have more impact

Table 6.6 Stream resource management strategy based upon habitat and water quality plus condition of aquatic macroinvertebrate community indicated by the Biotic Condition Index (*BCI*).

*Stream Condition**	*Management Strategy*	*Priority*
If habitat quality is high and *BCI* is high, assume water quality is also high	Maintain high stream quality, costs will be low in relation to stream quality	1
If habitat quality is moderate to low and *BIC* is high, assume water quality is high	Habitat improvement required, costs should be moderate to low in relation to stream improvement	2
If habitat quality is high but *BCI* is moderate to low, assume water quality is moderate to low	Water quality improvement required, costs will be moderate to high in relation to stream improvement	3
If habitat quality and *BCI* are low, assume water quality is moderate to low	Habitat and water quality improvement are required, costs will be high in relation to stream improvement	4

* High $BCI \geq 85$; Moderate *BCI* between 70 and 85; Low $BCI < 70$
Habitat quality to be determined by habitat method used and management goals.
Water quality to be determined in relation to position of stream in drainage and stream hydrogeology.

on macroinvertebrate community structure than changes in habitat do. It is generally less expensive to improve physical habitat than to significantly improve water quality, especially with nonpoint sources of pollution. A management strategy designed to give maximum return for unit of resource expenditure is given in Table 6.6. Priority is based upon a ratio between stream quality improvement per unit of management expenditure. Success of management is judged by measured physical habitat and return of actual macroinvertebrate community condition (*CTQa*) to the predicted potential condition (*CTQp*).

Example of Application of *BCI*

Two stations on the Provo River, one above and one below a proposed dam site, were located approximately 8 km apart. Measured water quality was almost identical for the two stations and habitat was also similar. Water sample analyses failed to show differences in water quality that would cause biotic community differences between the two stations. Macroinvertebrate density was less at the lower station both years of study (Table 6.7), as was standing crop measured as grams of dry weight. Dominance diversity (*H*) was also less at the lower stations both years.

Table 6.7 Summary statistics for analysis of macroinvertebrate samples from two sites on the Provo River, Utah, one station above and one below a proposed dam site.

	Mean Number/m²				
	Spring 1980		*Spring 1981*		
Taxa	*Above*	*Below*	*Above*	*Below*	*TQ*
Nematoda	22	129	—	43	108
Gastropoda	3	—	22	—	108
Hirudinea	—	22	11	—	108
Pelecypoda	151	—	54	38	108
Oligochaeta	1,657	—	172	578	108
Turbellaria	—	—	—	30	108
Hydracarina	280	323	1,302	129	108
Copepoda	86	—	—	—	108
Ostracoda	—	—	—	51	108
Ameletus	22	—	—	—	48
Baetis	3,269	4,024	2,120	1,627	72
Rhithrogena	65	—	194	—	21
Epeorus	861	—	947	30	21
Paraleptophlebia	172	129	151	8	24
Ephemerella grandis	301	22	581	11	24
E. doddsi	43	—	32	—	4
E. inermis	344	538	226	110	48
Taenionema	—	—	22	13	48
Pteronarcella badia	3	—	323	8	24
Skwalla parallela	—	—	22	3	18
Isoperla 1	—	86	43	51	48
Isoperla fulva	43	—	64	8	48
Chloroperlidae	194	172	86	169	24
Rhyacophila	22	—	43	—	18
Glossosoma	194	65	—	—	24
Protoptila	—	—	183	11	32
Hydropsyche	753	624	1,431	35	108
Cheumatopsyche	—	43	11	51	108
Arctopsyche	—	—	22	—	18
Hydroptila	—	—	—	8	108
Oecetis	—	43	—	19	54
Hesperophylax	—	—	—	3	108
Lepidostoma	86	—	54	8	18
Brachycentrus	3,185	473	4,347	32	24
Micrasema	—	—	—	11	24
Helicopsyche borealis	430	758	118	105	18
Elmidae	108	301	398	605	108
Antocha monticola	—	108	—	59	24
Holorusia grandis	—	3	—	—	72
Dicranota	—	—	—	30	24

Table 6.7 (*Continued*)

	Mean Number/m²				
	Spring 1980		*Spring 1981*		
Taxa	*Above*	*Below*	*Above*	*Below*	*TQ*
Eriocera	—	22	—	282	72
Blephariceridae	22	—	13	—	2
Simuliidae	689	65	484	48	108
Chironomidae	18,227	14,375	10,276	7,102	108
Ceratopogonidae	—	—	—	22	108
Atherix pachypus	129	5	269	3	24
Hemerodromia	—	—	22	19	95
Mean number/m²	31,328	21,829	24,041	11,360	
Standard Dev.	4,477	7,223	9,163	5,526	
Mean Dry Wt. gm/m²	19.5	8.9	19.9	1.8	
Number of Taxa	28	22	32	36	
H (Shannon-Weaver)	2.380	1.827	2.936	2.182	
CTQa	53	62	54	64	
CTQd	56	66	56	71	
CTQp	53	53	53	53	
BCI	100	85	95	83	

The *CTQa* was higher at the lower station, indicating relatively more tolerant taxa than at the upper station. There were twenty-eight taxa common to both stations and nineteen taxa found only at one or the other. Of the eleven taxa found only at the lower station, the average *TQ* was 73.6, compared with 40.9 for the eight taxa found only at the upper station.

The *BCI* indicated that the macroinvertebrate community at the upper station was between 95 and 100% of expected but only 83 to 85% at the lower station. The difference between *CTQa* and *CTQd* was only −2 to −3 at the upper station but −4 to −7 at the lower station. The decreased *H* values at the lower station was the result of dominance by the more tolerant taxa. The *BCI* values for the lower station were probably slightly high in relation to actual conditions. Since physical habitat was high and the *BCI* moderate at the lower station, it was assumed that the perturbation was water quality related (Table 6.6).

CONCLUSIONS

The methodology used in the *BCI:* (1) is sensitive to all types of environmental stress; (2) is applicable to various types of streams; (3) gives a linear assessment from unstressed to highly stressed conditions; (4) is independent of sample size providing the sample contains a representative assemblage of species; (5) is based upon water quality and physical habitat data readily available or easily acquired; and (6) meshes readily with existing stream habitat and water quality management programs.

REFERENCES

Bakke, B.M. 1977. Grazing is destroying our fish. *Conservation* Feb/Mar: 31–34.

Beaumont, P. 1975. Hydrology. In *River Ecology,* edited by B.A. Whitton, 1–38. Berkeley: Univ. of Calif. Press.

Beck, W.M. 1954. Studies in stream pollution biology: I. A simplified ecological classification of organisms. *Quatr. J. of Fla. Acad. of Sci.* 174(4):221–27.

______. 1955. Suggested method for reporting biotic data. *Sewage and Industrial Wastes* 27:1193–97.

Bormann, F.H., and G.E. Likens. 1979. *Pattern and Process in a Forested Ecosystem.* New York: Springer-Verlag.

Brillouin, L. 1960. *Science and Information Theory.* 2d ed. New York: Academic Press.

Bruce, J.R., and R.H. Clark. 1966. *Introduction to Hydrometeorology.* Oxford: Pergamon Press.

Cairns, J., Jr., and K.L. Dickson. 1971. A simple method for the biological assessment of the effects of waste discharges on aquatic bottom-dwelling organisms. *J. Water Poll. Contr. Federation* 43:755–72.

Cairns, J., Jr., K.L. Dickson, and G.R. Lanza. 1973. Rapid biological monitoring system for determining aquatic community structure in receiving systems. In *Biological Methods for the Assessment of Water Quality,* edited by J. Cairns, Jr. and K.L. Dickson, 148–63. ASTM STP 528. Philadelphia, PA: Amer. Soc. for Test. and Mater.

Chandler, J.R. 1970. A biological approach to water quality management. *Water Poll. Contr.* 69:415–22.

Chutter, F.M. 1972. An empirical biotic index of the quality of water in South African streams and rivers. *Water Res.* 6:19–30.

Colwell, R.K., and D.J. Futuyma. 1971. On the measurement of niche breadth and overlap. *Ecology* 52(4):567–76.

Cook, S.E. 1976. Quest for an index of community structure sensitive to water pollution. *Environ. Poll.* 11:269–88.

Cummins, K.W. 1974. Structure and function of stream ecosystems. *Bio. Sci.* 24(11):631–41.

______. 1975. The ecology of running waters; theory and practice. In *Proc. Sandusky River Basin Symposium,* edited by D.B. Baker, W.B. Jackson, and B.L. Prater, 227–93. 1976-653-346. Washington, D.C.: U.S. Government Printing Office.

______. 1977. From headwater streams to rivers. *Am. Biol. Teacher* 39:305–12.

Curry, R.R. 1977. Reinhabiting the earth: life support and the future primitive. In *Recovery and Restoration of Damaged Ecosystems,* edited by J. Cairns, Jr., K.L. Dickson, and E.E. Herricks, 1–23. Charlottesville: University Press of Virginia.

DeJong, T.M. 1975. A comparison of three diversity indices based on their components of richness and evenness. *Oikos* 26:222–27.

DeMarch, B.G.H. 1976. Spatial and temporal patterns in macrobenthic stream diversity. *J. Fish. Res. Bd. Can.* 33:1261–70.

Fisher, S.G., and G.E. Likens. 1973. Energy flow in Bear Brook, New Hampshire: An integrated approach to stream ecosystem metabolism. *Ecol. Monogr.* 43(4):421–39.

Fortescue, John A.C. 1980. *Environmental Geochemistry: A Holistic Approach.* New York: Springer-Verlag.

Gaufin, A.R. 1973. Use of aquatic invertebrates in the assessment of water quality. In *Biological Methods for the Assessment of Water Quality,* edited by J. Cairns, Jr., and K.L. Dickson, 96–116. ASTM STP 528. Philadelphia, Pa.: Amer. Soc. for Test. and Mater.

Golterman, H.L. 1975. *Physiological Limnology.* Amsterdam: Elsevier.

Grant, P.R., and R.J. MacKay. 1969. Ecological segregation of systematically related stream insects. *Can. J. Zool.* 47:691–94.

Harrel, R.C., and T.C. Dorris. 1968. Stream order, morphometry, physico-chemical conditions, and community structure of benthic macroinvertebrates in an intermittent stream system. *Am. Midl. Nat.* 80(1):220–51.

Hart, C.W., and S.L. Fuller, eds. 1974. *Pollution Ecology of Freshwater Invertebrates.* New York: Academic Press.

Hawkes, H.A. 1975. Biological aspects of river pollution. In *Aspects of River Pollution,* edited by L. Klein, 191–251. London: Butterworth.

Hendricks, A., D. Henley, J.T. Wyatt, K.L. Dickson, and J.K.G. Silvey. 1974. Utilization of diversity indices in evaluating the effect of a paper mill effluent on bottom fauna. *Hydrobiologia* 44:463–74.

Hill, A.R. 1975. Ecosystem stability in relation to stresses caused by human activities. *Canad. Geographer* 19(3)206–20.

Hilsenhoff, W.L. 1977. *Use of Arthropods to Evaluate Water Quality of Streams.* Wisconsin Dept. Nat. Resour. Tech. Bull. 100. Madison, WI.

———. 1982. *Using a Biotic Index to Evaluate Water Quality in Streams.* Wisconsin Dept. Nat. Resour. Tech. Bull. 132, Madison, WI.

Hoehn, R.C., J.R. Stauffer, M. Masnik, and C.H. Hocutt. 1974. Relationships between sediment oil concentrations and the macroinvertebrates present in a small stream following an oil spill. *Environ. Letter* 7(4):345–52.

Holling, C.S. 1973. Resilience and stability of ecological systems. *Ann. Rev. Ecol. and Systemat.* 4:1–23.

Horton, R.E. 1945. Erosional development of streams and their drainage basins: hydrophysical approach to quantitative morphology. *Bull. Geol. Soc. Am.* 56:275–370.

Huet, M. 1959. Profiles and biology of western European streams as related to fish management. *Trans. Am. Fish Soc.* 88:155–63.

Hynes, H.B.N. 1972. *The Ecology of Running Waters.* Toronto: Univ. of Toronto Press.

Kolkwitz, R., and M. Marsson. 1908. Ockologie der pflanzlichen Saprobien. *Berichte der Deutschen Botanischen Gesselschaft* 26:505–19.

———. 1909. Ockologie der tierischen Saprobien. *International Revue der Gesamten Hydrobiologie und Hydrographie* 2:126–52.

Macan, T.T. 1974. *Freshwater Ecology.* New York: John Wiley and Sons.

MacKay, D.W., P.G. Soulsby, and T. Poodle. 1973. *The Biological Assessment of Pollution in Streams.* London: Assoc. of River Anth. Yr. Bk. and Direct.

Margalef, D.R. 1958. Information theory in ecology. *General Systems* 3:36–71.

———. 1969. Diversity and stability: a practical proposal and a model of interdependence. In *Diversity and Stability in Ecological Systems,* edited by Brookhaven National Laboratory, 25–27. Brookhaven Symposia in Biology No. 22. Upton, NY: Brookhaven National Lab.

Masnik, M.T., J.R. Stauffer, C.H. Hocutt, and J.H. Wilson. 1976. The effects of an oil spill on the macroinvertebrates and fish in a small southwestern Virginia creek. *J. Environ. Sci. Health* A11(4&5):281–96.

Mathis, B.J. 1965. Community structure of benthic macroinvertebrates in an intermittent stream receiving oil field brines. Ph.D. thesis, Oklahoma State University, Stillwater.

______. 1968. Species diversity of benthic macroinvertebrates in three mountain streams. *Trans. III Okla. State Acad. Sci.* 61:171–76.

Minshall, G.W. 1978. Autotrophy in stream ecosystems. *BioScience* 28:767–71.

Mueller-Dombois, D., and H. Ellenburg. 1974. *Aims and Methods of Vegetation Ecology.* New York: John Wiley and Sons.

Narf, R.P., E.L. Lange, and R.C. Wildman. 1984. Statistical procedures for applying Hilsenhoff's Biotic Index. *J. Freshw. Ecol.* 2:441–48.

Nuttall, P.M., and J.B. Purves. 1974. Numerical indices applied to the results of a survey of the macroinvertebrates fauna of the Tamar Catchment (southwestern England). *Freshwater Biol.* 4:213–22.

Odum, E.P. 1969. The strategy of ecosystem development. *Science* 164:262–70.

Patrick, R. 1968. The structure of diatom communities in similar ecological conditions. *Am. Nat.* 102:173–83.

______. 1969. Some effects of temperature on freshwater algae. In. *Biological Aspects of Thermal Pollution,* edited by P.A. Krenkel, and F.L. Parker, 161–85. Nashville, TN: Vanderbilt University Press.

______. 1973. Use of algae, especially diatoms, in the assessment of water quality. In *Biological Methods for the Assessment of Water Quality,* edited by J. Cairns, Jr. and K.L. Dickson, 76–79. ASTM STP 528. Philadelphia, Pa.: Amer. Soc. for Test. and Mater.

Patten, B.C. 1962. Species diversity in net phytoplankton of Rarita Bay. *J. Mar. Res.* 20:57–75.

Pennak, R.W. 1971. Towards a classification of lotic habitats. *Hydrobiologia* 38(2):321–34.

Petersen, R.C., and K.W. Cummins. 1974. Leaf processing in a woodland stream. *Freshwater Biol.* 4:343–68.

Pielou, E.C. 1966. The measurement of diversity in different types of biological collections. *J. Theoret. Biol.* 13:133–44.

______. 1972. Niche width and niche overlap: a method for measuring them. *Ecology* 53(4):687–92.

Reice, S.R. 1977. The role of animal associations and current velocity in sediment-specific leaf litter decomposition. *Oikos* 29:357–65.

Resh, V.H., and J.D. Unzicker. 1975. Water quality monitoring and aquatic organisms: the importance of species identification. *J. Water Poll. Contr. Fed.* 47(1):9–16.

Ringler, N.H., and J.D. Hall. 1975. Effects of logging on water temperature and dissolved oxygen in spawning beds. *Trans. Amer. Fish. Soc.* 104(1):111–21.

Shannon, C.E., and W. Weaver. 1963. *The mathematical theory of communication.* Urbana, Ill.: Univ. of Ill. Press.

Shreve, R.L. 1966. Statistical law of stream numbers. *J. Geol.* 74:17–37.

Stoneburner, D.L. 1977. Preliminary observations of the aquatic insects of the Smokey Mountains: altitudinal zonation in the spring. *Hydrobiologia* 56(2):137–43.

Vannote, R.L., G.W. Minshall, K.W. Cummins, J.R. Sedell, and C.E. Cushing. 1980. The river continuum concept. *Can. J. Fish. Aquat. Sci.* 37:130–37.

Vitousek, P.M. 1977. The regulation of element concentrations in mountain streams in the northeastern United States. *Ecol. Monogr.* 47:65–68.

Vitousek, P.M., and W.A. Reiners. 1975. Ecosystem succession and nutrient retention: A hypothesis. *BioScience* 25:376–81.

Voshell, R.J., Jr. 1980. ECOSCAN: an interactive ecological classification system for analyzing benthic macroinvertebrate samples. Report for the U.S. Army Corp. of Engineers, Huntington, WV (DACW69-80-M-1663).

Weber, C.I., ed. 1973. *Biological Field and Laboratory Methods.* EPA-670/4–73–001. Cincinnati, OH: U.S. Environment Protection Agency.

Wiggins, G.B. 1977. *Larvae of the North American Caddisfly Genera (Trichoptera).* Toronto: University of Toronto Press.

Wilhm, J.L. 1965. Species diversity of benthic macroinvertebrates in a stream receiving domestic and oil refinery effluents. Ph.D. diss, Oklahoma State Univ., Stillwater.

———. 1967. Use of biomass units in Shannon's formula. *Ecology* 49(1):153–56.

———. 1970. Range of diversity index in benthic macroinvertebrate populations. *J. Water Poll. Contr. Fed.* 42(5):221–24.

Wilhm, J.L., and T.C. Dorris. 1968. Biological parameters for water quality criteria. *Bio. Sci.* 18:477–81.

Winget, R.N., and F.A. Mangum. 1979. *Biotic condition index: integrated biological, physical, and chemical stream parameters for management.* U.S. Department of Agriculture, Forest Service, Intermountain Region Bull. Provo, Utah.

Wurtz, C.B. 1955. Stream biota and stream pollution. *Sewage and Indust. Wastes* 27:1270–78.

Zand, S.M. 1976. Indices associated with information theory in water quality. *J. Water Poll. Contr. Fed.* 48(8)2026–31.

CHAPTER 7

Aquatic Community Response to Techniques Utilized to Reclaim Eastern U.S. Coal Surface Mine–Impacted Streams

Lynn B. Starnes*

Tennessee Valley Authority
450 Evans Building
Knoxville, Tennessee 37902

Coal mining and associated processing activities are currently required to use best available technologies to minimize impacts to air, water, and land. Mined areas must be reclaimed. Many older mined areas were poorly operated from an environmental standpoint and were left to be reclaimed by natural processes. This lack of environmental foresight culminated in such problems as acid mine drainage, mine subsidence, mine and refuse fires, and barren and highly erodable land. While these problems emanate from both abandoned strip and underground mines, this chapter will concentrate on methods to restore abandoned surface coal mines in the eastern United States.

Between 1930 and 1971 some 3.7 million acres were surface mined for coal and noncoal materials (such as sand, gravel, and minerals) (Paone et al. 1974). The U.S. Soil Conservation Service (1979) indicates that of almost 1.7 million acres of coal surface-mined land needing reclamation over 1.3 million acres (82%) are in the East (Alabama, Illinois, Indiana, Kentucky, Maryland, Ohio, Pennsylvania, Tennessee, Virginia, and West Virginia). While mine drainage problems also occur in western mining states, the overall problems there are not as extensive as in the eastern states.

Environmental impacts resulted in enactment of state and federal laws to

---

* Present Address: U.S. Fish and Wildlife Service, Division of Program Operations—Fisheries, Washington, D.C. 20240

control both mining and reclamation of surface-mined coal land. West Virginia (1939) was the first State to enact legislation followed by Indiana (1941), Illinois (1943), Pennsylvania (1945), Ohio (1947), Kentucky (1954), Maryland (1955), Virginia (1966), and Tennessee (1967) (Paone et al. 1974). As early as 1963, the Tennessee Valley Authority (TVA) recognized developing problems associated with coal surface mining and made recommendations in its regional report for provisions to be included in surface mining legislation (TVA 1963).

National concern for the detrimental impacts of surface mining was reflected in Section 205(c) of the Appalachian Regional Development Act of 1965 (Public Law 89–94) which called for the Department of the Interior, the Department of Agriculture, and the Tennessee Valley Authority to make an inventory and study of surface-mining operations and their effects on United States lands. Interest in mineral resources mining impacts on aquatic resources resulted in Section 233 of the River and Harbors, Flood Control Acts of 1970 (Public Law 91–611) authorizing the U.S. Army Corps of Engineers to review and study the effects of strip mining operations upon navigable rivers and tributaries. The U.S. Army Corps of Engineers (1974) surveyed all forms of surface-mining activities and their impacts on water resources with a plan [including a demonstration reclamation project in Cabin Creek, West Virginia (U.S. Army Corps of Engineers 1978)] to ameliorate water resource problems related to strip mine practices.

National involvement in surface coal mining and reclamation culminated in passage of the Surface Mining Control and Reclamation Act of 1977 (SMCRA), Public Law 95–87, 30 USC §§1202 *et seq.* (Supp. III, 1979). SMCRA establishes minimum federal standards applicable to the conduct of surface coal mining and reclamation operations. SMCRA and the regulations promulgated thereunder have the goal of regulating new mining to prevent unacceptable environmental change and, at the same time, providing for the reclamation of abandoned coal mines through fees imposed upon new coal produced (Abandoned Mine Reclamation Fund). Legislation culminating in SMCRA sought to require the reclamation of abandoned surface mines as well as to prevent future mining-related environmental impacts.

Approaches to stream reclamation discussed in this chapter are based both on previously unpublished studies and on literature. Historically there have been two basic approaches to stream restoration. The first emphasizes watershed reclamation through enhancement of vegetative cover and at-source treatment of problem drainage while the second focuses on instream habitat restoration. This chapter concludes with an example of a comprehensive approach that encompasses the watershed improvement approach and rehabilitation techniques designed to improve in-stream habitat.

ABANDONED MINES

Sedimentation and chemical pollution are typical water resource impacts from surface coal mining performed under laws with limited environmental insight.

Sedimentation is caused by erosion of disturbed surface features with inadequate plant cover. It is a universal problem and becomes especially severe when multiple mines are present in the same watershed and the assimilative capacity of the receiving streams is exceeded. Cumulative impacts on successively larger order streams also result with the inflow of this material from smaller streams. In the eastern United States chemical pollution results when the products formed by the oxidation of pyritic materials associated with the coal strata are mixed with water and form acid. The initial products of the oxidation of pyritic materials are ferrous sulfate and sulfuric acid:

$$2FeS_2 + 7O_2 + 2H_2O \longrightarrow 2FeSO_4 + H_2SO_4 \tag{7.1}$$

Ferrous sulfate may be further oxidized to ferric sulfate which can oxidize additional pyritic material:

$$4FeSO_4 + O_2 + H_2SO_4 \longrightarrow 2Fe(SO_4)_3 + 2H_2O \tag{7.2}$$

When ferric sulfate is further oxidized, ferric hydroxide (a precipitate sometimes called *yellowboy*) and sulfuric acid are produced:

$$Fe_2(SO_4)_3 + 6H_2O \longrightarrow 2Fe(OH)_3 + 3H_2SO_4 \tag{7.3}$$

A relatively recent discovery has been the consistent isolation of sulfur- and iron-oxidizing bacteria from acid mine drainage. Experiments conducted to assess the relative importance of various microorganisms to acid formation are summarized in Boyer et al. (1978). It is apparent bacteria can increase the ferric-ferrous ratio; however, their relative importance (in relation to chemical processes) has yet to be determined.

Three types of surface mining (area, contour, and mountain top removal) are the principal forms of mining in the eastern United States. Contour strip mining is a technique used extensively in mountainous terrain. Its use is common in northern Alabama, eastern Kentucky, western Pennsylvania, eastern Tennessee, western Virginia, and West Virginia. This area is commonly referred to as the Appalachians or Appalachia (Figure 7.1). More reclamation problems are encountered with contour than area mining because of steeper topography, higher precipitation, and associated erosion. The configuration and terminology associated with contour mining is illustrated in Figure 7.2.

Contour mining techniques create a bench along the same elevation (contour) of the mountainous area from which first the material over the coal, then the coal itself is removed. In abandoned mines, the material over the coal—the overburden or spoil—was stacked on the outer edge of the bench or pushed entirely off the bench and down the outslope. This spoil material was generally unstable (depending on slope) and tended to erode quickly and vegetate slowly. In addition, as it absorbed moisture its weight increased and it became susceptible to sliding or creeping down slope, causing further problems (such as landslides, or erosion

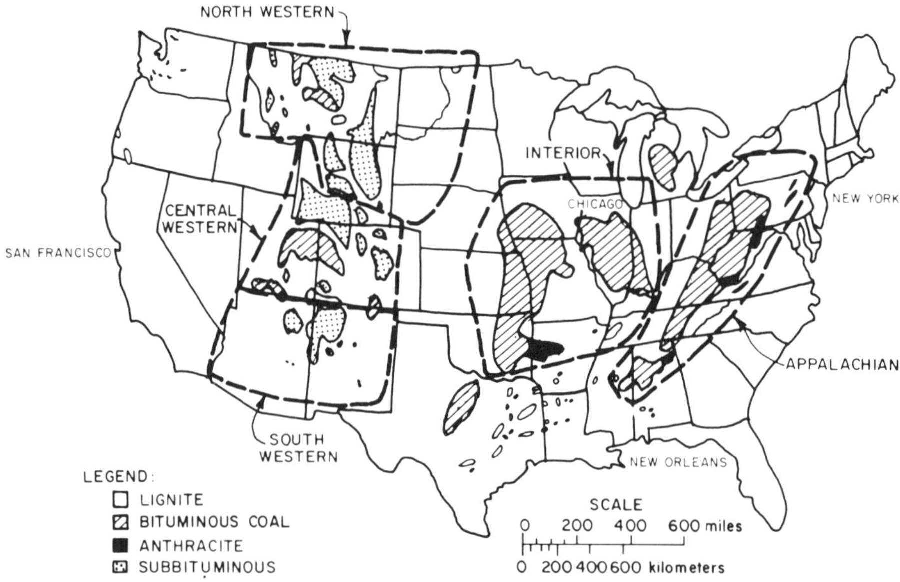

Figure 7.1 Map of the continental United States with major coal producing regions indicated. Adapted from Rowe 1979.

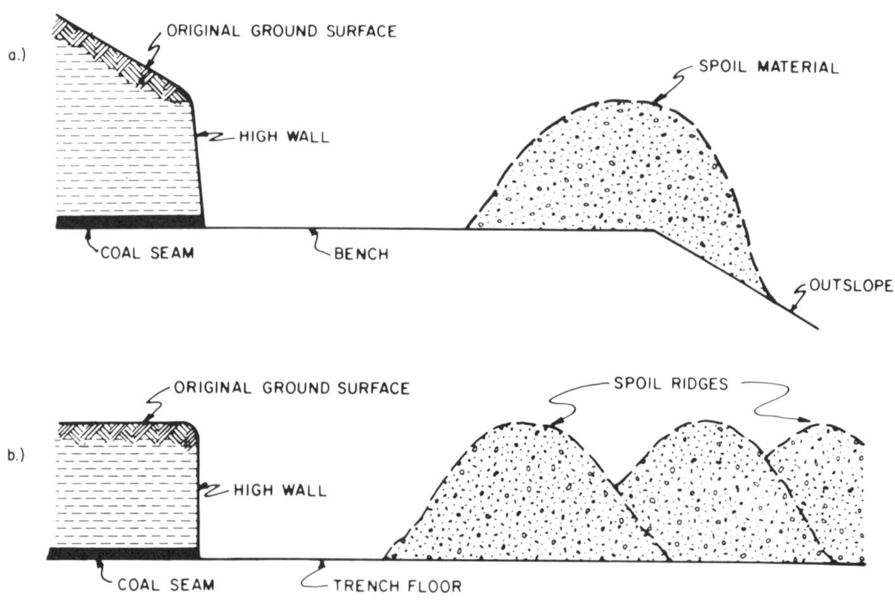

Figure 7.2 Schematic of (a) contour strip mining and (b) area mining illustrating appropriate terminology.

damaging homes and property). Spoil material sometimes reaches from these mountain side mines to valley floors (Figure 7.3). Today according to SMCRA the overburden or spoil must be replaced to approximate the original contour.

The *area mining* technique is generally used in relatively flat terrain where the coal seam runs essentially parallel to the surface. In the eastern United States, it is used commonly in Illinois, Indiana, and western Kentucky, or the area referred to as the Interior (Figure 7.1). An area mine begins with a vertical cut. The spoil material from that first cut is placed to one side forming a parallel ridge. Spoil material from each successive parallel cut or trench is placed in the preceding trench. The configuration of such a mine is illustrated in Figure 7.2. The last trench will be bounded by a highwall on one side and spoil material on the other. This final pit normally fills with water, forming a lake. Abandoned area mines created environmental problems primarily from the manner of spoil layering. The last material placed on the ridges was normally in closest association with the coal. Therefore, pyritic materials were frequently exposed to be readily oxidized. The production of acid mine drainage prevented vegetative cover on the ridges and produced acidic ponds in the trenches. In abandoned mines, the ridges were left intact but under SMCRA must be regraded to the approximate original contour with spoil burial.

Mountain top removal involves following a coal seam across the mountain and removing the overburden or mountain peak. In abandoned mines, spoil material

Figure 7.3 Abandoned contour coal mine with spoil reaching from bench down slope to valley. Photo, with permission, Tennessee Valley Authority.

was stacked on the outer edge of or pushed off the bench creating environmental problems similar to abandoned contour mines. Current reclamation techniques for mountain top removal are similar to those for area mining. The result is a relatively stable flat-top mountain resembling a western butte.

IMPACTS TO AQUATIC ECOLOGY

A major watershed disturbance, such as surface coal mining, can have significant impacts on aquatic systems with the extent of impact primarily related to spoil handling and success of terrestrial reclamation. The abandoned mine drainage problem is extensive since it is estimated that 1.7 million acres are involved (U.S. Soil Conservation Service 1979). Where polluted abandoned mine waters have entered streams, aquatic habitats are altered and sometimes permanently destroyed. Sedimentation and acid mine drainage are the two major components of mine drainage problems. These may degrade aquatic habitat from the actual mine site downstream for several miles.

Sedimentation

Hynes (1974) reported two adverse effects of inorganic sediments on aquatic ecosystems. Silt increases turbidity which, in turn, decreases light penetration. Thus, overall stream productivity is decreased. The second effect is related to the sediments covering the stream substrate. The impacts of sediments to fish and aquatic invertebrates was reviewed by Cordone and Kelly (1961). Effects can be direct by causing abrasive injuries to delicate external organs such as gills, protective mucal coverings, fins, etc., or by smothering eggs and nests. Indirect effects can range from elimination of a preferred food source to elimination of preferred reproductive habitat for fish and aquatic invertebrates.

Acid Mine Drainage

Johnson and Miller (1979) estimated that acid mine drainage impacted over 10,000 miles of streams. Spaulding and Ogden (1968) reported that 31% of the streams in mining areas had yellowboy on the substrate and 37% had iron-colored water. With active mining now regulated by SMCRA, abandoned mines cause the worst water quality problems.

Principally, precipitated iron hydroxide (yellowboy) can smother aquatic invertebrates and reduce habitat diversity. With reduced pH, metals such as aluminum and manganese can reach lethal levels or a combination of mineral acids and ions of iron and sulfur can cause mortality to organisms. Roback and Richardson (1969) reported that nonbenthic Hemiptera and Coleoptera were least affected while the Odonata, Ephemeroptera, and Plecoptera were most severely affected.

Within Trichoptera (*Ptilostomis*), Megaloptera (*Sialis*), and Diptera (*Chironomus*), some species proved tolerant. Several authors, including Parsons (1952) and Riley (1960), have found acid mine drainage to drastically reduce or eliminate aquatic invertebrates. The combination of decreasing pH, precipitating iron compounds, and low dissolved oxygen severely alters aquatic communities.

In general, sensitive aquatic species are eliminated soon after pollution enters the stream. Only increasingly tolerant species with unspecialized diets and reproductive requirements persist after the acid drainage continues or increases. Both reproductive capability and available reproductive habitat are reduced. Growth rates in tolerant fish species in mined watersheds have been shown to be slower than for individuals in unmined watersheds (Bettoli 1979).

ECOLOGICAL RECOVERY

Recovery rates for aquatic communities impacted by abandoned mines vary with geology, topography, and mining technique. Talak (1977), Tolbert and Vaughan (1980), and Vaughan et al. (1978) reported full natural (no remedial or extra reclamation) benthic community recovery after twenty-four years in nonacid mine drainage situations. This is in contrast to acid mine drainage situations where seasonal fluctuations in water quality is great and annual recovery is imperceptible, making reasonable prediction of recovery impossible. Herricks and Cairns (1975) found recovery time was variable, but with short-term stress full community recovery would occur within two to four weeks. Zarger et al. (1979) described remedial reclamation of Ollis Creek, an acid mine situation. Separate subwatersheds varied, but depending on techniques utilized and the percent of the watershed disturbed, we have found reclamation efforts decrease the expected time of recovery. This study will be examined in greater detail in the Watershed Reclamation section.

RECLAMATION RATIONALE

Most of the land disturbed by coal mining and subsequently abandoned can be rehabilitated for productive uses. Reclamation projects should be evaluated and designed around threats to public health and safety, technical feasibility, environmental degradation, and surrounding land use and site constraints. Where acid drainage pollutes public water supplies or where water impoundments threaten homes with inadequate and/or deteriorating dams/spillways, back-to-contour reclamation may be the only way to totally eliminate severe threats. The amount of reclamation to be done at a site varies but should attempt to eliminate all problems associated with a particular abandoned mine.

For the 1.7 million acres needing reclamation, total restoration to premining conditions or original contour is not practical due to the tremendous costs, and may not even be environmentally desirable since such intensive reclamation efforts re-disturb areas partially vegetated, possibly re-exposing toxic materials and creating new unvegetated areas. Thus, two management techniques that attempt to ameliorate existing problems may be employed. The first technique attempts to prevent or possibly treat mine drainage before reaching water courses. In this alternative, the mine drainage is detained onsite or near the mine, but before the drainage is discharged to adjacent streams. In water courses or impoundments where mine drainage, especially sediments, has already impacted and damaged the channel, the second, or corrective action, technique is employed. Depending on the length of time and severity of the problem since mining, these two techniques, *prevention of additional impacts* and *correction of existing damages,* may be used in concert to effect aquatic rehabilitation.

Research activity into reclamation of abandoned coal-mined lands has yielded a voluminous body of literature. With passage of SMCRA emphasis has changed from documentation of abandoned mine damage to development of techniques not only to revegetate mines but also to enhance fish and wildlife (Leedy 1981; Leedy and Franklin 1981). Since this chapter does not provide a complete literature review, specific information on reclamation of surface-mined land and coal mine drainage may be obtained from these early bibliographies: Linstrom (1953), Funk (1962), and Kieffer (1972). Additional, more recent, bibliographies have been published by EPA [Gleason (1979, 1980)] and the U.S. Department of Interior (1977–1981). These contain lists and abstracts of literature dealing with mine drainage and reclamation research.

WATERSHED RECLAMATION

Orphan Land Reclamation Demonstration Program

The Tennessee Valley Authority sponsored abandoned mine land reclamation programs in Alabama, Kentucky, Tennessee, and Virginia during a four-year period from 1976 to 1980 (Vines 1979). The purpose of the program was to alleviate offsite environmental impacts by reclaiming orphan strip mines in a regional demonstration of techniques and administrative arrangements that could be applied elsewhere in Appalachia. Funds were provided by congressional appropriations to reclaim abandoned mine lands agreed to between TVA, the cooperating state, and landowner. In total $7.8 million was spent in treating 14,500 acres (5,830 ha) of abandoned surface mines and haul roads in a 38-county area of the southern Appalachians. Basic reclamation or minimum disturbance techniques were applied to correct problems of surface water flow, active erosion, and acid mine drainage with the goal of returning treated areas to productive forest and wildlife uses and enhancing aesthetic values. Reclamation costs averaged $537/acre.

Data collected on mines were then analyzed and work priorities assigned based on the severity of environmental problems emanating from the abandoned site and the landowner's interest in cooperating in the project. The local unemployment problems also helped determine initial operating areas to provide work opportunities. The mines were treated on subwatershed basis or a hydrologic unit.

At each mine reclamation techniques were designed and utilized to treat site-specific problems. The kinds and placement of erosion control structures considered topography, soil characteristics, and onsite vegetation patterns, as well as drainage patterns and water courses. Soil erosion control practices were designed to reduce both the amount and the velocity of runoff by creating check dams or sediment traps as close to the source as possible. Also, runoff was routed to natural drainage channels where possible instead of attempting to stabilize or to create new drainage ways. Where gullies or larger scale runoff problems existed, sediment control devices (such as log-fabric dams) were designed to catch and hold sediments. Finally, a vegetation plan was developed for each site with plant species chosen on the basis of mine site characteristics.

The description of construction details for devices utilized during the program are contained in Nicholson and Snyder (1939) and Wilder and Rains (1976). Table 7.1 evaluates the strengths and weaknesses of the particular techniques. The effectiveness of each device was limited by care in placement and construction. Care is essential to prevent washout or erosion around ends of the devices. The devices most successful in controlling erosion were placed close to the source of the problem. The most successful design, particularly in steep slope situations, utilized a series of structures rather than one large structure or pond.

As a part of the project, TVA planned to determine the effectiveness of its efforts through a comprehensive program monitoring terrestrial and aquatic systems associated with the reclamation of orphaned mines in selected watersheds. This plan was abandoned with one year of pre-treatment and one year of post-treatment data. During that monitoring phase, no aquatic changes (water quality, aquatic macroinvertebrate, or periphyton) were evident (Taylor and Nicholson, TVA unpublished results 1979). However, I noted improvements in water quality and biota in subsequent visits to these stations.

Watersheds to be monitored were selected from those abandoned mines where the watershed had a permanent stream with unaffected reaches above the mine to be used as a natural control. Six stations in two Tennessee watersheds (Figure 7.4), Jones Branch and Kent Hollow, were monitored monthly for water quality, aquatic macroinvertebrates, and periphyton. Characteristics of these watersheds were high acidity, low pH, increased levels of turbidity and silt, and high concentrations of heavy metals (iron and manganese). In all affected stations, aquatic macroinvertebrate communities were reduced. Jones Branch, Scott County, Tennessee, had creek chubs, *Semotilus atromaculatus,* while Kent Hollow, Campbell County, Tennessee, had no fish. These two watersheds were revisited in 1979 and 1980 to examine obvious aquatic changes.

The Kent Hollow watershed was remined from 1979 to 1980. The spoil material on the outslope, which had relatively good vegetation, was removed during remining and disposed of in a pit along with new mining coal-associated strata. The mining operation, subject to SMCRA and its implementing regulations, appar-

Table 7.1 Comparison of advantages and disadvantages of soil erosion and sediment control practices utilized in orphan land reclamation projects

	Soil Erosion Control Practices		
	Sandbags	*Hay Bales*	*Rock*
Advantages	—Can use onsite material, not just sand —Relatively permanent —In series effective in slowing runoff velocity	—Effective filter to trap sediment —Seed in hay will germinate and grow	—Limestone can be utilized to help neutralize acid drainage —Can be used to stabilize and line drainage channels —Can utilize available onsite rock (not just limestone)
Disadvantages	—Manpower required to fill bags	—Relatively temporary/will disintegrate —Flood can move staked bales	—Can be costly and involve additional transportation (if onsite material unavailable)
	Sediment Control Devices		
	Log-fabric Dam	*Bag Structure*	*Silt Fence*
Advantages	—Most effective containment device —Can be made with onsite materials —Well suited for deeper gullies that fill in with trapped sediments	—Quick and easy to install —Relatively effective at slowing and filtering limited quantities of water	—Quick and easy to install —Good filter —Successfully slows water velocity —Fence/cloth combination provides reinforcement for filling behind structures —Can extend linearly long distances
Disadvantages	—Requires careful, labor-intensive construction (logs anchored into banks, sand bags lining front *and* back of structure	—Relatively short life span as sun ages cloth/plastic —Bag/cloth can be expensive	—Size/strength preclude use in high velocity or sediments —Relatively limited storage

Source: Wilder and Rains (1976); Jack Muncy, TVA, 1982, personal communication.

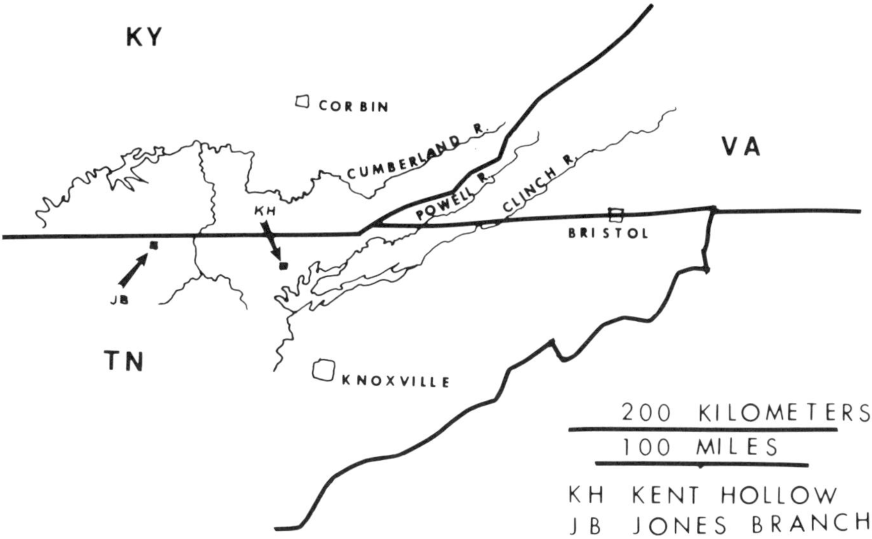

Figure 7.4 Location of aquatic monitoring stations in Orphan Land Reclamation Demonstration Program. Adapted from Taylor and Nicholson 1979.

ently eliminated the previous acid drainage problem while new sediment ponds prevented additional sedimentation problems allowing natural stream recovery. In 1980, the stream substrate was clean with winter flows having flushed all silt from pool and riffle areas. The pH (5.5 to 5.7) and metal levels ($<$ 1 mg/1) were similar to those previously reported for the upstream control. Aquatic macroinvertebrates had responded with increases in density and diversity per square foot. No fish were present.

By 1979 the Jones Branch site was being altered by a natural method. The pool areas which had silted in were becoming colonized with aquatic and semiaquatic macrophytes. The result was a braided stream channel which developed in several former pools of the stream. In 1980 there did not appear to be any fresh input of sediments. Between 1979 and 1980 the water quality also improved, perhaps augmented by filtering through the new "wetlands" within the stream. The pH increased from 5.6 to 5.8 from 1979 to 1980 and total suspended solids ranged from 4 to 17 mg/1.

While this monitoring effort did not prove conclusive, it does appear that the Orphan Land Reclamation Demonstration Program activities are ameliorating watershed problems.

Piney Creek Watershed Reclamation Demonstration

This joint demonstration was undertaken by the University of Tennessee, the Tennessee Valley Authority, the Tennessee Departments of Conservation, Corrections, and Public Health, and various landowners. The project interim report (Byerly et al. 1978) described reclamation activities and initial water quality responses which are summarized in the following discussion.

The study watershed is located within Van Buren County, Tennessee, and the affected stream, Piney Creek, drains into Fall Creek Falls State Park (Figure 7.5). Acid mine drainage and siltation problems emanated from sixteen underground mines, 226 acres (92 ha) of pre–Tennessee reclamation law mines, and 203 acres (82 ha) of post-law mines with insufficient ground cover (Byerly et al. 1978).

Reclamation activities were essentially manual and involved revegetation activities combined with erosion control devices, silt structures, and sealing underground mines. While hydroseeding and hydromulching were utilized, the best cover of grass was obtained where the area was limed, scarified, and mulched by hand. Over two tree- and shrub-planting seasons, survival approached 80%. Erosion and silt control structures were similar to those utilized by the Orphan Land

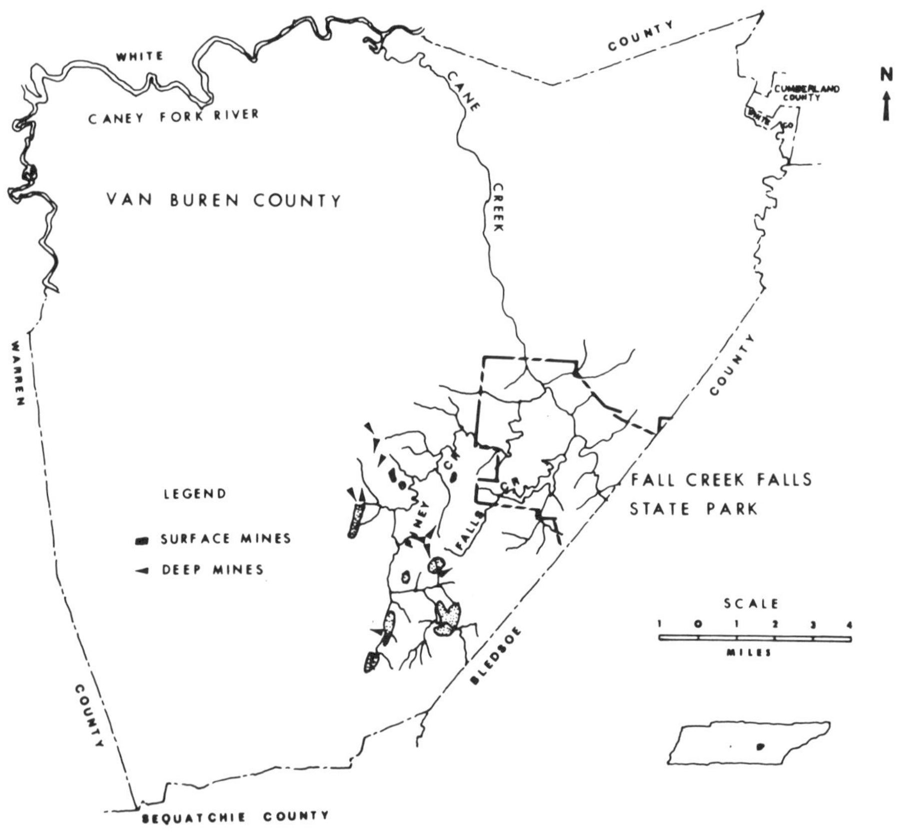

Figure 7.5 Location of surface and deep mines in Piney Creek watershed, Van Buren County, Tennessee. Adapted from Byerly et al. 1978.

Table 7.2 Annual average of water quality and benthic communities in two Ollis Creek tributaries, Thompson Creek and Unnamed Tributary, from 1975 through 1980.

	Thompson Creek					
	1975	*1976*	*1977*	*1978*	*1979*	*1980*
Water quality						
pH	4.2	4.5	4.2	4.2	4.2	4.5
Fe mg/p	1.1	0.5	0.2	0.4	0.2	0.1
SO_4 mg/p	75.7	88.5	72.0	96.5	85.2	60.0
Mn mg/p	1.5	1.9	1.4	2.5	1.9	1.0
Macroinvertebrates						
Total taxa	10.0	18.0	18.0	15.0	19.0	15.0
Mean number	3.0	15.0	18.0	26.0	43.0	29.0
	Unnamed Tributary					
	1975	*1976*	*1977*	*1978*	*1979*	*1980*
Water quality						
pH	3.9	4.7	4.2	4.5	4.2	4.4
Fe mg/p	9.3	5.2	3.6	3.5	4.2	2.3
SO_4 mg/p	610.0	497.0	386.7	467.2	445.0	383.0
Mn mg/p	13.6	8.4	6.9	18.1	13.5	12.6
Macroinvertebrates						
Total taxa	12.0	10.0	14.0	8.0	10.0	10.0
Mean number	2.0	2.0	2.0	3.0	4.0	4.0
Number of collection dates	3	12	4	4	4	3

It is interesting that aquatic macroinvertebrates responded with such large-scale changes when pH, for example, was only marginally acceptable.

Unnamed Tributary

Remedial reclamation activities began in 1974 with the placement of a large silt pond across this tributary. Erosion and sediment control structures similar to those used in the Orphan Land Reclamation Demonstration Program were placed on bench and outslope areas.

Watershed revegetation efforts were initiated in 1976. As can be seen from the annual averages in Table 7.2, this watershed had significantly poorer water quality than Thompson Creek at the initiation of treatments.

In 1975 the silt pond was working well with sediments being largely collected by the structure; however, water quality was changing and yellowboy coated the substrate. By 1977, this structure had filled and the tributary was flowing over and infiltrating through the collected sediments. No heavy equipment was available to properly clean and dispose of the sediments so it was allowed to remain in place. Revegetation activities, as well as silt and sediment control structures, had greatly decreased incoming sediments but water quality was dominated by the leaching of materials from the collected sediments. In 1978, a flood topped the structure. The tributary began cutting through collected sediments resulting in heavy downstream siltation.

This scenario explains water quality and aquatic biota recovery rates. Water quality at this station significantly improved between 1975 and 1980; however, resultant concentrations remained above the Thompson Creek initial concentrations. This indicates the length of time remaining until recovery at this tributary. Aquatic communities remained depauperate with no change from 1975 to 1980. With water quality still exceeding U.S. EPA (1977a) water quality standards and substrate covered with precipitate or silts, this result is expected. Using regression analysis, I calculated that an additional 59 to 82 years will be required at this tributary for water quality to reach acceptable standards for manganese and iron respectively (Zarger et al. 1982) if present recovery trends continue.

The overall success of intensive reclamation techniques has been demonstrated with several developments in the watershed. The 1980 vegetation surveys documented an overall 78% vegetation cover (in contrast to 24% in 1974) (Figure 7.7). The Tennessee Wildlife Resources Agency (TWRA) stocked largemouth bass, *Micropterus salmoides;* bluegill, *Lepomis macrochirus;* and catfish, *Ictalurus punctatus,* in the water supply reservoirs in fall 1979 and spring 1980 where previously there had been fishkills. The TWRA transplanted beaver populations into upper Thompson Creek where they have reproduced and spread to other tributaries. A side-effect of the introduced beaver population is that their impoundments appear to further trap silts and act as biologically maintained settling basins. The once acidic ponds on the bench were stocked in 1979 and 1980 by TVA and TWRA with bluegill, green sunfish (*Lepomis cyanellus*), and largemouth bass. A study of stocking success was conducted by Summers (1981), who found that growth and reproduction were good in 1980. In 1981 there was no bass reproduction and before the end of the year forage fish were stunted. Aquatic vegetation dominating the ponds may have been responsible for some of the fisheries response.

STREAM RESTORATION

Various structural techniques have been utilized within stream channels to manipulate water quality problems in the eastern United States. Many of these structures such as clay liners, limestone riprap, or flow deflectors are constructed routinely in abandoned or new mine reclamation. While water quality evaluations are frequently performed, biotic evaluations are rarely conducted. The studies selected for discussion were chosen based on aquatic evaluations and overall project scope.

The choice of technique is governed by water velocities, installation costs, and the kinds of effects desired. Devices can, for example, create pools and slow

Figure 7.7 Vegetative comparisons in a 24-acre plot, Ollis Creek drainage **(a)** at initiation of remedial reclamation in 1976 and **(b)** in 1980 after five growing seasons. Photos, with permission, Tennessee Valley Authority.

flow to act as settling basins or can accelerate flows to keep substrate swept clean. While it must be realized that indiscriminate use of instream devices can be detrimental to aquatic life, with careful planning and placement these tools can encourage rapid aquatic recovery and development. Structures can ameliorate water quality problems as well as provide habitat diversity.

Limestone Barriers—Installation and Evaluation

Pearson and McDonnell (1975) and Yocum (1976) reported on the installation and performance of limestone barriers in Trough Creek in south-central Pennsylvania. Use of limestone to neutralize acid mine drainage typically is of limited success due to a coating of both iron hydroxide, which develops from chemical reactions, and silt from unvegetated abandoned mines in the watershed. Pearson and McDonnell (1975) utilized models to rationalize barrier design. Comparisons of model predictions with field measurements indicated that adequately designed barriers can restore streams to acceptable water quality with some shortcomings. Yocum (1976) reported the results of a gaging and sampling program designed to evaluate limestone barriers during instream usage. A schematic (Figure 7.8) shows the configuration of four limestone barriers. Efficiency was related to total width regardless of whether the width was divided between multiple structures or was in one structure. Efficiency was greater for single structures No. 1, 4, and 5 than for No. 2 which was comprised of small, multiple barriers. Structures were most efficient in low flow when the water had maximal contact with the limestone but were relatively inefficient when flows were high and most water was not in contact with the limestone.

If water quality were improving from other watershed treatment measures, these structures would accelerate recovery. On the other hand, where the source of acid mine drainage and sediments is not eliminated, the barriers will become ineffective. While they improve water quality, such structures make major stream habitat alterations. As such, they require routine maintenance and eventual removal, as conditions improve, in order to allow complete biological recovery.

Stream Channel Reconstruction

In new and abandoned mine reclamation, streams are frequently relocated. In abandoned mine reclamation, it is often easier to move the stream away from mine spoils, subsidence areas, or refuse piles and thus prevent sedimentation or acid mine drainage than it is to treat these problems after formation. While numerous reclamation projects include stream reconstruction as a facet of environmental repair, we will look in detail at studies by Klingensmith et al. (1976) and U.S. EPA (1977b).

Klingensmith et al. (1976) described stream reconstruction in Pennsylvania in Catawissa Creek Project, Beech Creek Project (Little Sandy Run), and Tioga

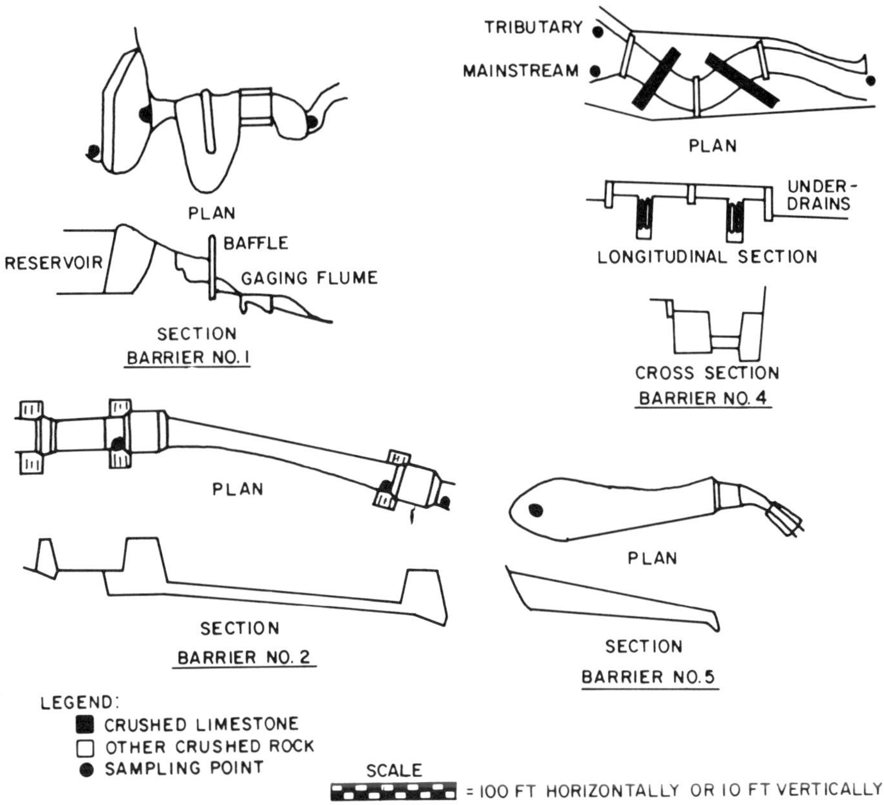

Figure 7.8 Schematic of four limestone barriers. Adapted from Pearson and McDonnell 1975.

River Project (Morris Run). Rationale for and details of reconstruction differed for each project.

Catawissa Creek Project

A 1,700-foot section of alternate channel was cleared of mine spoils and debris and deepened. An alternate channel had originally been created to allow for mining through the original stream channel. The stream was subsequently diverted to a deep mine to allow mining near the alternate channel, but never returned to either the original or alternate channels. Renovation of the alternate stream channel did allow for reduced acid load to Catawissa Creek by 1,830 lbs/day, but increases in iron concentration emanating from the deep mine resulted in iron remaining at the same concentration.

Beech Creek Stream Project

Subsidence and drainage into deep mines dictated reconstruction of 1,000 feet of Little Sandy Run. The new channel was excavated to eliminate slow flow areas. Sand and bentonite were layered in the channel and riprap (crushed rock layer) was added to protect the bentonite seal and provide stability. Stream channel reconstruction reduced acid and iron loads by 624 and 156 lbs/day respectively.

Tioga River Project

Site I of this project involved reclamation of a 16-acre abandoned mine and subsequent relocation of the stream across the reclaimed mine. To make the bed impermeable, a clay liner, protective filter, and finally quarry stone were layered in the channel. Site II involved reclamation of 80 acres of surface mines and construction of an infiltration ditch downhill from a 4-acre wastewater sludge test plot. Reclamation of strip mines at Sites I and II and reconstruction of stream bed at Site I reduced acid and iron loadings 8,480 and 550 lbs/day respectively.

Elkins Mine Drainage Control Demonstration Project

This project was conducted between 1964 and 1968 on Roaring Creek and Grassy Run Watersheds near Elkins, West Virginia. Scott et al. (1972) detail reclamation measures and costs. Approximately 400 ha of surface mines and 1200 ha of underground mine works were the subject of mine seals, water diversion, burial of spoil and refuse, and revegetation. Monitoring results in Grassy Run indicated some reduction in acidity but little change in flow, sulfate, or iron. In Roaring Creek, the diversion had impact on flow. Acidity and sulfate, however, did not decrease. Biological monitoring from 1964 to 1968 documented little change in fauna in either stream. Information being collected during report preparation (1976) was described by U.S. EPA (1977b). A fishery has returned to the Tygart Valley River from above the mouth of Roaring Creek and Grassy Run down to Tygart Reservoir. This followup evaluation indicated a delay in recovery but overall success in reclamation efforts utilized on this site.

STREAM RESTORATION

Neither energy development nor stream channel realignment need be synonomous with environmental degradation. To stabilize and protect stream diversions or altered stream channels, natural energy dissipators (such as bends, vegetation, logs, and boulders) are created or simulated by placing structures in the stream channel. In addition to providing natural physical changes, these structures provide habitat necessary for fish and other aquatic life. The absence of energy dissipators increases the velocity of flow and intensifies erosion and sedimentation at areas further downstream.

While stream channelization is prohibited by SMCRA, stream diversions during new mining are permitted. Tourbier and Westmacott (1980) developed guidelines for use in the Office of Surface Mining's Small Coal Operator Program. Their recommendations apply to rehabilitation of channelized streams as well as restoration of new stream diversions. Streams should have sinuosity or meandering pattern, alternating riffles and pools, and both longitudinal profile and cross section that approximate premining or nearby unaltered streams.

Rehabilitation of altered streams in the eastern United States is being researched to determine what conditions can be ameliorated. Perhaps the most common rehabilitation practice involves creation of structures within the stream to alter flow (Saunders and Smith 1962). Structures can be utilized to increase, decrease, or divert flows which will create riffle-pool areas (Gee 1952). Brusven et al. (1974) evaluated the sediment removal capabilities of different structures. Instream structures, gabion deflectors, channel diversion, and log jam removal all resulted in increased sediment transport. Downstream of surface mines, where control of sedimentation is the focal point of reclamation efforts, structure success and life span will be dictated to the extent the structure can accommodate sedimentation. If the structure design facilitates removal of recently deposited silt, it will help the stream return to a more stable as well as more natural configuration without dredging, channelization, or complete loss of aquatic life. Barton and Winger (1973) found structures in altered channels provided for aquatic invertebrate recovery which enhanced rapid fish recovery. This is in marked contrast to White and Fox (1980) who found limited biotic recovery even twenty-one years after channelization with no restoration practices in South Carolina streams.

While the previous studies apply to stream channelization, Thompson (1980) reported similar results for stream relocations associated with surface mining. Fish and aquatic macroinvertebrate monitoring from 1977 to 1979 on four intermittent streams at three Illinois surface mines indicated significant aquatic community recovery within two to three years. Greatest recovery was associated with older stream relocations and with creation of habitat diversity (such as riffles and pools).

Stream restoration techniques have been implemented into western mining and reclamation as discussed by Gore in Chapter 4 and Wesche in Chapter 5 of this book. While stream restoration principles are similar in both the eastern and western United States, climate, geology, and hydrology dictate that specific techniques implemented in the two regions are different. Whether in the East or the West, success of stream restoration measures will be contingent on the selection of structures that are permanent, relatively maintenance free, and designed for site-specific conditions.

In the following discussion of structures, descriptions will include advantages, disadvantages, and potential problems with their use in altered channels associated with coal surface mining. White and Brynildson (1967) and Maughan et al. (1978) described and evaluated these structures in detail as they were utilized in trout management. While these structures alone will not solve water quality problems, when used in a watershed restoration scenario, they can effectively stabilize stream channels and enhance aquatic recovery. Detailed plans for an abandoned mine watershed reclamation project including stream channelization and restoration will be discussed in the Summary section.

Bank Cover

Bank cover is effective in providing relatively permanent habitat in areas where banks are smooth and unvegetated as typically occurs in areas of stream channelization, relocation, or diversion. The structure is most effective on the "current-swept" outside bank of a stream bend. The structure will simultaneously stabilize banks and provide protection for aquatic species. If anchored securely into the bank and if all wood is submerged, life expectancy of the structure is good, with little maintenance required.

Deflectors

Made of concrete, logs, rocks, and/or gabions (a mesh container used to hold rocks) deflectors divert flows from unprotected banks. Used individually or with a structure such as previously described, deflectors can protect eroding banks or direct flows toward more desirable channels. If paired, they can be used to restrict flows and depth, an asset in low-flow periods. Used in series and alternating banks, they can produce sinusoidal stream flows. Below abandoned mine areas, these structures should have low profile to prevent channel filling. Sediments will accumulate behind the structures. Gabions or riprap (broken rock placed on a stream bank or in the stream channel to protect soils from water action) are effective bank protectors but provide very limited aquatic habitat. Structures must be designed and installed carefully to prevent erosion around the structure or removal by floods.

Cascades

Cascades are as numerous as construction materials: K-dams, rock dams, log dams, log drop structures, gabion brace-log dams, log-shell, rock-filled dams, etc. The dams are constructed in uniform shallow riffle areas where limited aquatic habitat exists. Cascades are constructed to create a pool below the dam. To be effective, a structure should have no wood above the water surface and should be anchored securely into both stream banks. Thus, rock dams are often cheapest to build but may be more difficult to retain or anchor. The K-dam has the advantage of other dams but allows flow through the structure, permitting silt to pass downstream and fish to swim upstream.

Rock Boulders

Use of boulders is perhaps the cheapest and easiest way to provide habitat diversity. Large, irregularly shaped rocks are placed in the stream channel. Care must be taken in placement, as current can be deflected toward unprotected banks and too many boulders will block flow. Boulders cannot be anchored and are, therefore, susceptible to movement. Cracks and indentations provide aquatic invertebrate habitat. Fish will utilize low-flow areas behind the boulders.

SUMMARY

The Ollis Creek (Tennessee) and Elkins (West Virginia) projects indicated that aquatic recovery may occur at mine reclamation sites, but even with remedial or intensive reclamation complete recovery will be slow. Recovery is greatly accelerated over natural rates which may be very slow. Tolbert and Vaughn (1980) and Talak (1977) reported twenty-four years to full biological recovery in nonacid drainage situations. For abandoned mines, this is probably the fastest possible natural recovery; and unfortunately, problems such as acid or sedimentation would extend recovery into additional decades. For aquatic resources, this is unacceptable when total watershed reclamation schemes can not only ameliorate water quality problems but also enhance aquatic biological recovery.

To conclude and summarize the principles discussed thus far, this chapter will examine a project proposal developed by TVA for Commonwealth of Virginia, Department of Conservation and Economic Development, Division of Mined Land Reclamation; and for U.S. Department of Interior, Office of Surface Mining Reclamation and Enforcement. The multiphase and multidisciplinary project was designed to develop short- and long-term solutions to the problems of degraded stream channels and flood plains resulting from adverse impacts of coal mining practices in the watershed. While not all portions of the proposal have been implemented, the design is a model for these kinds of projects. Only aspects directly affecting stream restoration will be examined.

St. Charles, Virginia, Watershed Reclamation Project (TVA 1979)

The St. Charles project area is located in northeast Lee County, Virginia, northwest of Pennington Gap. Figure 7.9 shows the location of the Straight Creek watershed, subwatersheds (which will be described shortly), and location and type of environmental problems existing in the watershed. Typical of the Appalachian coal fields, a mixture of problems such as abandoned surface and deep mines, haul roads, and gob piles (coal refuse from earlier deep mines) exist. The project area consists of approximately 8,000 acres (3,237 ha). Approximately 800 acres (324 ha) have been disturbed by surface-mining activities.

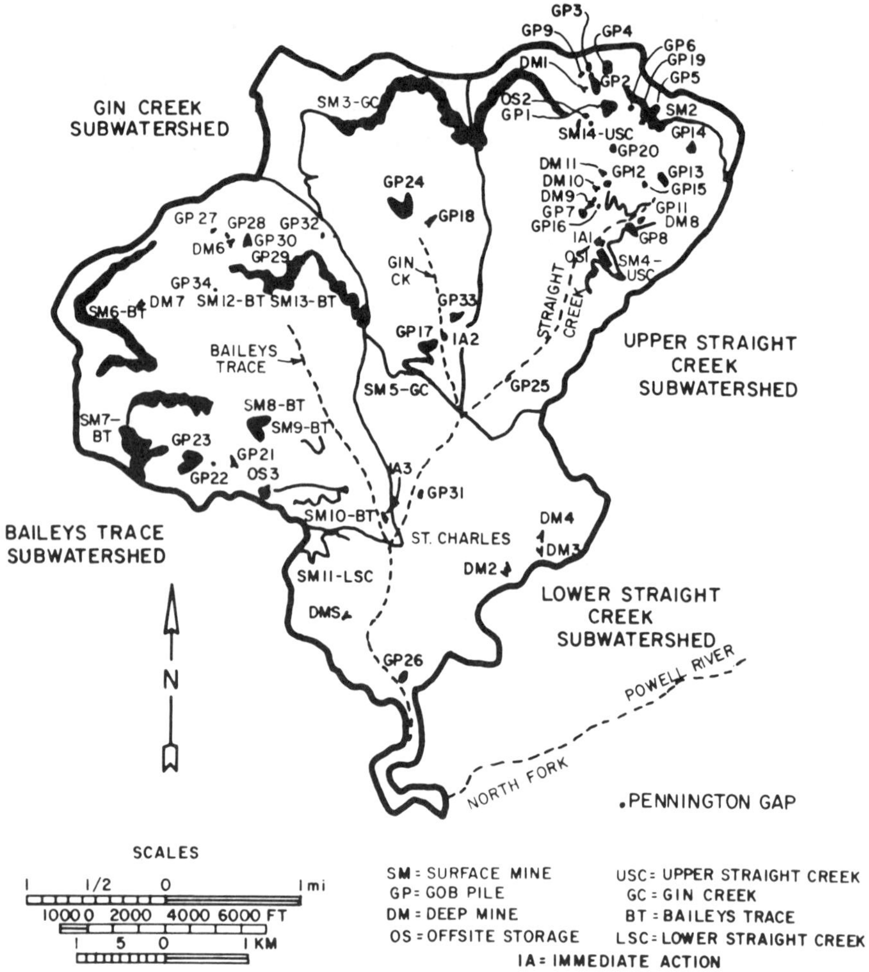

Figure 7.9 The Straight Creek watershed showing subwatersheds, surface mines, deep mines, and refuse piles. Adapted from TVA 1979.

St. Charles (Straight Creek watershed) was designated by the Commonwealth of Virginia as the number one priority in the State under the Office of Surface Mining's Abandoned Mine Reclamation Fund. (Established under Section 401 of SMCRA, the Fund is financed through fees collected from active coal-mining operations. Priority is given to alleviating dangers to public health and safety.) This area received this priority based primarily on severe flooding that endangered public health, safety, and property. Flooding was largely the result of lost carrying capacity of streams. In 1978, the U.S. Office of Surface Mining Reclamation and Enforcement released funds to Virginia Division of Mined Land Reclamation for initiation of work on the project. The initial work was conducted in two parts (TVA 1979). Part 1, with plans prepared by TVA, reclaimed three gob piles (waste coal, rock pyrites, or other material of no commercial value) in juxtaposition to watershed streams and removed sediment and debris from selected reaches of affected streams. This work was coordinated and supervised by the Virginia Division of Mined Land Reclamation. Part 2 was prepared by TVA in coordination with Virginia Division of Mined Land Reclamation and involved development of the long-term plan for reclamation and flood reduction as well as monitoring to assess treatment success. While the project was multidisciplinary in design and implementation, this chapter examines only aquatic aspects.

Stream Restoration

Watershed reclamation (terrestrial) began with Upper Straight Creek, identified as having the worst watershed problems, and proceeded to Gin Creek, Baileys Trace and, lastly, Lower Straight Creek (Figure 7.9), which had the fewest problems even though it was affected by upstream improvement or degradation. As designed, channel improvements would span the entire project life. While gob piles and abandoned surface mines were reclaimed and deep mine drainage sealed or treated, selected portions of streams were deepened and the channel slopes increased. This was followed by bank stabilization efforts. While activities have been interrupted at this point, it was planned that if these efforts did not reduce flooding, levees or walls would be constructed. Part of the plan was to relocate or improve the hydraulic efficiency of artificial obstructions. Models would be utilized to locate stream sections where potential aggradation would occur following Part 1 channelization. Channel maintenance would be performed as necessary. The final aspect to be implemented after drainage improved or ceased and sedimentation was slowed would be the fisheries restoration plan.

Straight Creek is approximately six miles (9.6 km) long. Between the headwater tributaries [elevation 2,300 feet (701 m)] and its confluence with the North Fork Powell River, Straight Creek loses 350 feet (107 m) in elevation. Identified stream problems were extreme stream flow fluctuations, ranging from floods to near-zero flow, increased turbidities, heavy sedimentation with channel filling including loss of pools and riffles, and degraded water quality.

Aquatic biological surveys (TVA 1979) conducted from November 1978 to April 1979 revealed three fish species (stoneroller, *Campostoma anomalum;* blacknose dace, *Rhinichthys atratulus;* and northern hog sucker, *Hypentelium nigricans*) present in Straight Creek. Due to its stream size and confluence with the North Fork Powell River, more species were expected. Also, fish collected were always associated with deep holes or bridge abutments, making it appear that cover was a limiting factor (Strant Colwell, 1982, TVA, personal communication). Aquatic macroinvertebrate sampling from November 1978 through April 1979 indicated a relatively restricted population. Over the six months of sampling a relatively large number of species (thirty-six) were collected. Thus, for this watershed, planned fisheries improvement devices would have relatively rapid and significant benefit for both fish and aquatic macroinvertebrates. A total of forty-eight fisheries improvement devices were designed for the four subwatersheds.

To illustrate the treatment recommendations, Baileys Trace subwatershed will be examined in detail. Because Baileys Trace is the least affected tributary to Straight Creek it is the area where stream restoration should begin. The location of eleven proposed structures is illustrated in Figure 7.10. At sites 1 and 2 there would be a series of stair-step rock dams and revegetated banks. Both sites 3 and 4 involve installing *K*-dams, riprapping banks and returning the stream to original channel. The oil tank at site 4 would need to be removed or wing deflectors utilized to divert stream flow around the obstruction. At sites 5 and 6 bank cover–wing devices would be utilized to stabilize banks and provide habitat and cover. A *K*-dam or rock dam would be installed at site 7. At sites 8 and 9 small wing devices combined with boulders would enhance the bridge abutment pool. Riprapping of steep (channelized) banks at sites 10 and 11 and possibly wing deflectors would help protect banks. The overall purpose is to place stream devices so as to create productive pools with cover, maintain good riffle-pool ratios, and create refuges in low-flow conditions.

In a watershed, such as St. Charles where mining and even the town itself has altered the natural channel, flow patterns, and flood plain, the purpose of

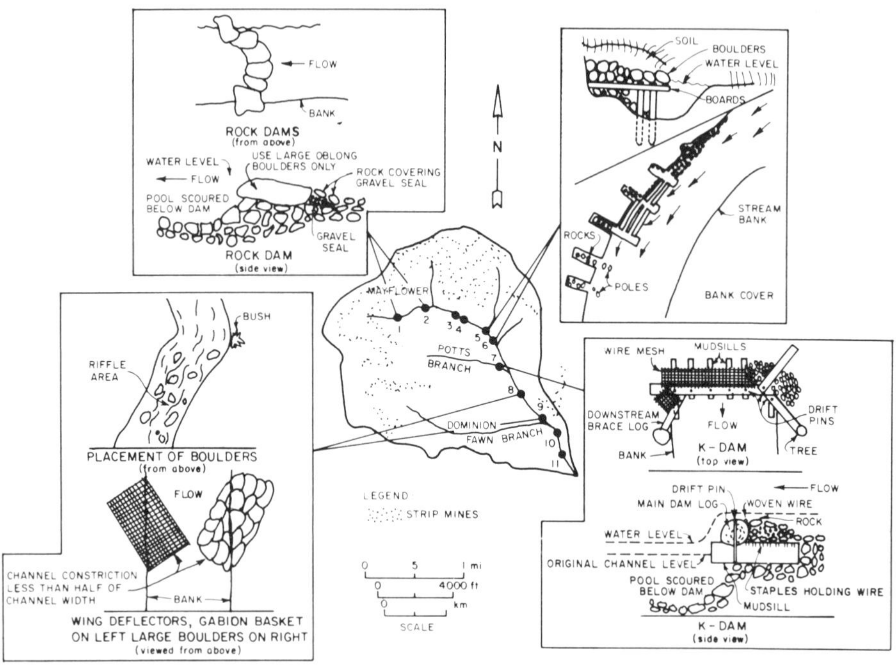

Figure 7.10 Baileys Trace subwatershed with location of four different types of structures proposed for habitat improvement at eleven locations. Adapted from TVA 1979.

in-stream restoration is to return stability to the watershed and, at the same time, return recreation. Structures must be designed to return the stream to its natural or an equally permanent channel. The structure of the stream community, once studied to determine the limiting factor(s), may be enhanced or returned by implementing the proper corrective measures.

CONCLUSION

The greatest ecological recovery in the receiving streams within those projects evaluated in this chapter has occurred when reclamation methodology has simultaneously treated both drainage and vegetation problems and was based on careful planning of site-specific problems. While this chapter examined the exception, not the rule, for the majority of projects (especially in the East) reclamation efforts have been concentrated at the abandoned mine with little regard for amelioration or elimination of in-stream or secondary impacts. The tacit assumption that successful revegetation indicates successful aquatic recovery may not be true. Matter and Ney (1981) state, "It is apparent that terrestrial reclamation does not guarantee lotic reclamation and recovery . . . acceptable vegetative cover may still result in substantial erosional impacts to streams." Procedures for aquatic restoration, while neither intricate nor expensive, do enhance biological recovery—the true measure of reclamation success.

REFERENCES

Barton, J.R., and P.V. Winger. 1973. *A Study of the Channelization of the Weber River Summit County, Utah.* Final Res. Rept. Utah Div. Wildl. Res. and Utah State Dept. of Highways. p. 188.

Bettoli, P.W. 1979. *Age, Growth and Food Habits of Fishes Common to the New River, Tennessee.* Nashville, TN: U.S. Army Corps of Engineers. p. 74.

Boyer, J.F., Jr., C.T. Ford, V.E. Gleason, and A. Price. 1978. Assessment of research and development needs and priorities for acid mine drainage abatement. Bituminous Coal Research Inc. Report L-822. Prepared for Bureau of Mines. Washington, D.C., p. 162.

Brusven, M.A., F.J. Witts, R. Leudtke, and T.L. Kelly. 1974. *A Model Design for Physical and Biotic Rehabilitation of a Silted Stream.* Research Tech. Completion Report A-032-1DA. Water Resources Research Institute. p. 96.

Byerly, D.W., J.B. Maddox, C.S. Fletcher, and D.T. Eagle. 1978. Reclamation of mined land in the Piney Creek watershed—an interim report. In *Proceedings of a Symposium: Surface Mining and Fish/Wildlife Needs in the Eastern United States,* edited by E.E. Samuel et al., p. 267–75. FWS/OBS-78/81. Washington, D.C.: U.S. Fish and Wildlife Service.

Cordone, A.J., and D.W. Kelly. 1961. The influence of inorganic sediment on the aquatic life of streams. *Calif. Fish and Game.* 47:189–229.

Funk, D.T. 1962. *A revised bibliography of strip-mine reclamation.* U.S. Department of Agriculture, For. Serv. Cent. States, For. Exp. Stn. Misc. Release. Columbus, Ohio. p. 35.

Gee, M. 1952. *Fish Stream Improvement Handbook.* Washington, D.C.: U.S. Dept. of Agric., U.S. Forest Serv. p. 21.

Gleason, V.E. 1979. *Bibliography on Mined-land Reclamation.* EPA-600/7-79-102. Washington, D.C.: U.S. Dept. of Interior. p. 373.

______. 1980. *Mine Drainage Bibliography 1929–1981.* EPA-600/7-80-113. Washington, D.C.: U.S. Dept. of Interior. p. 184.

Herricks, E.E., and J. Cairns, Jr. 1975. *Rehabilitation of Streams Receiving Acid Mine Drainage.* Va. Water Resources Center Bulletin 66. Blacksburg, Va.: Va. Polytech. Inst. and State Univ. p. 284.

Hynes, H.B.N. 1974. *The Biology of Polluted Water.* 5th ed. Toronto: University of Toronto Press. p. 202.

Johnson, W., and G.G. Miller. 1979. *Abandoned Coal-mined Land—Nature, Extent and Cost of Reclamation.* Washington, D.C.: U.S. Bureau of Mines. p. 29.

Kieffer, F.V. 1972. *A Bibliography of Surface Coal Mining in the United States.* Columbus, Ohio: Forum Associates. p. 71.

Klingensmith, R.S., A.F. Miorin, and J.R. Saluinas. 1976. At source control through the application of several abatement techniques. In *Sixth Symposium on Coal Mine Drainage Research,* Washington, D.C.: Nat. Coal Assoc., p. 270–84.

Leedy, D.L. 1981. *Coal Surface Mining Reclamation and Fish and Wildlife Relationships in the Eastern United States.* Volume I. FWS/OBS-80/24. Washington, D.C.: U.S. Dept. of Int., Fish and Wildlife Service. p. 75.

Leedy, D.L., and T.M. Franklin. 1981. *Coal Surface Mining Reclamation and Fish and Wildlife Relationships in the Eastern United States.* Volume II. FWS/OBS-80/25. Washington, D.C.: U.S. Dept. of the Interior Fish and Wildlife Service. p. 169.

Linstrom, G.A. 1953. A bibliography of strip-mine reclamation. U.S. Department of Agriculture, For. Serv. Cent. States, For. Exp. Stn. Misc. Rel. 8. Columbus, Ohio. p. 25.

Matter, W.J., and J.J. Ney. 1981. The impact of surface mine reclamation on headwater streams in southwest Virginia. *Hydrobiologia* 78:63–71.

Maughan, O.E., K.L. Nelson, and J.J. Ney. 1978. *Evaluation of Stream Improvement Practices in Southeastern Trout Streams.* Bulletin 115. Blacksburg, Va.: Va. Water Resources Research Center. p. 67.

Nicholson, J.H., and J.E. Snyder. 1939. *Manual for Soil Erosion Control in the Tennessee Valley.* Division of Land and Forest Resources. Norris, Tenn.: Tennessee Valley Authority. p. 124.

Paone, J., J.L. Morning, and L. Giorgetti. 1974. Land utilization and reclamation in the mining industry, 1930–71. U.S. Dept. of the Interior, Bureau Mines Inf. Circ. 8742. p. 61.

Parsons, J.W. 1952. A biological approach to the study and control of acid mine pollution. *J. Tn. Acad. Sci.* 24(4):304–10.

Pearson, F.H., and A.J. McDonnell. 1975. Limestone barriers to neutralize acidic streams. *J. of the Environmental Engineering Division, ASCE* 101(EB2):425–40. Proc. Paper H382, June 1975.

Riley, C.V. 1960. The ecology of water areas associated with coal strip-mined lands in Ohio. *Ohio J. Sci.* 60(2):106–21.

Roback, S.S., and J.W. Richardson. 1969. The effects of acid mine drainage on aquatic insects. *Proc. Acad. Nat. Sci. Phil.* 121:81–107.

Rowe, J.E., ed. 1979. *Coal Surface Mining: Impacts of Reclamation.* Boulder, Colo.: Westview Press. p. 490.

Saunders, J.W., and M.W. Smith. 1962. Physical alteration of stream habitat to improve brook trout production. *Trans. Am. Fish. Soc.* 91(2):185–88.

Scott, R.B., R.C. Wilmouth, and R.D. Hill. 1972. Cost of reclamation and mine drainage abatement, Elkins demonstration project. *AIME Trans.* 252:187–93.

Spaulding, W.M., Jr., and R.D. Ogden. 1968. *Effects of Surface Mining on the Fish and Wildlife Resources of the United States.* Washington, D.C.: U.S. Fish and Wildlife Service, Bureau of Sport Fisheries and Resources. Publication 68. p. 51.

Starnes, L.B., and J.B. Maddox. 1978. Effects of remedial reclamation treatments on terrestrial and aquatic ecosystems—A progress report. In *Proceedings of a Symposium: Surface Mining and Fish/Wildlife Needs in the Eastern United States,* edited by E.E. Samuel et al., p. 276–86. FWS/OBS-78/81. U.S. Fish and Wildlife Service.

Summers, P.B., Jr. 1981. *An ecological assessment of twenty-one sediment ponds, Ollis Creek Mine, Campbell County,* TN. Master's thesis, Tenn. Tech. University. Cookeville. p. 242.

Talak, A. 1977. *The recovery of stream benthic insect communities following coal strip mining in the Cumberland Mountains of Tennessee.* Master's thesis, University of Tenn., Knoxville. p. 82.

Taylor, E.A., and C.P. Nicholson. 1979. *Orphan Land Reclamation Demonstration Program Monitoring Project.* Division of Water Resources. Final Report. Norris, Tenn.: Tennessee Valley Authority. p. 78.

Tennessee Valley Authority. 1963. *An Appraisal of Coal Strip Mining.* Knoxville, Tenn. p. 13.

______. 1979. *St. Charles, Virginia. Watershed Reclamation Project.* Knoxville, Tenn. p. 545.

Thompson, C.S. 1980. Effect of stream relocations on benthic macroinvertebrates and fishes in surface mine areas. Presented at Spring 1980 Technical Session, Indiana Chapter of AFS. Ball State University. p. 35.

Tolbert, V.R., and G.L. Vaughan. 1980. Strip-mining as it relates to benthic insect communities and their recovery. *WV Acad. Sci.* 51(3):168–81.

Tourbier, J.T., and R. Westmacott. 1980. *A Handbook for Small Surface Coal Operators.* Water Resources Center. Dover, DE: Univ. of Del. p. 130.

U.S. Army Corps of Engineers. 1974. *The National Strip Mine Study.* Volume I. Washington, D.C.: Department of the Army. p. 92.

______. 1978. *Demonstration Reclamation Project Cabin Creek, West Virginia.* Final Environmental Statement. Huntington, W.Va.: Department of Army. p. 73.

U.S. Department of the Interior. Quarterly 1977–1981. *Surface Environment and Mining Alert to Current Literature (SEAMALERT).* Washington, D.C.: Office of Surface Mining. Division of Information and Records Mgmt.

U.S. Environmental Protection Agency. 1977a. *Quality Criteria of Water.* Washington, D.C.: Office for Water and Hazardous Materials. p. 256.

______. 1977b. *Elkins Mine Drainage Pollution Control Demonstration Project.* EPA-600/7-77-090. Cincinnati, Ohio: Industrial Environmental Research Laboratory. p. 297.

U.S. Soil Conservation Service. 1979. *The Status of Land Disturbed by Surface Mining in the United States; Basic Statistics by State and County as of July 1, 1977.* Publication SCS-TP-158. Report No. 1979–631–344/2860. Washington, D.C.: U.S. Gov't Printing Office. p. 124.

Vaughan, G.L., A. Talak, and R.J. Anderson. 1978. The chronology and character of recovery of aquatic communities from the effects of strip mining for coal in east Tennessee. In *Proceedings of a Symposium: Surface Mining and Fish/Wildlife Needs in the Eastern United States,* edited by E.E. Samuel et al., p. 119–25. FWS/OBS-78/81. U.S. Fish and Wildlife Service.

Vines, C.A. 1979. *Reclaiming Mined Lands: An Update—Tennessee Valley Perspective.* Knoxville, Tenn.: TVA. p. 19–23.

White, R.J., and O.M. Brynildson. 1967. *Guidelines for Management of Trout Stream Habitat in Wisconsin.* Bulletin 39. Madison: Wis. Department of Natural Resources. P. 65.

White, T.R., and R.C. Fox. 1980. *Recolonization of Streams by Aquatic Insects Following Channelization.* Clemson Univ. Water Res. Res. Inst. Tech. Rept. No. 87. Vol. I. Clemson, S.C. p. 120.

Wilder, C.A., and M.T. Rains. 1976. *Site Improvement Techniques on the Y-LT: A Status Report.* Forest Service. Oxford, Miss.: U.S. Department of Agriculture. p. 12.

Yocum, S.C. 1976. *Trough Creek Limestone Barrier Installation and Evaluation.* EPA-600/2-76-114. Washington, D.C.: U.S. EPA, Office of Research and Dev. p. 93.

Zarger, T.G., J.B. Maddox, L.B. Starnes, and W.M. Seawell. 1979. *Ecological Recovery After Reclamation of Toxic Spoils Left by Coal Surface Mining. Phase I.* EPA-600/7-79-209. Washington, D.C.: U.S. Environmental Protection Agency. p. 77.

Zarger, T.G., D.H. Scanlon, C.P. Nicholson, S.R. Brown, L.B. Starnes, and W.D. Harned. 1982. *Ecological Recovery After Reclamation of Toxic Spoils Left by Coal Surface Mining. Phase II.* Submitted by TVA to EPA October 1981. Knoxville, TN: Tennessee Valley Authority.

CHAPTER 8

Some Effects of Stream Habitat Improvement on the Aquatic and Riparian Community of a Small Mountain Stream

Stephen A. Burgess*

André Marsan et Associés Inc.
1210 Sherbrooke St. W.
Montreal, Quebec, Canada
H3A 1H7

The province of Quebec is well known as a region offering opportunities for sport fishing for such species as northern pike, *Esox lucius;* walleye, *Stizostedion vitreum;* and lake trout, *Salvelinus namaycush.* The most sought-after fish species in the province is undoubtedly the brook trout, *Salvelinus fontinalis.* Brook trout are found throughout Quebec, except for the northern tip of the Ungava peninsula, and because of its preference for small lakes, rivers, and streams, and since it is easily caught on simple gear, it is the most readily accessible salmonid species in the province. This accounts for its wide popularity as a sport fish.

In general, the majority of the angling effort for brook trout is concentrated on small lakes and ponds, since larger fish tend to be found in those areas, and because stream fishing requires considerably more effort than lake fishing. Streams can support important numbers of trout, however. Saunders and Power (1970) found that during the summer months 86% of the total brook trout population of Matamek Lake resided in spawning streams representing approximately 50% of the standing stock of the lake/stream ecosystem. The lake population was composed principally of trout aged 2+ to 7+, while trout obtained a maximum age of 4+ in the streams, due to a general downstream movement of fish out of the spawning streams after their second year.

Although stream-dwelling trout may be generally smaller than those in lakes, stream fishing can offer a good alternative during the summer months when the

* Present address: Environment and Right-of-Way Branch, National Energy Board, 473 Albert Street, Ottawa, Ontario, Canada, K1A 0E5

lakes warm up and trout become difficult to catch. That is particularly true in small lakes in the southern part of the province where trout often seek refuge in tributary streams during that time of the year.

Unfortunately, many streams, because of overcrowding or a lack of adequate habitat for adult fish, tend to be unproductive and support populations of small fish which are of little interest from a sport fishing standpoint. As a result, these streams are often underexploited by sport fishermen. In many cases, fishing in these streams could be significantly improved if habitat improvement to favour either larger fish or increased trout production was carried out.

Numerous authors have studied the critical environmental parameters affecting stream trout populations. Of those, water velocity, water depth, and cover have been shown to be important factors regulating populations (Cooper and Wesche 1976; and Wesche, Chapter 5, this book). Probability-of-use curves for brook trout (Bovee 1978) indicate preferred flow velocities and depths for spawning are 0.15–0.41 m/sec and about 0.15 m respectively. Adults prefer velocities less than 0.15 m/sec and depths greater than 0.24 m. Thompson (1972) suggested that ideal rearing habitat should have a riffle-pool ratio near 50:50, with riffle velocities of 0.30–0.45 m/sec and pool velocities of 0.10–0.25 m/sec, with approximately 60% of the riffle area of sufficient water depth for fish passage. In order for a stream to achieve its maximum potential for trout production, it must offer a diversity of flow velocities and depths to fulfill the habitat requirements for the various life stages of a productive trout population.

As discussed by Cooper and Wesche (1976), cover also plays an important role in determining stream trout production. Wipperman (1969) found that a 40% reduction in discharge below normal summer flow did not significantly affect the amount of available cover, and that cover was probably largely responsible for maintaining trout populations during dewatering periods. Boussu (1954) showed a direct relationship between cover, including overhanging brush and undercut banks, and biomass of trout over 18 cm in length. By increasing the amount of available cover in four experimental sections, he increased trout biomass by over 258%.

The natural pattern of stream flow in Quebec is characterized by extremely high spring discharge, a gradual reduction to base flow levels in late summer, followed by a second, minor peak in the autumn and then a significant reduction in flow during the winter months. That pattern is particularly evident in small trout streams that depend primarily on runoff as their principal source of water. As a result of extreme fluctuations in seasonal rates of discharge, trout populations are often limited by low flow conditions. For example, Côté (1970), investigating the ecology of a population of brook trout in a southern Laurentian stream, found that although fry production was extremely high, winter fish mortality was common due to insufficient flows and a lack of deep pools in which trout could overwinter. The lack of pools and scarcity of in-stream or streambank cover also reduce stream habitat quality, thus limiting trout production. In such streams, artificial modifications to create pools should improve habitat quality and help to protect trout populations from the effects of low flows. The addition of cover should also help to improve habitat in streams where inadequate cover is available.

Habitat improvement to increase trout populations is not a new concept. One of the first studies dealing with this subject was published by Hubbs et al. in 1932. Since that time, numerous studies have shown that stream improvement will result in increases in both numbers and biomass of trout (Shetter et al. 1946; Warner and Porter 1960; Saunders and Smith 1962; Hunt 1969, 1971; White 1975). Few attempts have been made to determine the secondary effects that habitat improvement might have on nontarget organisms such as stream invertebrates or trout predators. One would expect that any changes in stream habitat characteristics would affect other stream organisms as well as trout, and that any increases in prey availability would lead to increased predator use of the improved area.

Of the potential trout predators commonly occurring in the southern Laurentians of Quebec, the mink, *Mustela vison* is the most abundant. Numerous studies have shown that aquatic prey, and fish in particular, constitute a variable, but often important part of the mink's diet. There is some disagreement, however, as to the importance of salmonids in the diet. Alexander (1976) stated that trout populations in Michigan streams would benefit from mink population control because of their heavy dependence on trout populations in that area. Day and Linn (1972), on the other hand, found that mink prefer coarse fish, even in areas where salmonids abound.

The purpose of this study was to determine the effectiveness of a relatively simple habitat improvement program in increasing trout biomass in an experimental section of a small mountain stream. The intention was to use relatively simple techniques with low cost and labor requirements. In addition to monitoring the effects of habitat improvement on the trout population, the responses of other members of the aquatic and riparian community, notably crayfish and mink, were also investigated.

STUDY AREA

The study area was located 105 km northwest of Montreal, Quebec (lat 46° 09′, long 74° 29′) in an area of the mid-Laurentian series and at an elevation of about 300 m above sea level (Figure 8.1). The study area was situated in the midst of a heavily used recreational region. Angling pressure in many lakes and streams is intense, and in many, the quality of trout fishing has diminished over the last fifteen years due to overfishing and habitat degradation.

The stream under investigation is an unnamed, spring-fed mountain stream with an artificial pond having limited storage capacity located at the headwaters. The mean gradient of the study stream is 1.25%, although most of the drop in elevation occurs over the stream's lower 2 km. The study area was located about 5.5 km from the stream's headwaters. Less than 100 m below the study area,

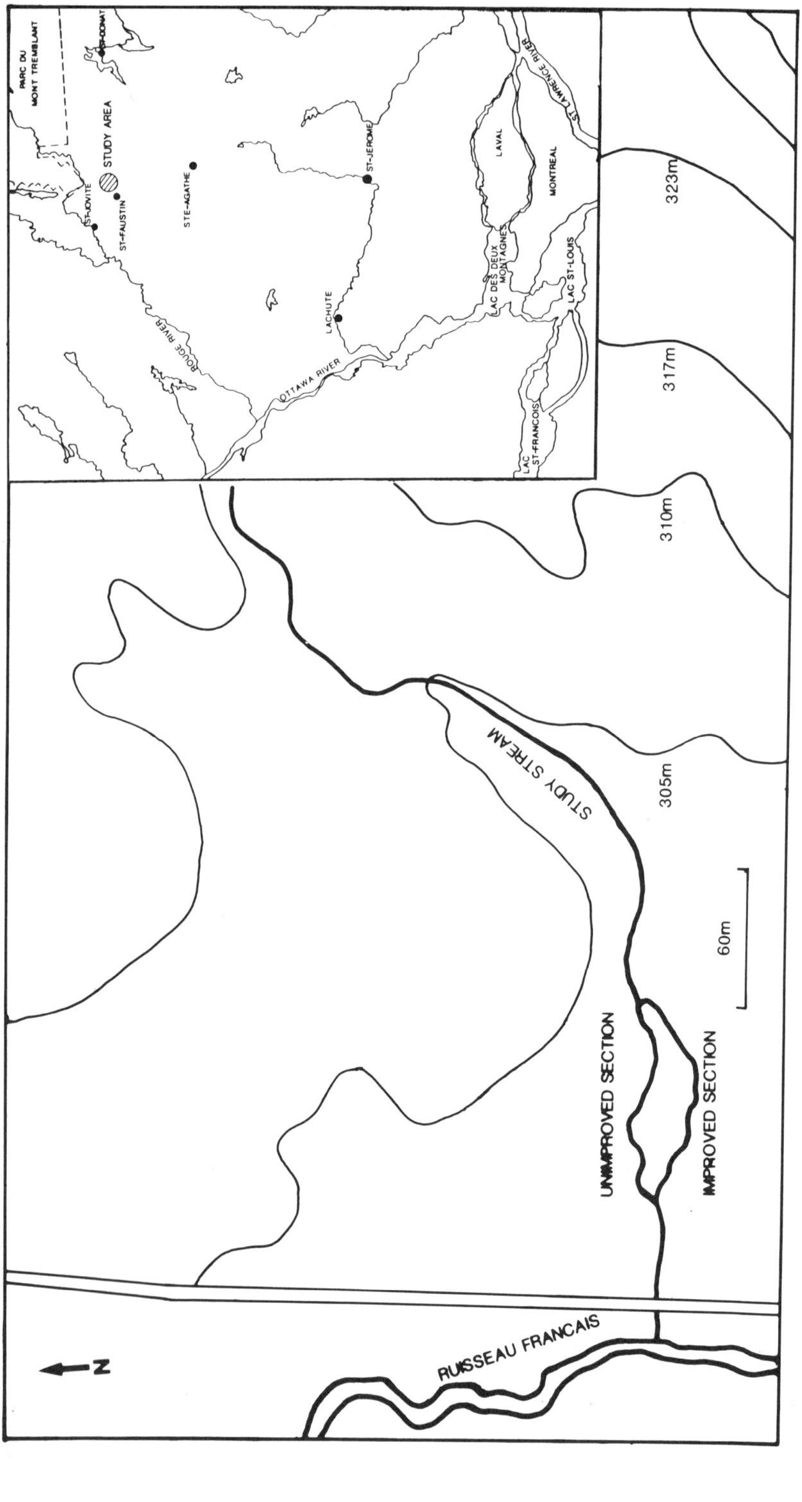

Figure 8.1 Map of the study area.

the study stream discharges into the Ruisseau Français, a slow-moving warm-water creek with very little potential trout spawning habitat.

Prior to habitat improvement, the stream was characterized by extensive riffle areas, with a mean depth of less than 25 cm and a width of approximately 3–5 m. The stream bed was composed primarily of gravel and cobble with some accumulation of sand, silt, and organic detritus in a few zones of reduced current. Occasional fallen trees or large boulders had created flow restrictions which led to deepening of the channel and the creation of small pools in those areas. No physical obstructions to trout movement were found within the study area or between the study area and the Ruisseau Français.

The mean September flow velocity in the area prior to stream management was calculated by Côté (1970) to be 55 cm/sec, with a maximum of 84 cm/sec and a minimum of 9 cm/sec. Due to the small size of the drainage basin, discharge was strongly affected by runoff, so that daily variations in flow were significant. Mean summer water temperature (June–September) during the two years of the study were 12° and 13° C respectively, with a minimum observed temperature of 6.0° C and a maximum of 20.5° C.

Approximately 2.5 km above the study area, the stream flowed through cleared pasture land. The dominant fish species in that area were yellow perch, *Perca flavescens;* pumpkinseed, *Lepomis gibbosus;* and brown bullhead, *Ictalurus nebulosus.* The fish community in the study area was dominated by brook trout. Other species were seldom observed in that area, except during midsummer when occasional downstream movements of pumpkinseeds were observed.

Riparian vegetation in the study area consisted primarily of speckled alder, *Alnus rugosa,* although other species such as white spruce, *Picea glauca;* balsam fir, *Abies balsamea;* white ash, *Fraxinus americana;* and willow, *Salix sp.* were also common. Relatively little overhanging cover was provided by stream bank vegetation.

TROUT POPULATION ECOLOGY

During the summer of 1968, Côté (1970) studied the ecology of the brook trout population in the study stream. Between the time of Côté's work and the time stream improvement was carried out in 1976, certain changes in stream habitat and trout populations had occurred.

The major difference in stream characteristics between this study and that of Côté (1970) is that at the time of his work in 1968, the stream consisted of only one channel in the study area. Côté built a dam near the upstream end of the channel to provide a means whereby flow could be diverted through a second parallel channel which was dry under normal conditions. The diversion permitted drainage of the main stream channel to allow the collection of fish for census purposes. This dam, however, constituted an insurmountable barrier to upstream trout movements, and resulted in crowding of fish below the dam during the summer months. At the time stream improvement was carried out in 1976, the dam had fallen into disrepair, and both the main channel and the diversion channel had water flowing through them, thus allowing unrestricted trout movement upstream through the study area. For the purposes of this study, the original main channel served as the unimproved, or control section, and the diversion channel was improved.

Daily Activity Patterns

Côté examined activity patterns of trout using counting fences placed at various locations in the stream. The resulting data showed that trout movements take place primarily at night, with peaks of activity occurring from 20:00–24:00 hours and from 03:00–05:00 hours. Because feeding takes place mostly during the day—and drift feeding, which requires little movement from a fixed station, is the most common feeding strategy of stream-dwelling trout populations—Côté believed that nocturnal movements of fish are primarily for purposes of habitat exploration or migration.

During the day, trout tend to select feeding stations under cover, or in shaded areas, close to zones of bright light where food particles are visible (Gibson and Keenleyside 1966). During the day, therefore, trout are fairly secure from predation by "plunge-type" predators such as mink, which generally observe a potential prey item from above water and then plunge into the water in attempt to capture it before it has time to react. At night, however, when trout exploratory movements increase, and they actually tend to become slightly photopositive (Gibson and Keenleyside 1966), they are likely to be more susceptible to nocturnal predators such as mink.

Seasonal Activity Patterns

Côté noted two types of trout activity patterns within the study area. As a rule, trout exhibited a low, fairly constant level of activity which consisted basically of exploratory, or circulatory movements, within a fairly restricted area. Two periods of more intense activity were observed, however, which were a result of movements of fish into the study stream from the Ruisseau Français. The first of those periods occurred between 14 and 22 June, while the other was observed between 11 and 21 July.

The reasons for the June migration were unknown since water temperature in the Ruisseau Français was close to optimum at that time, but were probably related to a search for food or other favorable habitat conditions. The July migration was likely triggered by high water temperatures in the Ruisseau Français. Temperatures there reached 21° C at that time while conditions in the study stream were several degrees cooler and more favorable for trout.

Because of its generally shallow depth and the severity of the winters in the region, overwintering conditions in the study stream tend to be extremely rigorous and winter mortality can be high. As a result, a general downstream movement of fish into the Ruisseau Français evidently occurs after spawning is completed, either in late fall or during the winter. Trout probably overwinter in the Ruisseau Français and move back into the study stream during the following summer due to the more favorable habitat conditions found there.

Age and Growth

Growth of brook trout in the study stream is relatively slow, particularly during the first two years of life. The growth rates of males and females are similar, although males tend to mature earlier. In general, males become mature at a length of about 121 mm, usually in their third year, while females mature at 139 mm, in their third or fourth year. The sex ratio is 1 female to 1.6 males. Mean lengths of fish at the end of the growing season as calculated by Côté (1970) are given in Table 8.1.

The maximum age of fish in the study area was found to be 3+ years. It is at that age that mortality due to angling becomes significant, probably accounting for the virtual elimination of fish older than three years.

The trout population in the study area was found by Côté to be 269 fish, yielding an estimate of 6215 trout per hectare and a biomass estimate of 106 kg/ha. The age structure of the trout population in the area was 7.4% age 0+, 53.2% age 1+, 29% age 2+, and 10.4% age 3+ (Table 8.1). Recruitment of young fish into the area was from an important spawning and rearing area located upstream, while older fish were generally migratory in nature, as discussed previously. Survival rates were calculated for the entire population, and were 6% for age class 0, 34% for age class 1, 70% for age class 2, and 20% for age class 3 (Table 8.1).

Côté stated that environmental factors that account for the slow growth rates of trout in the stream, particularly during the first two years of life, as well as the high mortality rates of older fish, were the lack of pools, combined with the short growing season and severe winters in the area. Excessive fishing pressure also accounted for the high mortality of fish of angling size. Intraspecific competition

Table 8.1 Characteristics of the brook trout population in the study stream. (Adapted from Côté 1970.) Population and biomass estimates are for the study area only. Mean lengths and mortality rates are for the entire stream.

Age Class	*Mean Length (mm)*	*Mortality Rate (%)*	*Population*	*Biomass (kg/ha)*
0+	51	94	20	0.59
1+	79	66	143	17.52
2+	135	30	78	45.86
3+	203	80	28	54.59

probably played a minor role in regulating trout growth and mortality in this stream.

METHODS

Habitat Improvement

In order to test the value of habitat improvement as a means of increasing trout biomass, a section of the study stream that divided to form two parallel sections of approximately 100 m in length was selected for study. One section was selected for habitat improvement while the other was left unmanaged and served as a control. The two sections were separated by a maximum distance of 25 m. Prior to habitat improvement in May of 1976, both sections were similar in terms of streambed characteristics, water quality, flow volumes, and cover characteristics. It was therefore assumed that trout populations in the two sections were comparable and no preimprovement inventories were carried out.

In order to ensure control of discharge through the two sections, flow control structures (Figure 8.2) were constructed at the head of each. As a result, flows through the two sections could be equalized, or flow through one section could be cut off completely to allow the collection of fish and crayfish, *Cambarus bartoni*, from the drained section. The check dams were relatively simple, low-head structures intended solely for flow regulation. The difference in water levels upstream

Figure 8.2 View of a flow control structure, with the water level gauge located in the pool upstream. In the background are a sand transect and a counting fence.

and downstream of the dams was 15–25 cm and water depths below them were greater than 40 cm. It was therefore unlikely that they constituted barriers to upstream movements of trout, other than perhaps first-year fish (Stuart 1962).

Numerous techniques are available for improving stream habitat for trout. In streams such as the one studied here, which are characterized by extensive riffle areas and relatively little trout cover, those include such devices as flow deflectors and low dams, which constrict flow, causing the formation of pools upstream and scouring of the stream bank or bed downstream. Cover structures often consist of artificial bank overhangs or other in-stream cover devices.

Habitat improvement in the study area involved the construction of small rock dams and deflectors. Rocks used for dam construction were usually at least 30 cm in diameter, and dams and deflectors were triangular in cross section, with the base of the dam being roughly three times the height. Those specifications were adopted to ensure stability of the structures. Maximum height of the structures from the streambed was less than 1 m. That resulted in the creation of pools up to 1 m deep upstream of the structures and scouring of the stream bed downstream, which also resulted in the deepening of the stream channel (Figure 8.3). Before habitat improvement, less than 10% of the stream channel consisted of pools. Afterwards, the riffle: pool ratio in the improved section was approximately 1:1. In-stream cover, in the form of logs, as well as rafts of alders lashed together (Saunders and Smith 1962), was introduced into the improved section, usually in pools and near areas of high food availability. Temporary cover devices

Figure 8.3 An example of a rock deflector in the improved stream section directing most of the stream flow against the far bank, promoting bank undercutting and bed scour, and providing excellent trout habitat.

were used to allow quick removal to facilitate capturing fish for census purposes.

Habitat improvement was accomplished using only materials available on-site and involved no expenses for equipment or materials. Labor requirements for improvement of approximately 100 m of stream channel were three man-days. Maintenance of the improvement devices during the first summer of operation was negligible. The following spring, minor repairs to certain cover structures were undertaken, which required approximately one man-day of labor mainly involved repositioning cover devices that had been displaced in the spring flood. If more permanent types of cover structures had been used, more initial effort would have been required, but the need for periodic maintenance would have been eliminated thus reducing long-term labor requirements.

Angling in the study area was restricted to eliminate any bias that might result due to that activity. Unfortunately, angling in the Ruisseau Français could not be controlled so that numbers of fish migrating into the study area from Ruisseau Français may have been reduced and size classes skewed in favor of smaller fish. It was unlikely, however, that this had any impact on the relative numbers of fish entering the improved or control sections.

Stream stage height was continually monitored during both summers of the study using a portable water level gauge installed in the pond upstream from the check dams. Water temperatures were also measured in the pond upstream from the check dams, and at the outlet of each of the study channels. Temperatures were recorded once in the morning and once in the evening during the summers of 1976 and 1977, and daily averages were calculated. Daily mean values were analyzed to determine whether stream improvement had any influence on water temperatures.

Trout and Crayfish Populations

Trout Movements

The effects of habitat improvement on trout movements were determined through the use of counting fences placed at each end of both the improved and control channels. The counting fences consisted of double-ended traps with funnel entrances which were designed so that fish moving upstream were segregated from those moving downstream. The entrances were made of ¼ in (0.64 cm) wire mesh. That was found to be the smallest mesh size which could be used without constantly becoming fouled by drifting debris, and it was 100% efficient for fish older than age 0+. First-year fish were often caught, but others were observed escaping from the traps. As a result, our counting fence data tended to underestimate captures of first-year fish, particularly early in the summer. Although the counting fences were opened on several occasions due to flood conditions, the data obtained should be fairly reliable. Elwood and Waters (1969) and Hoopes (1975) found that flooding caused relatively little movement of trout.

Traps were located in the deepest section of the stream channel and rocks were placed in the bottom to act as ballast, to reduce flow velocities in the trap, and to provide hiding places for captured fish. Steel rods driven into the streambed anchored the traps in place. Once the trap was in position, fences, also constructed of ¼ in mesh, were erected between the trap and the streambanks, so that fish passage past the counting fence was impossible under normal conditions. The counting fences were in operation from 5 June to 9 September 1977, and traps were checked once a day throughout the sampling period, except for five short periods when flood conditions due to heavy rainfall required opening the traps to prevent them from being washed out. All fish captured were recorded, their fork lengths measured, and their direction of movement noted. Fish were fin-clipped according to trap location and direction of movement so that movements of individual fish could be followed based on recapture records. All fish were released in the direction they were headed when captured.

Trout and Crayfish Biomass

Trout population density and biomass, as well as crayfish biomass, were determined in September 1976 and 1977 by alternately draining the unimproved channel, and then the improved channel, and collecting all individuals that became trapped in the isolated pools that formed as a result of drainage. Fish were counted and their fork lengths measured. Trout were not weighed in order to minimize handling, but weights were estimated using a length-weight conversion taken from Côté (1970) for the population studied here. Crayfish were not counted but were weighted collectively. In 1977, counting fences remained in operation during the drainage period in the event that reduction of flows resulted in movements of fish out of the study area. No such emigration was observed.

Mink Activity

Mink movements in the study area were monitored during 1976 and 1977 using a modified sand-transect technique (Bider 1968; Burgess and Bider 1980). Data obtained were used to describe certain characteristics of mink behavior, and in particular, the response of mink to stream habitat improvement. Transects consisted of styrofoam sheets measuring 30 cm × 122 cm. A thin layer of fine sand was spread over the styrofoam which was protected from rain by a clear polyethylene canopy. Transects were placed on the streambanks, as close to the water's edge as possible along the improved section, the control section, and the Ruisseau Français near its confluence with the study stream.

Aspects of mink activity examined included their seasonal pattern of activity and their response to stream habitat improvement and fluctuating water levels. All vertebrate activity on the sand-transects was recorded twice daily, shortly after dawn and just prior to sunset, in an effort to separate nocturnal and diurnal activity. Because a number of potential terrestrial prey species were common in the study area and the availability of those could have influenced mink activity

(Gerell 1970), the temporal and spatial activity patterns of these species were examined in order to determine if observed mink activity could be related to the availability of terrestrial prey.

During the second summer of the study, mink began using the sand-transects as feeding platforms and defecating sites. A total of forty mink scats were thus collected and analyzed to provide information on mink diet during that time (Burgess and Bider 1980). Dorsal guard hairs and teeth were used for identification of mammalian remains, scales and bones for fish, and feathers for birds. Exoskeletal remains of crayfish and insects were easily identified. All food items were classified according to their percent occurrence and percent of the total scat volume.

RESULTS

Temperature

During the two summers of the study, no changes in water temperatures resulting from habitat improvement were detected. During the summer of 1976, the mean temperature in both sections was 13° C, while in 1977 the mean temperature was 12° C. Extreme water temperatures were also the same, with a minimum of 7.5° C and a maximum of 18.5° C observed in 1976. Minimum and maximum temperatures in 1977 were 6.0° C and 20.5° C respectively.

That is not to imply, however, that water temperatures cannot be affected by stream habitat improvement. The similarity in water temperatures between the stream sections in this study is not surprising and was likely due to the fact that the sections were short ($\simeq$ 100 m in length) and that the overhanging tree canopy was relatively dense. In certain situations, however, where tree cover is more open, water temperatures could be increased as a result of the creation of pools. The magnitude of such an increase would depend upon how much the flow through the managed area was slowed and how much the area was exposed to insolation. On the other hand, significant increases in the amount of overhanging bank cover could cause water temperatures to be reduced in improved streams where this type of management is adopted.

Trout Population Density and Biomass

Trout population response to habitat improvement occurred rapidly. By September 1976, three months after improvement, the population in the improved channel was approximately 107 fish, while the control channel contained 75 individuals. By the end of the second summer, the populations were 162 and 77 trout respectively (Burgess and Bider 1980). Subsequent population counts in the improved section indicated that the increased population in that section appears to be relatively stable, despite the fact that angling in the area is no longer controlled. Population estimates varying from 96 to 119 ($\overline{X} = 110$) have been obtained for the improved section as a result of four stream surveys carried out between 1977 and 1981 (Doucet 1981, personal communication). Unfortunately, no population estimates for the control section were available for comparison.

Brook trout biomass in the control section was 630 g in 1976 and 388 g in 1977, while in the improved section we obtained values of 1114 g and 1105 g respectively. Conversion of those values resulted in mean biomass estimates of 36.53 kg/ha and 53.51 kg/ha for the control and improved sections respectively during the two years of the study. Those estimates were for trout populations during the month of September, at a time when densities in the study area were at a maximum as a result of immigration from the Ruisseau Français. Over the two years of this study, trout biomass in the improved area averaged 218% more than in the unimproved section, while on a per unit area basis, the improved section was 145% more productive.

Côté (1970) estimated trout biomass in the study area to be approximately 106 kg/ha, at a time when only the control channel was active. The improved section, at the time of Côté's study, was used only as a diversion channel during trout biomass estimate studies. At the time of Côté's study, upstream trout movements were blocked by a flow control structure located at the upstream end of the study area, probably causing trout moving upstream to concentrate below the structure, resulting in abnormally high numbers of fish in that area. During the course of this study, trout movement through the area was not restricted, so that no artificial increases in trout biomass occurred.

While trout size-class distributions were not significantly altered as a result of habitat improvement (Figure 8.4), the number of angling-size fish (>15 cm) was greater in the improved section by an average of 6.5 trout/100 m of stream channel. That corresponded to an average increase of 10.4 kg/ha in angling-size fish in the improved section. While angling-size trout accounted for 62% of the observed biomass increases in the improved area, habitat improvement also resulted in a mean increase of 170% in the number of trout less than 15 cm in length over the 2 years of the study. It is thus evident that habitat improvement resulted in increases in all age-classes of fish, rather than of a single size-class only.

Because a large percentage of the trout population in the study area immigrated from the Ruisseau Français during the summer, the age-class distributions of both the improved and control sections were largely representative of this migratory segment of the population. It is difficult to say whether a similar phenomenon would be observed in a stream supporting a more sedentary trout population, although with the greater diversity of habitats present in an improved stream, a broader age-class structure would be expected.

Trout Movements

Immigration of trout into the study area from the Ruisseau Français was known to be an important phenomenon (Côté 1970), and it was anticipated that habitat improvement might influence trout movements and thus population levels in the study area. Examination of the counting fence data (Figure 8.5) generally supported this, since by the end of the summer of 1977, immigration, defined as net immigration at the downstream traps minus net emigration at the upstream traps, had resulted in net increases of 48 and 71 trout in the control and improved sections

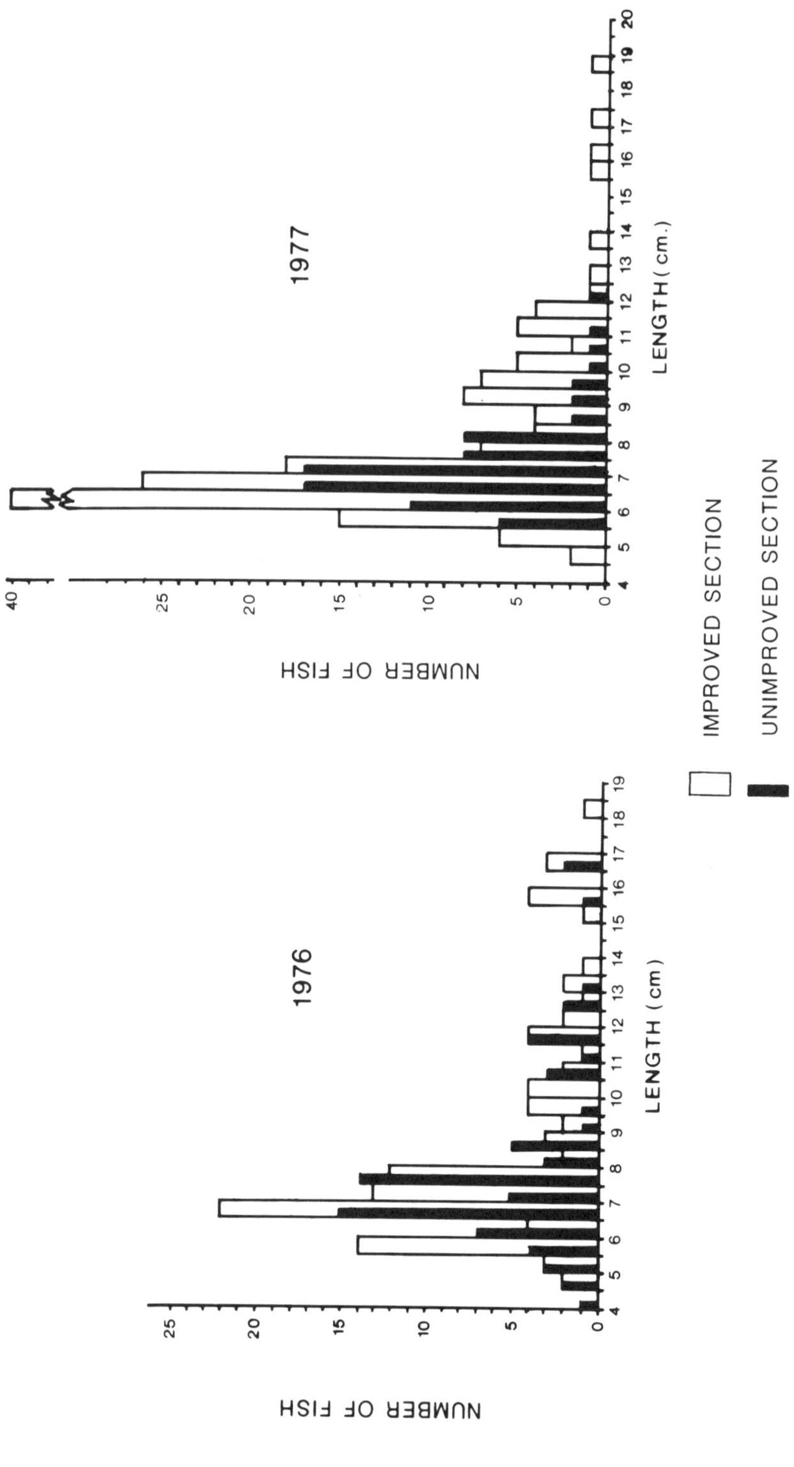

Figure 8.4 Comparison of the size class distribution of brook trout in improved and unimproved sections.

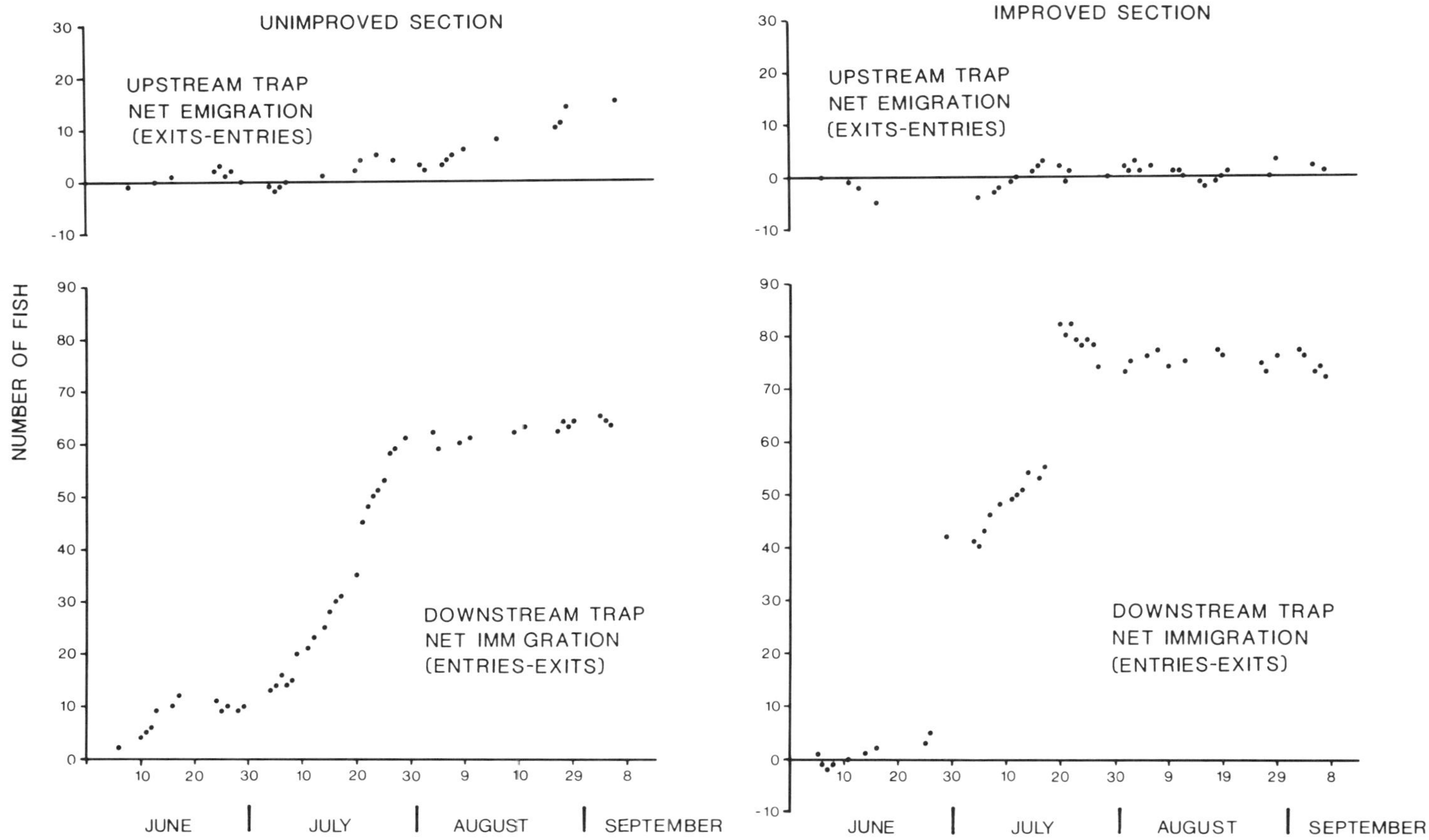

Figure 8.5 Summary of counting fence capture data, showing net immigration at the downstream ends, and net emigration at the upstream ends of the study sections.

respectively, thus accounting for approximately 62% and 44% of the trout populations in those areas by the end of the summer. The results show that a greater proportion of the population in the control section consisted of immigrants but that more fish were attracted into the improved section, indicating that the carrying capacity of the improved section was increased both during winter and summer. The reason trout preferentially selected the improved section during their upstream movements is unknown.

The seasonal pattern of trout movements in the study area was similar to that observed by Côté (1970). Several aspects of trout movements were influenced by habitat alteration, however. The general pattern of upstream movement into the study area was such that the bulk of immigration had occurred by the end of July. During August and early September, upstream movement into the area was negligible and most counting fence captures consisted of local, random movements of fish. Interestingly, two major influxes of fish into the improved section were observed during the course of the summer, whereas only one was observed in the unimproved channel. The first influx occurred during the nights of 26 and 28 June, when a total of 36 fish moved into the improved section. During that same period, no noticeable immigration into the unimproved channel occurred. The second major influx occurred on 21 July, when 27 trout moved into the improved area and 10 fish moved into the unimproved section. The reasons for the observed upstream movements are likely related to high water temperatures or other unfavorable habitat characteristics in the Ruisseau Français, but as stated above, the mechanism whereby trout were able to select the improved section remains unclear. It is evident, however, that the lack of immigration into the control section at the end of June was largely responsible for the difference in numbers of immigrants between the two sections.

Another difference between the two sections was that upstream emigration from the unimproved section was substantially greater than from the improved section. By the end of the summer, there had been a net upstream emigration of 15 trout from the unimproved section, while only one fish moved upstream out of the improved channel. During the months of June and July, upstream movement out of the unimproved section was negligible. Most of the upstream movement out of that section occurred during August.

The above data suggest that the carrying capacity of the unimproved section had been attained or exceeded as a result of immigration by the end of July, and the population there was in the process of restabilization during August. On the other hand, because no emigration from the improved section was evident, it appears that that section had not reached its carrying capacity.

Crayfish Biomass

Because crayfish are an important component of the stream community and are also a potential prey item for mink, all crayfish observed during the stream surveys were collected to obtain a biomass estimate. Collection took place at night, when crayfish were most active and most easily observed. Although every crayfish in the stream could not be captured, a relatively intense collection program was carried out so that values given here should be close to the actual crayfish biomass present.

As with trout, the crayfish population also benefitted from stream habitat improvement. In 1976, 652.9 g (46.85 kg/ha) of crayfish were collected from the control section while 1699.7 g (81.67 kg/ha) were taken from the improved channel. In 1977, the corresponding values were 713 g (51.16 kg/ha) and 1568 g (75.34 kg/ha) respectively.

Assuming that trout and crayfish represent nearly all of the stream biomass potentially available to mink, our results showed that crayfish represent over 50% of that total. Erlinge (1969) found that, in Sweden, mink prefer areas of high crayfish and small mammal availability. Given the importance of crayfish in the total biomass of the study stream and the higher densities present in the improved section, it was not surprising for mink, and probably racoons, to respond to that resource.

Mink Activity

Seasonal Distribution of Mink Activity

The pattern of mink activity observed indicated that the Ruisseau Français was a preferred habitat for denning and rearing of young. When the young were able to accompany the female on hunting trips, they restricted their activity to the vicinity of the den site, but as they became more independent, they began to explore other areas. The study area seemed to be a habitat used on a temporary basis, primarily as an activity corridor for young animals during the dispersal period.

A comparison of mink activity between the Ruisseau Français and the study stream (Figure 8.6) indicates that the latter was used primarily by young animals or family groups after the first week of August. Mink were active in the area of the study stream a total of 17 days in August. Prior to that time, the study area was used only sporadically by individual males. The Ruisseau Français, on the other hand, was used heavily by females and family groups throughout July (10 days) and August (11 days), after which time mink activity in that area decreased.

Spatial Distribution of Mink Activity

Based on transect data collected during the two years of the study, mink activity averaged 53% greater along the banks of the improved section as compared to the control section, supporting the hypothesis that greater food availability in the improved section would result in increased use of that area by mink. It was not evident that mink were responding to increased trout availability. As discussed previously, crayfish biomass also increased in the improved section, and other

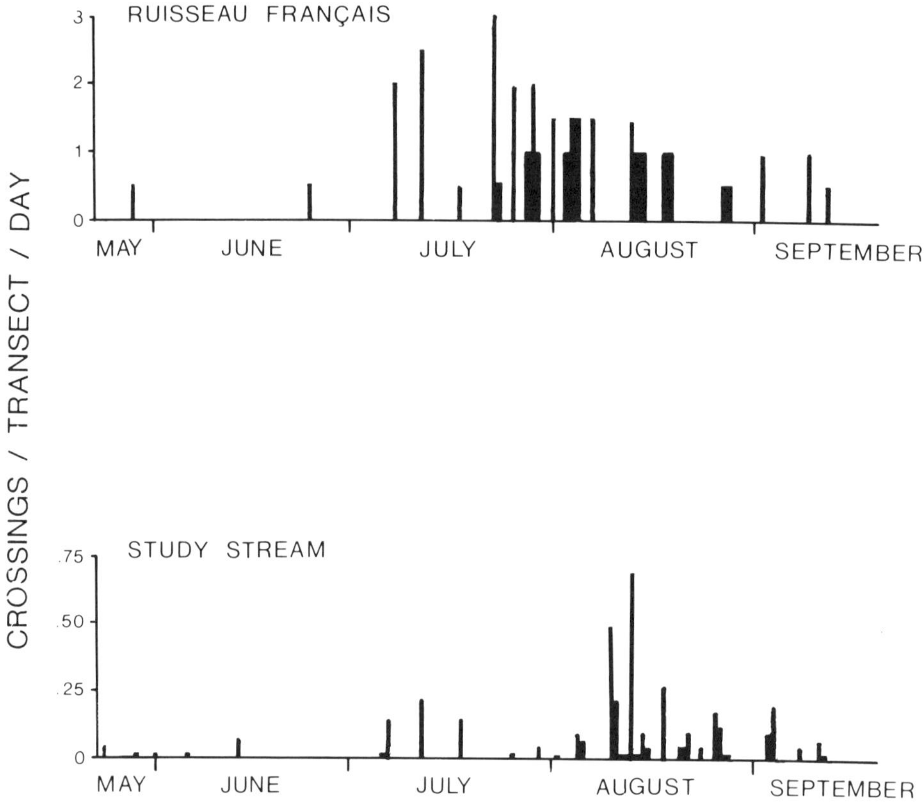

Figure 8.6 Comparison of the seasonal distribution of mink activity in the vicinity of the Ruisseau Français and the study stream.

potential prey species such as small mammals and amphibians were abundant in the area. The availability of those species could also influence the distribution of mink activity.

During the summer of 1977, between 11 July and 9 October, mink commonly used the sand-transects as defecation sites, so that a total of forty scats were collected and analysed. Most of those (75%) were from the area of Ruisseau Français. No other mink scats were encountered in the study area. Results of the fecal analysis show that small mammals, crayfish, frogs, red squirrels, *Tamiasciurus hudsonicus,* fish and aquatic insects contributed in order of importance, to the majority of the mink diet in the study area (Burgess and Bider 1980). Of the scats collected in the area of the study stream, 50% contained crayfish remains, 33% contained shrews (Soricidae), 17% contained deer mice, *Peromyscus maniculatus,* and 17% contained meadow voles, *Microtus pennsylvanicus.* No trout remains were present in any scats found, although yellow perch, and minnows (Cyprinidae) were present in some scats collected in the vicinity of the Ruisseau Français.

These results indicated that trout played a minor role in the diet of mink in the study area, while crayfish and small mammals were considerably more important. If mink activity was a function of prey availability, it should have been related more to the availability of those species than that of trout.

During the course of the summer, an important shift in habitat utilization was exhibited by the mink population, Consisting of a shift from aquatic food sources to the exploitation of terrestrial prey species to demonstrate the seasonal change in habitat use. This was evident from an analysis of the fecal material. The scat sample was divided into two equal subsamples. The first subsample included scats collected between 11 July and 17 August, while the second consisted of those collected between 18 August and 9 October. The subsamples were analyzed separately, and the percentage of scats containing food items from either an aquatic environment (crayfish, fish, frogs, and aquatic insects) or a terrestrial environment (mice, shrews, squirrels, and birds) was compared between time periods. The results indicated that early in the season, aquatic prey were dominant, while later in the season, terrestrial prey became more important (Table 8.2). The reasons for this are not clear, but may be related to an increased availability of terrestrial prey later in the season or an inability on the part of young mink to capture aquatic prey.

The shift in habitat use was also reflected in the relationship between stream flow and mink activity. During the first part of the summer, when stream flows may have been a factor governing the availability of aquatic prey, mink activity in the vicinity of the study stream was significantly related to water levels. Mink

Table 8.2 Seasonal variation in mink diet as determined by fecal analysis

	11 July–17 August		*18 August–9 October*	
Prey origin	*% Occurrence*	*% Volume*	*% Occurrence*	*% Volume*
Aquatic				
Crayfish	30.0	30.0	10.0	10.0
Frogs	20.0	14.5	10.0	9.5
Fish	15.0	14.8	5.0	0.5
Aquatic insects	20.0	10.7	15.0	1.8
Total	85.0	70.0	40.0	21.8
Terrestrial				
Mice, voles & shrews	15.0	15.0	50.0	47.2
Squirrels	5.0	5.0	15.0	15.0
Unidentified mammal	10.0	10.0	10.0	5.5
Birds	—	—	10.0	10.0
Vegetable matter	—	—	5.0	5.0
Total	30.0	30.0	90.0	82.7

displayed a distinct preference for low-water conditions at that time (Figure 8.7). As the season progressed, this relationship was no longer observed, which would be expected if dependence on aquatic prey was reduced. Based on the results, it seems that before mid-August the mink population, which consisted mostly of adult animals, depended most heavily on aquatic prey, and crayfish in particular. After that time, when young mink comprised the majority of the population, and populations of small mammals were at their highest levels, mink preferred terrestrial prey.

Because terrestrial prey were an important component of the mink diet the influence of terrestrial prey availability on the spatial distribution of mink activity in the study area was studied (Burgess and Bider 1980). An analysis of the spatial activity patterns of all potential prey species revealed that in 1976, only chipmunks, *Tamias striatus,* were more active in the vicinity of the improved section, whereas northern two-lined salamanders, *Eurycea bislineata,* meadow voles, red squirrels and deer mice were more active in the unimproved area. In 1977, salamanders, red squirrels, deer mice, frogs, *Rana clamitans, R. sylvatica,* and American toads, *Bufo americanus,* were more active in the improved area, while snowshoe hares, *Lepus americanus,* meadow voles, and shrews were more active in the unimproved area. All other species exhibited a random activity pattern across the study area.

Of the terrestrial prey species present in the area, small mammals comprised the major proportion of the mink's diet and were also quite abundant throughout the summer. While it seemed that this was one terrestrial prey group to which mink would respond on a regular basis, no correlation could be demonstrated between the spatial distribution of mink activity and the activity patterns of deer mice, meadow voles, or shrews (Burgess and Bider 1980). It thus appeared that mink predation on terrestrial prey was largely opportunistic in nature.

Raccoon Activity

Another mammalian predator commonly occupying riparian habitats is the raccoon, *Procyon lotor.* While raccoons are not particularly adept at capturing trout, crayfish are often considered an important component of that species' diet. Given the increased availability of crayfish in the improved section, one might also anticipate that raccoons would also prefer that area. While tracking data revealed a 358% increase in raccoon activity along the improved section, the seasonal activity pattern of raccoons was different from that of mink. Raccoons were most active in the study area during the early part of the summer. Their occurrence became sporadic after mid-June, with the exception of a brief period of increased activity at the beginning of September. The reasons for the observed shift in habitat use are unclear, but may be related to seasonal changes in food availability. It does seem likely, however, that the spatial distribution of raccoon activity in the study area was influenced by the increased availability of crayfish in the improved section.

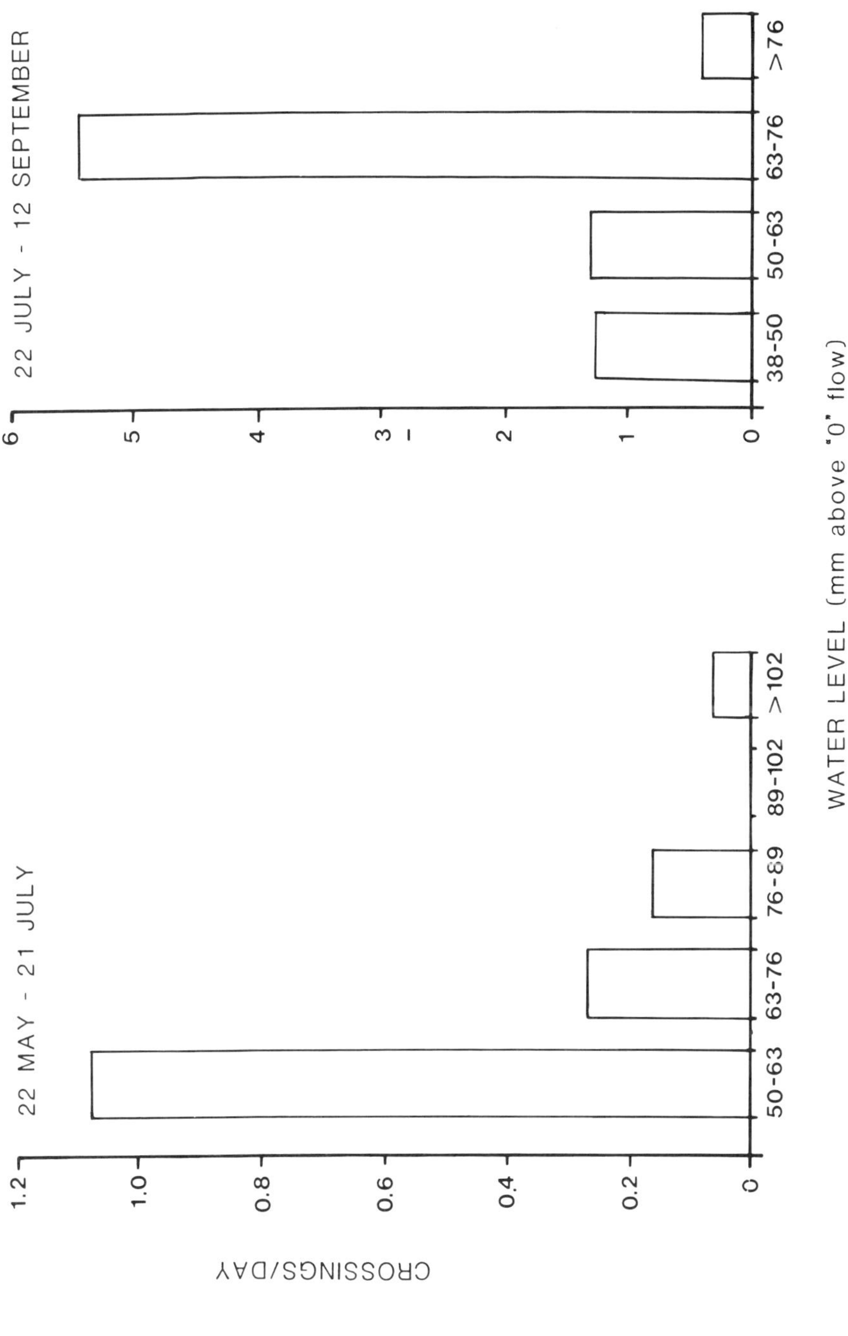

Figure 8.7 The seasonal relationship between mink activity and stream flow.

MANAGEMENT IMPLICATIONS

The results of this study demonstrated that stream habitat improvement can benefit trout populations in small mountain streams of southern Quebec. This is particularly interesting due to the increased density of angling-sized trout which occurred as a result of the management program. However, the results of this study do not allow the prediction of long-term effects of stream improvement on trout productivity or population dynamics, due to the very restricted size of the area that was managed and the migratory nature of the population studied. Given the limited amount of improved habitat in this study, it is unlikely that the overall trout population in the study stream will be significantly affected by this project. In other situations, where more extensive habitat improvement might be carried out, it is likely that trout populations would benefit due to the creation of suitable habitat for angling-sized trout, as well as improved overwintering habitat. Although stream habitat improvement shows promise as a means of increasing stream trout populations in the southern Laurentians of Quebec, overfishing and general degradation of fish habitats will limit its effectiveness as a management tool unless it is combined with other management practices.

At the present time, angling regulations for brook trout in this region of Quebec define a daily limit of fifteen trout, with no size restrictions. Personal observations in the study area and in the vicinity of the Ruisseau Français indicate that most angling in those streams is done with worms as bait, resulting in high hooking mortality. The daily limit is often not respected and is rarely enforced, and very small trout of 10 cm or less are often creeled. As a result of extreme fishing pressure from local residents and cottagers, the quality of trout fishing has greatly declined in recent years, since a large number of potential spawners are taken by anglers before they have an opportunity to reproduce.

Stream habitat improvement will not correct problems related to overfishing, but combined with other management techniques, has excellent potential for improving trout fishing in small streams. In areas where fishing can be controlled, such as in provincial parks, fishing clubs, or other similar areas, this technique would be most effective in increasing the potential of streams with poor habitat characteristics.

A number of management options are available for streams in areas where access is open. Those include reduced daily creel limits, institution of minimum size limits, restriction of the use of live bait, stream habitat improvement, stocking, or any combination of the above.

Because of the enforcement problems likely to be involved in prohibiting the use of live bait, that is probably not a viable management option. The high mortality of released fish caught on live bait would render the establishment of a minimum size limit at best ineffective, and more likely detrimental to the trout fishery, as mortality of released fish would probably significantly increase total fishing mortality.

Due to the presence of a relatively extensive road network in southern Quebec, and the fact that most fishing pressure occurs in the vicinity of bridges or other road access points, habitat improvement in the vicinity of access areas, combined with stocking on a put-and-take basis, as well as a possible reduction of daily limits holds promise as being the most effective means of improving the brook trout fishery and preserving natural stocks. Of course, every effort should be made to preserve existing trout habitat and any management work that is undertaken should be preceded by a detailed study of the stream in question and its trout population. That would allow the management scheme to be tailored to suit the needs of the stream and allow an optimization of management efforts.

SUMMARY

This study demonstrated that in addition to increasing trout biomass, stream habitat improvement also affected populations of nontarget organisms. Crayfish populations increased substantially in the improved section, which likely resulted in the increased use of that area by mink and raccoons. In this case, no significant loss of trout biomass occurred as a result of increased use of the area by mammalian predators. In areas where no alternate prey species are available, some trout might be lost as a result of predation. It is unlikely, however, that predation by mink would negate the value of habitat improvement as a management tool.

The management techniques employed in this study had several advantages. The structures were simple to build, using readily available materials. Labor requirements were low and all work was accomplished using hand tools. As a result, the total cost of habitat improvement was relatively low. In addition, because they were constructed of materials available on site, the structures preserved the natural aspect of the stream. This is an important consideration, since preserving the natural or aesthetic quality of the environment is often critical to maintaining the overall enjoyment of the angling experience.

REFERENCES

Alexander, G.R. 1976. *Diet of Vertebrate Predators on Trout Waters in North Central Lower Michigan.* Lansing, MI: Mich. Dep. Nat. Resour. Fish. Res. Rep. No. 1839.

Bider, J.R. 1968. Animal activity in uncontrolled terrestrial communities as determined by a sand transect technique. *Ecol. Monogr.* 38:269–308.

Boussu, M.F. 1954. Relationship between trout populations and cover in a small stream. *J. Wild. Manage.* 2:229–39.

Bovee, K.D. 1978. *Probability-of-Use Criteria for the Family Salmonidae.* Instream Flow Information Paper No. 4. Fort Collins, Colo.: Cooperative Instream Flow Service Group. P. 68–70.

Burgess, S.A., and J.R. Bider. 1980. Effects of stream habitat improvements on invertebrates, trout populations, and mink activity. *J. Wildl. Manage.* 44(4):871–80.

Cooper, C.O., and T.A. Wesche. 1976. *Stream Channel Modifications to Enhance Trout Habitat under Low Flow Conditions.* Water Resources Series No. 58. Laramie, Wyo.: Water Resources Research Institute, Univ. of Wyoming.

Côté, Y. 1970. Etude ecologique de l'omble de fontaine (*Salvelinus fontinalis,* Mitchell) d'un Ruisseau des Laurentides. Masters thesis, McGill University, Montreal, Quebec.

Day, M.G., and I. Linn. 1972. Notes on the food of feral mink *Mustela vison* in England and Wales. *J. Zool, Lond.* 167:463–73.

Elwood, J.W., and T.F. Waters. 1969. Effects of floods on consumption and production rates of a stream brook trout population. *Trans. Am. Fish. Soc.* 98(2):253–62.

Erlinge, S. 1969. Food habits of the otter, *Lutra lutra* L., and the mink, *Mustela vison* Schreber, in a trout water in southern Sweden. *Oikos* 20:1–7.

Gerell, R. 1970. Home ranges and movements of the mink, *Mustela vison* in southern Sweden. *Oikos* 21:160–73.

Gibson, R. J., and M.H.A. Keenleyside. 1966. Responses to light of young Atlantic salmon (*Salmo salar*) and brook trout (*Salvelinus fontinalis*). *J. Fish. Res. Bd. Canada* 23:1007–24.

Hoopes, R.F. 1975. Flooding as the result of Hurricane Agnes and its effect on a native brook trout population in an infertile headwater stream in central Pennsylvania. *Trans. Am. Fish. Soc.* 104(1):96–99.

Hubbs, C.L., J.R. Greeley, and C.M. Tarzwell. 1932. Methods for the improvement of Michigan trout streams. Inst. for Fisheries Research, Bull. No. 1. Ann Arbor: Univ. of Michigan Press.

Hunt, R.L. 1968. Effects of habitat alteration on production, standing crops and yield of brook trout in Lawrence Creek, Wisconsin. pp. 281–312. In: T.G. Northcote, ed., *Symposium on Salmon and Trout in Streams.* Univ. of British Columbia Vancouver.

______. 1971. Responses of a brook trout (*Salvelinus fontinalis*) population to habitat development in Lawrence Creek. Madison, WI: Wis. Dep. Nat. Resour. Tech. Bull. 48.

Saunders, J.G., and M.W. Smith. 1962. Physical alteration of stream habitat to improve brook trout production. *Trans. Am. Fish. Soc.* 91(2):185–88.

Saunders, L.H., and G. Power. 1970. Population ecology of the brook trout, *Salvelinus fontinalis,* in Matamek Lake, Quebec. *J. Fish. Res. Bd. Canada.* 27:413–24.

Shetter, D.S., O.H. Clark, and A.S. Hazzard. 1946. The effects of deflectors in a section of a Michigan trout stream. *Trans. Am. Fish Soc.* 76:248–78.

Stuart, T.A. 1962. *Leaping behavior of salmon and trout at falls and obstructions.* Freshwater and Salmon Fisheries Research, Report No. 28. Dept. of Agric. and Fisheries for Scotland.

Thompson, K. 1972. Determining stream flows for fish life. In Proceedings of the Instream Flow Requirement Workshop, 31–50. Vancouver, Wash.: Pacific N.W. River Basins Commission.

Warner, K., and I.R. Porter. 1960. Experimental improvement of bulldozed trout stream in northern Maine. *Trans. Am. Fish Soc.* 89(1):59–63.

White, R.J. 1975. Trout population responses to streamflow fluctuation and habitat management in Big Roche-a-Cri Creek Wisconsin. *Verh. Internat. Verein. Limmol.* 19:2469–77.

Wipperman, A.H. 1969. *Southwest Montana Fishery Study: Effects of Dewatering on a Trout Population.* Montana Fish and Game Dept. Fed. Aid Project F-9-R-17, Job Completion Report. Helena, MT: Montana Fish and Game Dept.

CHAPTER 9

Enhancement of Urban Water Quality through Control of Nonpoint Source Pollution: Denver, Colorado

Robert D. Judy, Jr.*

Engineering-Science, Inc.
10 Lakeside Lane
Denver, Colorado 80212

This chapter focuses on my proposed methodology to quantify pollutant loadings in the South Platte River, Denver, Colorado, resulting from storm event generated nonpoint source pollution. Nonpoint source pollution is defined as pollution originating from many different sources, such as streets, parking lots, industrial and residential developments, atmospheric deposition, etc. Nonpoint source pollution is not presently governed by the National Pollution Discharge Elimination System or any type of wet weather water quality criteria.

I refer the reader to Herricks and Osborne (Chapter 2, this book) for a more descriptive presentation of nonpoint source pollution, its sources, types, and possible severity of impact to receiving waters.

Stream restoration is significantly dependent on controlling nonpoint sources of pollution. Restoration explicitly implies control, and to control only point sources neglects large quantities of toxic and stream habitat-degrading materials such as lead, cadmium, and suspended sediments. Urban area nonpoint source pollutants are significantly different in nature from mined or timbered area generated pollutants. The urban area is a complex web of public and private property zones, multiple land uses, and various intensities of pollutant generation. Mined and timberland areas however are relatively discrete, produce fewer numbers of contaminants, and can be more effectively regulated through existing state and federal statutes.

---

* Present address: USDI, Office of Surface Mining, Mining Analysis Division, Brooks Towers, 1020 15th Street, Denver, Colorado 80202

National, state, and local attention has focused, until recently, on point source pollutants as the most significant cause of water quality degradation in and around urban areas. This focus was particularly apparent in passage of PL 92–500 and subsequent amendments that established the National Environmental Policy Act (NEPA) and the United States Environmental Protection Agency (EPA) in 1968. Sections 201 and 205(g) specifically instructed the EPA to establish the construction grant program to assist funding of local sewage treatment plants, either for upgrading or expansion. Section 208, however, instructed the establishment of Areawide Planning Agencies (APA), certified by the governor of each state and approved by the EPA. The function of the APAs was to estimate population growth levels over twenty years and to design regionally tailored Clean Water Plans (CWP). These plans included all sources of pollution, especially Publicly Owned Wastewater

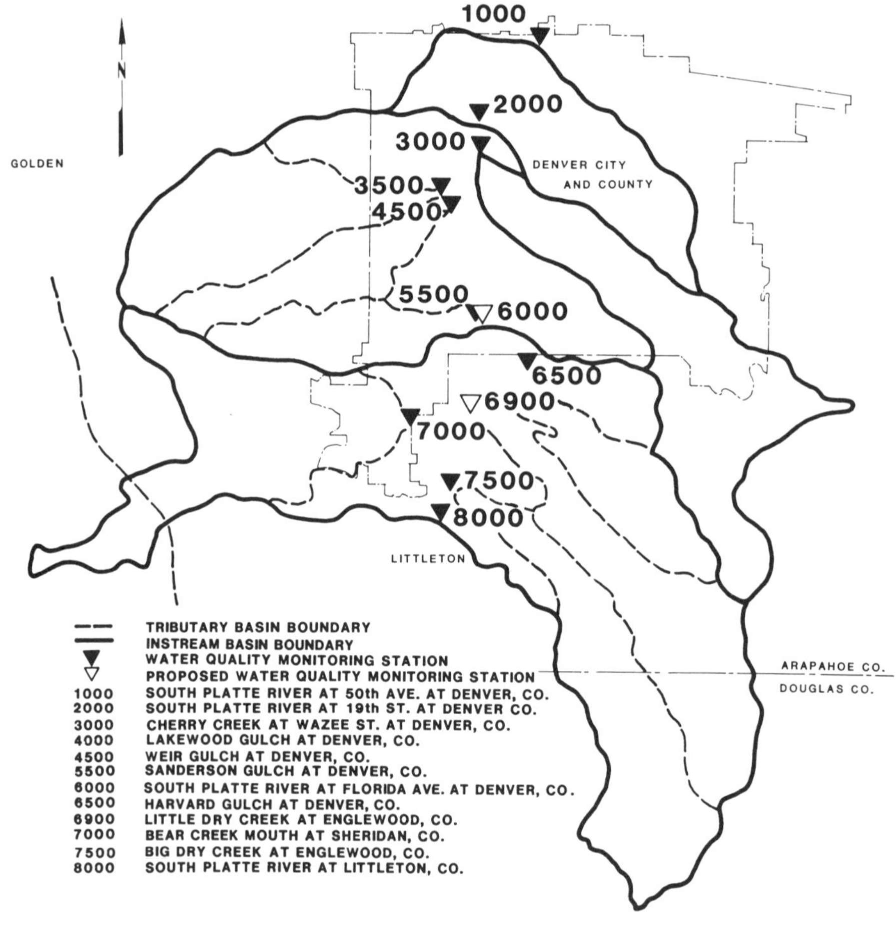

Figure 9.1 Map of DRURP study area showing in-stream and tributary basins.

Treatment Works (POTWs) and industrial effluents, and less recognizable sources of pollutants which can collectively be considered under the term nonpoint sources (NPS). Little emphasis, however, was placed on actual data collection, i.e., effluent monitoring, in-stream storm-water quality monitoring, in-stream ambient quality monitoring, or suspected sources of pollutants entering the waterways.

The Denver Regional Council of Governments (DRCOG) accepted a grant awarded from the EPA to conduct the Clean Water Plan study in 1976–77 and used the Hydrologic Simulation Program Version X (DRCOG 1977) to model all pollutants entering the South Platte River Basin. The planning area extended from Chatfield Reservoir in the southwest of Denver to the Town of Henderson in the northeast and included the major tributaries of Big Dry Creek, Little Dry Creek, Harvard Gulch, Weir Gulch, Sanderson Gulch, Cherry Creek, Clear Creek, Bear Creek, and Sand Creek (Figure 9.1).

The results of this previous study indicated that the NPS pollution accounted for at least as much pollution entering the South Platte River Basin as from all point sources (DRCOG 1977). Note, however, that very limited water quality samples were taken with which to calibrate the simulation model. The DRCOG CWP was accepted and conditionally approved by the governor in 1977. These conditions included extension of the CWP to include all of Douglas County (to the south of the study area) and a better estimation of the causes and severity of NPS pollution in the planning area.

In 1979 DRCOG received a grant award from the EPA to investigate the nature, causes, severity, and opportunities for control of NPS pollution in the Denver region. This study was funded under the Nationwide Urban Runoff Program (NURP) described by Heaney et. al. (1981) and is known as the Denver Regional Urban Runoff Program (DRURP).

This chapter focuses on tentative results of the DRURP and proposes a methodology for estimating basin response to storms in the Denver region using a lump sum parameter model. Additionally, this chapter concentrates on Best Management Practices suitable for implementation in this region. The Best Management Practices that receive the greatest attention are:

1. Temporary and permanent controls on new construction sites, e.g., utilization of temporary detention via straw bale dikes, grass-lined swales, and contour planning.
2. Percolation pits for the control of storm water in stabilized developments and recharge of ground water.
3. Protection of existing wetlands and installation of artificial wetlands in suitable areas.
4. Retrofitting of existing flood detention ponds with structures designed to retain storm water for water quality improvement.

Prior to a discussion of either the model or urban water quality improvements through Best Management Practices, a brief explanation of the climate, soils, and pollution problems in the Denver region is warranted.

Denver, Colorado

Climate

Denver, Colorado, sits on the High Plains at an elevation of 5,280 feet above mean sea level in close proximity to the Rocky Mountain Cordillera. Precipitation is scarce, averaging 14–15 in/yr with the overall climate being described as semiarid to arid. Relative humidity is low, averaging 15–20% and potential evapotranspiration averages 35 in/yr.

The precipitation in the region can be characterized by three major types of events: snowfall (November–March); upslope rainfall covering a broad area (March–May, October–November); and high-intensity, localized convective thunderstorms covering relatively small areas (May–September). This pattern of precipitation added complexity to my attempts to quantify nonpoint source pollution or control its effects on the receiving water, the South Platte River. Additionally, artificial irrigation contributes significant quantities to the flow of all tributaries and alluvial ground water levels.

Soils

The soils of the Denver region are generally highly calcareous clays with low infiltration rates except in alluvial areas where sands and sandy loams are predominant. An additional difficulty exists, however, in classifying soils from existing records because the U.S. Soil Conservation Service has not mapped the soils in the City and County of Denver.

Colt (1980, 1981) has mapped the soils of Adams County to the northeast of Denver and has defined areas suitable for percolation pit construction. These conditions also exist in alluvial areas of Denver, Douglas, and Arapahoe counties. The U.S. Army Corps of Engineers has mapped areas of existing wetlands in the contiguous lower forty-eight states and these maps have been procured for the DRCOG planning area. These wetland inventories have several limitations. The satellite image interpretations depend on the time of year the photograph was taken and the resolution of wetland determination is greater than or equal to 1.2 acres. Therefore, many small wetlands in the region were not mapped. Wetlands generally are formed in areas of constant water availability but they can withstand brief periods of drought. The predominant plant species in the wetlands of the region are the cattail, *Typha* spp. Rushes and sedges also occur, but in less abundance. Wetland soil types have not been adequately described in the region and, for the purposes of this chapter, are presumed to be similar to those in areas that have been.

Pollution Problems in the Denver Region

Point sources of pollution in the planning area can be described as primarily industrial and municipal with some minor illegal connections (personal observation).

All major point sources of pollution are regulated by the Colorado Department of Health's Water Quality Control Division. Regulations on beneficial uses in the South Platte River are set by the Colorado Water Quality Control Commission whose members are appointed by the governor and confirmed by the Colorado Senate. These dischargers come under the National Pollution Discharge Elimination System (NPDES). Colorado statutes require self-monitoring by the dischargers and enforcement monitoring by the U.S. EPA Surveillance and Analysis Division and the Colorado Department of Health or its designated agencies, i.e., City and County of Denver Department of Health and Hospitals. Point sources, then, are relatively well controlled and quantifiable.

Nonpoint sources of pollution in the Denver region are tremendously diffuse and difficult to quantify. Sources vary from lawn fertilizers and pesticides applied in excess of manufacturer's recommendations, to oil and grease on impervious areas deposited by automobiles, to leaf litter and other organic carbon sources originating from atmospheric fallout, tree fall, pet droppings, and so on. Additionally, NPS loads of total suspended solids (*TSS*) originate from atmospheric fallout and are related to areas undergoing construction. Quantification of NPS pollution is dependent on land use types, ability to measure all storm-generated runoff, and the ability to accurately measure localized precipitation.

The DRURP established nine automatic sampling stations in seven small urban basins representing the predominant land use types in the Denver region. These land use types were natural grassland, single-family residential, multifamily residential, mixed commercial and multifamily, and a commercial shopping center. The monitoring stations recorded five-minute rainfall, continuous discharge, and included automatic water quality sampling devices. These devices took water quality samples at five minute intervals and were triggered by preset changes in stage.

Concurrently, we manually sampled the South Platte River at four stations (1980) and five stations (1981) to determine ambient and storm-generated pollutant loads. The sampling network was designed with an upstream control station representing little urbanization and moved downstream through the heavily urbanized sections of Denver. Additionally, the major tributaries were sampled simultaneously with the small sites and in-stream stations during three major storm events to allow an estimation of total basin response to storm-generated pollutants.

The preliminary results of the small site, in-stream, and tributary sampling program will be discussed first, followed by a discussion of Best Management Practice pollutant removal efficiencies.

METHOD AND MATERIALS

The DRURP sampled ten major storm events on the South Platte River during 1980–81. Three events were sampled during 1980 at four locations. These occurred 11–12 July, 14–15 August, and 9–10 September. The 14–15 August storm received additional attention as all major tributaries were sampled in addition to the mainstream river stations and small urban sites.

Small Sites

The small urban sites were instrumented with automatic flow recorders, multiple intake water samplers, tipping-bucket type rainfall recorders, and wetfall/dryfall atmospheric deposition collectors which were placed in operation in March 1980. Table 9.1 is a description of these sites, areas, percent total imperviousness (% I_a), percent effective imperviousness (% I_e), and other pertinent information.

Table 9.1 Urban monitoring site descriptions

USGS Station Identifier	*Name of Monitoring Site*	*Latitude Longitude*	*Drainage Area/ Acres*	*% I_a*	*% I_e*
06710225	Big Dry Creek Tributary at Easter St. near Littleton, CO (MF)[1]	39°35′17″ 104°57′20″	30.0	*	41.3
06710610	Rooney Gulch at Rooney Ranch near Morrison, CO (NG)[2]	39°41′27″ 105°11′32″	405	0.6	0.6
06711585	Upper Asbury Park Storm Drain at Tejon St., Denver, CO (SF)[3]	39°40′52″ 105°00′42″	121	32.2	22.2
06711586	Lower Asbury Park Storm Drain at Asbury Avenue, Denver, CO (SF)[3]	39°40′51″ 105°00′41″	127	31.3	21.5
06711635	Upper N. Ave. Storm Drain at Denver Fed. Center, Lakewood, CO (M)[4]	39°43′21″ 105°07′47″	68.7	59.9	50.0
06711637	Lower N. Ave. Storm Drain at Denver Fed. Center, Lakewood, CO (M)[4]	39°43′22″ 105°07′46″	79.7	54.7	46.0
06713010	Cherry Knolls Storm Drain Denver, CO (MF)[1]	39°38′58″ 104°52′47″	57.1	63.6	37.5
06720420	116th & Claude Ct. Storm Drain, Northglenn, CO (SF)[3]	39°54′23″ 104°57′34″	167	37.3	23.9
394236105042400	Villa Italia Shopping Center Storm Drain, Lakewood, CO (C)[5]	39°42′36″ 105°04′24″	73.5	91.2	91.2

[1] MF = Multifamily
[2] NG = Natural grassland
[3] SF = Single family
[4] M = Mixed commercial, single family
[5] C = Commerical shopping area
* Not calculated

Rainfall/Runoff Relationships

Total rainfall and total runoff volumes were measured at each site during the 1980–81 sampling seasons. Total storm runoff can be calculated as a function of total storm rainfall and % I_e without regard to additional factors (Seahurn and Aronson 1974). These additional factors include antecedent moisture conditions, infiltration rate, intensity of percipitation, length of storm, soil and air temperatures, canopy cover, evapotranspiration, and storm sewerage. I have presented a method for the estimation of total storm runoff as a function of % I_e (Judy 1981).

Each aerial photograph was planimetered to determine the total impervious surface area (I_a), i.e., rooftop area, street, driveway, and walkway surface areas, and parking lots. Percent I_e was calculated followed by field verification to determine the areas that directly contributed to surface runoff via impervious connections. Table 9.1 gives these results. Note that for some of the basins no difference is apparent between I_a and I_e.

Total rainfall and total runoff were plotted for each basin. The relationship that best described the predictability of storm runoff was of the form:

$$y = bx^m \tag{9.1}$$

where

y = estimated runoff (inches)
x = total storm rainfall (inches)
b = intercept
m = slope

Figure 9.2 shows the relationship between total storm rainfall, total storm runoff, and % I_e for a series of highly probable storms in the Denver regions. These relationships are probably not linear but r-values ranged from 0.93 for low-volume, highest probability storms to 0.71 for high-volume, low probability storms. Additionally, there were not enough data to accurately test the value of nonlinear relationships. Regardless of the linearity or nonlinearity of the relationships, this set of functions shows the increase in runoff with increase in % I_e.

Pollutant Load/Runoff Relationships

Data for each site were analyzed to determine if a predictable relationship occurred between unit-area pollutant loads and unit-area runoff volumes based upon a combination of convective and upslope storm conditions. Table 9.2 gives the results of this analysis for the sites with sufficient data. Linear regression coefficients are included in the table for each site, to enable quick calculation of pollutant loads for any runoff value within the range of storms measured.

Figures 9.3 and 9.4 show these relationships graphically for water quality constituents of particular concern. Suspended solids are of importance from both quality and quantity standpoints. In particular, phosphorous is of prime concern

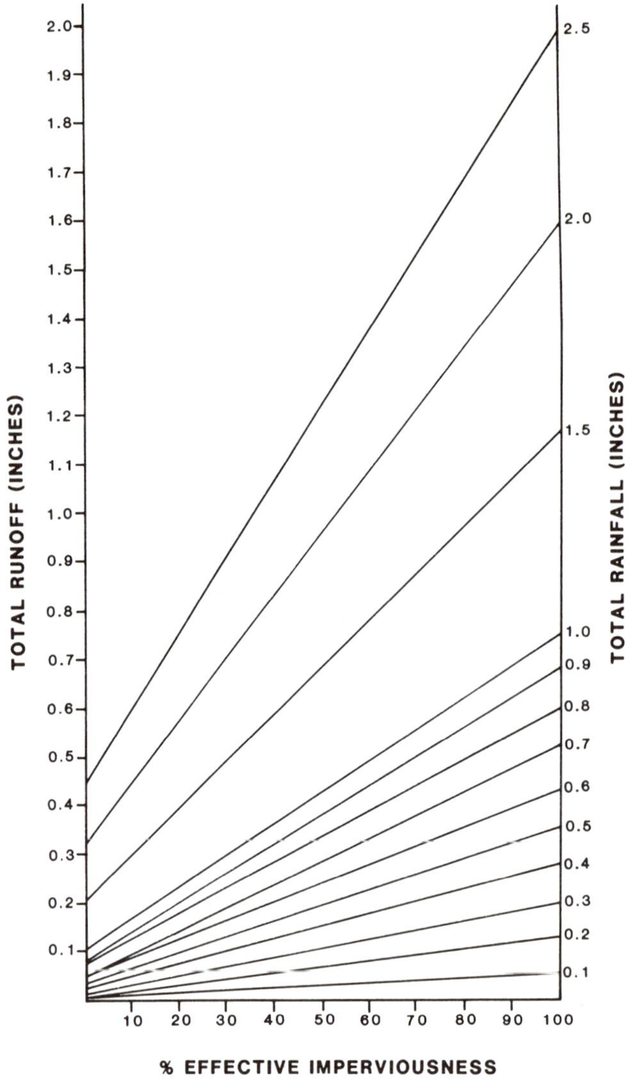

Figure 9.2 Relationships between total rainfall (inches), percent effective imperviousness, and total runoff (inches) derived from urban monitoring data.

because of its effects on downstream reservoirs and its implication in the process of eutrophication (Hutchinson 1975; Hynes 1970).

Zinc is associated with acute and chronic toxicity to aquatic life. Judy and Davies (1977) examined the effect of calcium addition to zinc toxicity of fathead minnows. We found that the LC-50 values of static bioassay for zinc increased

Table 9.2 Total runoff amounts (inches), pollutant loads for each constituent (pounds/acre), and the empirical relationships for selected urban basins[1]

Runoff (Inches)		Pollutant Loads Pounds/Acre (lb/a)														
		TP	*DP*	*O-PO*$_4$	*TOC*	*DOC*	*SS*	*Pb*	*Zn*	*Cu*	*Mn*	*TN*	*TKN*	*NH*$_3$	*NO*$_2$*NO*$_3$*N*	*COD*
Rooney Gulch at Rooney Ranch (NG)[2], I_e *= 0.6%*																
0.02		0.002	0.0006	0.0006	0.0803	0.0442	1.7746	0.0001	0.0005	0.00007	0.0015	0.0234	0.0213	0.00001	0.0021	0.2732
0.37		0.0312	0.0084	0.0137	1.3011	0.6633	31.6802	0.0014	0.0049	0.0011	0.0195	0.2117	0.1744	0.0029	0.0372	3.9955
0.02		0.0013	0.0004	0.0006	0.0683	0.0547	1.0389	0.0008	0.0005	0.0001	0.0011	0.1127	0.0092	0.0017	0.0021	0.3539
0.06		0.0033	0.0009	0.0009	0.3461	0.2368	3.2249	0.0003	0.0007	0.0015	0.0032	0.0258	0.0246	0.0009	0.0012	0.6556
Coeffi-	$b =$	0.0006	−0.0001	−0.0005	0.0474	0.0529	−0.8624	0.0003	0.0001	0.0005	0.0002	0.0406	0.0033	0.0007	−0.0015	0.0773
ents of	$m =$	0.0856	0.0229	0.0382	3.4172	1.6756	87.5922	0.0029	0.0128	0.0020	0.0521	0.4496	0.4606	0.0060	0.1036	10.5725
equations	$r =$	0.9981	0.9978	0.9960	0.9939	0.9803	0.9983	0.8428	0.9977	0.4779†	0.9998	0.8544	0.9965	0.8237	0.9908	0.9996
116th and Claude Court, Northglenn, CO., (SF)[3], I_e *= 23.9%*																
0.34		0.026	0.0149	0.0182	0.9303	0.6510	12.2145	0.0085	0.0086	0.0008	0.0081	0.2005	0.1559	0.0397	0.0446	4.6272
0.09		0.021	0.0030	0.0034	1.1277	0.3046	13.3590	0.0166	0.0089	0.0010	0.0087	0.1468	0.1325	0.0117	0.0143	5.6622
0.04		0.006	0.0027	0.0030	0.3854	0.2463	0.9936	0.0013	0.0011	0.0001	0.0010	0.0429	0.0314	0.0110	0.0115	1.0388
0.08		0.009	0.0022	0.0026	0.2395	0.1703	2.4343	0.0041	0.0028	0.0004	0.0028	0.0654	0.0443	0.0040	0.0211	1.8911
Coeffici-	$b =$	0.0077	−0.0004	−0.0008	0.4660	0.1382	3.3220	0.0062	0.0028	−0.0014	0.0028	0.0509	0.0428	0.0014	0.0081	2.1817
ents of	$m =$	0.0564	0.044	0.0550	1.4886	1.4900	28.5695	0.0102	0.0187	0.0275	0.0173	0.4580	0.3507	0.10102	0.0073	8.1682
equations	$r =$	0.81	0.99	0.99	0.48	0.96	0.61	0.21*	0.64	0.99	0.61	0.86	0.77	0.95	0.97	0.51*
Upper North Avenue Storm Drain at Denver Federal Center, Lckewood, CO(M)[4], I_e *= 50.0%*																
0.12		0.019	0.0032	0.0056	0.8293	0.2216	17.9255	0.0061	0.0075	0.0014	0.0140	0.1057	0.0623	0.0199	0.0434	3.1224
0.03		0.005	0.0006	0.0004	0.6087	0.2945	4.7771	0.0024	0.0046	0.0004	0.0011	0.0526	0.0268	0.0053	0.0259	1.8691
0.23		0.018	0.0027	0.0039	1.1914	0.4706	17.6164	0.0077	0.0096	0.0015	0.0135	0.1328	0.0878	0.0061	0.0450	4.2048
0.04		0.009	0.0028	0.0016	0.6607	0.3731	4.6293	0.0037	0.0050	0.0005	0.0048	0.0829	0.0605	0.0102	0.0224	2.5985
0.33		0.20	0.0093	0.0085	1.6859	1.2708	10.9176	0.0071	0.0122	0.0013	0.0090	0.2061	0.1501	0.0374	0.0560	7.3644
0.03		0.006	0.0005	0.0007	0.3612	0.1223	5.2794	0.0024	0.0024	0.0003	0.0036	0.0255	0.0193	0.0006	0.0061	2.1021
Coeffici-	$b =$	0.0067	0.0003	0.0005	0.4048	0.0869	6.2948	0.0027	0.0033	0.0004	0.0042	0.0375	0.0210	0.0026	0.0164	1.4969
ents of	$m =$	0.0473	0.0222	0.0228	3.7286	2.8612	29.9695	0.0168	0.0278	0.0036	0.0270	0.4881	0.3596	0.0816	0.1285	15.7436
equations	$r =$	0.86	0.86	0.89	0.98	0.86	0.59	0.89	0.97	0.81	0.63	0.95	0.95	0.76	0.88	0.96

Table 9.2 *(Continued)*

Runoff (Inches)	*Pollutant Loads Pounds/Acre (lb/a)* TP	DP	O-PO_4	TOC	DOC	SS	Pb	Zn	Cu	Mn	TN	TKN	NH_3	NO_2NO_3N	COD
Villa Italia Storm Drain, Littleton, CO., (C)[5], I_e = 91.2%															
0.37	0.025	0.0139	0.0069	3.3880	2.2718	9.4969	0.0139	0.0242	0.0017	0.0197	0.3891	0.3041	0.0660	0.0851	13.1117
0.05	0.018	0.0106	0.0071	1.3415	1.2515	4.3360	0.0098	0.0099	0.0009	0.0085	0.1842	0.1362	0.0383	0.0480	1.4420
1.44	0.072	0.0398	0.0110	6.2052	3.9496	43.8024	0.0528	0.0658	0.0053	0.0053	0.7841	0.5476	0.1681	0.2365	22.4638
0.30	0.019	0.0066	0.0060	1.7643	1.2674	3.8493	0.0069	0.0096	0.0008	0.0054	0.1854	0.1192	0.0462	0.0662	6.4638
0.38	0.020	0.0157	0.0105	1.9106	1.4345	2.9270	0.0052	0.0136	0.0015	0.0086	0.2474	0.1893	0.0878	0.0581	8.6274
Coeffici- b =	0.0093	0.0055	0.0068	1.1264	1.0045	−3.0776	−0.0001	0.0028	0.0003	0.0112	0.1286	0.1025	0.0333	0.0261	3.3894
ents of m =	0.0423	0.0233	0.0030	3.5346	2.0285	31.4171	0.0352	0.0430	0.0034	−0.0034	0.4516	0.3085	0.0944	0.1430	13.8432
equations r =	0.98	0.96	0.70	0.95	0.95	0.97	0.95	0.97	0.98	−0.31*	0.96	0.94	0.97	0.98	0.94

† Value of regression significantly different from relationships for other sites due to high loading of Cu (0.0022 lb/a) for low runoff (0.06")

* Value significantly different from other sites

[1] Basins not shown lacked significant data to evaluate pollutant loading relationships

[2] NG = Natural grassland state

[3] SF = Single family residential

[4] M = Mixed commercial and residential

[5] C = Commercial

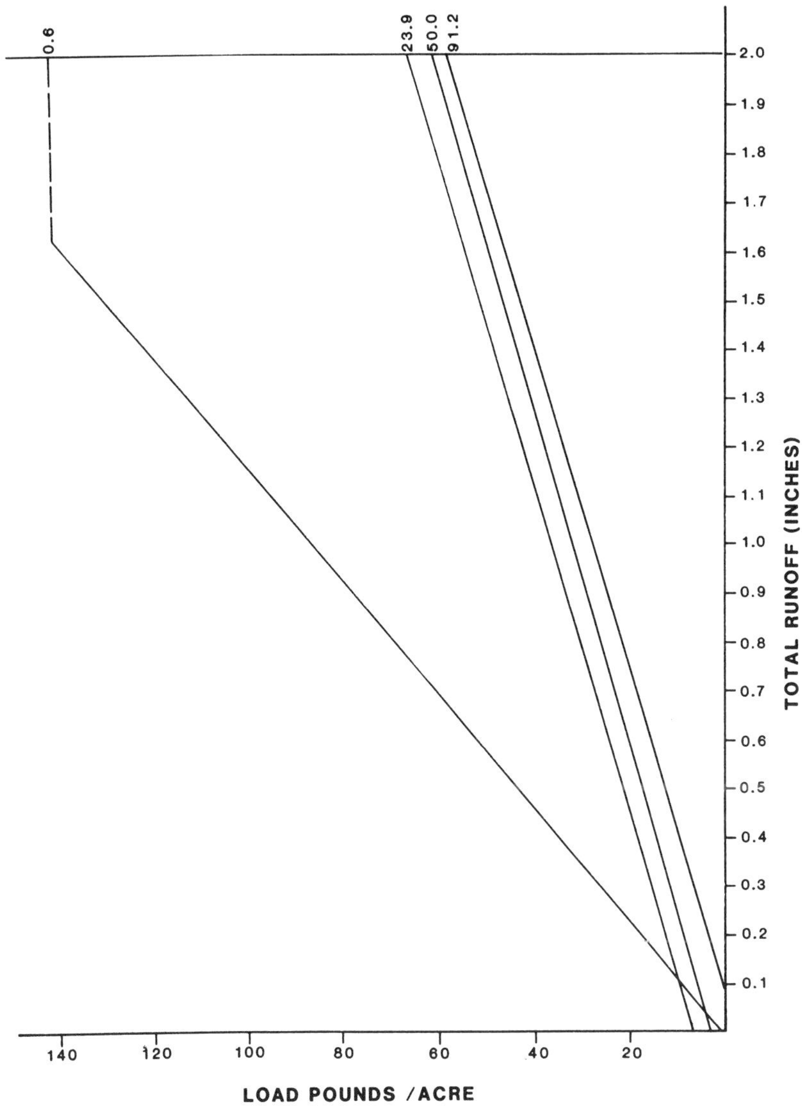

Figure 9.3 Suspended solid load as a function of runoff and percent effective imperviousness.

as the total hardness increased and that slight variations in temperature and pH had no effect on the test results.

Water quality standards for the South Platte River basin are set, in part, based on the hardness levels found in the natural system. The important aspects of these relationships are the increase or decrease in unit loading as a function of the land use and its associated % I_e. Total suspended solids load decreased

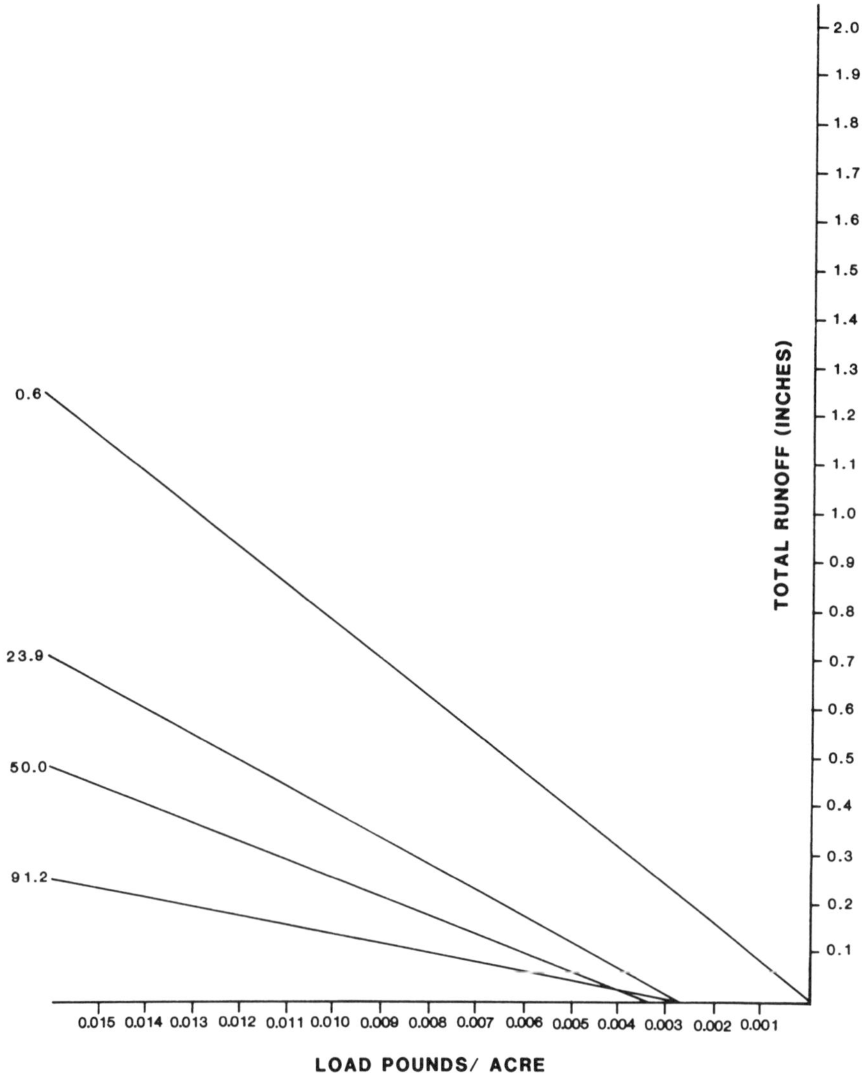

Figure 9.4 Zinc load as a function of runoff and percent effective imperviousness.

as the surface became more impervious. This is not in accordance with past studies (Ellis and Alley 1979) and may be a result of unusual conditions in the basin. Additionally, there is a large difference between the event-based load and the seasonal load. The natural grassland site generated less runoff on an annual basis than the developed sites. The other sites show little absolute difference between unit loads.

Orthophosphate load also decreased as the % I_e increased. Orthophosphate is found in soils and is used extensively in the Denver region as a major component of lawn fertilizer. The least impervious open space area also had the highest absolute storm unit-load for the range of storms measured. It was apparent, however, that a convergence occurred between the line of best fit for the natural (I_e = 0.6%) and single-family areas (I_e = 23.9%), near 2.0 inches runoff. This may be the result of inadequate data.

Zinc loading followed the logical pattern of increasing as the impervious area increases due to human influences. Zinc is a widely used industrial additive to hydrocarbon lubricants and tire material and is ubiquitous in the urban environment (Shaheen 1975; U.S. EPA 1980). Shaheen (1975) calculated a deposition of 0.039 grams (person/day) zinc on street surfaces as a result of automobile travel.

Planning Tool Development

The results of the rainfall/runoff and the pollutant load/runoff relationships were outlined in a series of nomographs to allow quick interpretation and estimation of pollutant loads originating from developed areas or prediction of loads from predevelopment plans. Figures 9.5, 9.6, and 9.7 illustrate the combination of these functions into a single nomograph.

Tributary Storm Event Data

The major tributaries of the South Platte River were sampled on 14–15 August 1980; 4 March 1981; and 28 May 1981. These storms represented a combined convective and upslope event, post snowfall event, and a severe convective event, respectively. Additionally, the tributaries were sampled once-monthly during May, June, July, and August 1981 to estimate ambient tributary loads. These data are not presently available but will be made a part of the model at a later date.

In-Stream Loads

In-stream samples were taken at four stations on the South Platte River during 1980 for three storms. The 1980 sampling season was during an extended drought period which is reflected in the low number of storm samples. These storms occurred 11–12 July, 14–15 August, and 8–19 September 1980. Table 9.3 presents the total load at each of the four stations for each water quality constituent of concern.

Small-Site to Large-Site Modeling

All of the small urban sites selected for monitoring reflect the worst case condition for runoff-generated pollution in the Denver region. This also applied to the sites containing detention ponds before retrofitting of the outlet structures to detain small storms. Therefore, all values for the small urban sites with respect to unit-loads and unit-runoff volume were expected to reflect no water quality improvement.

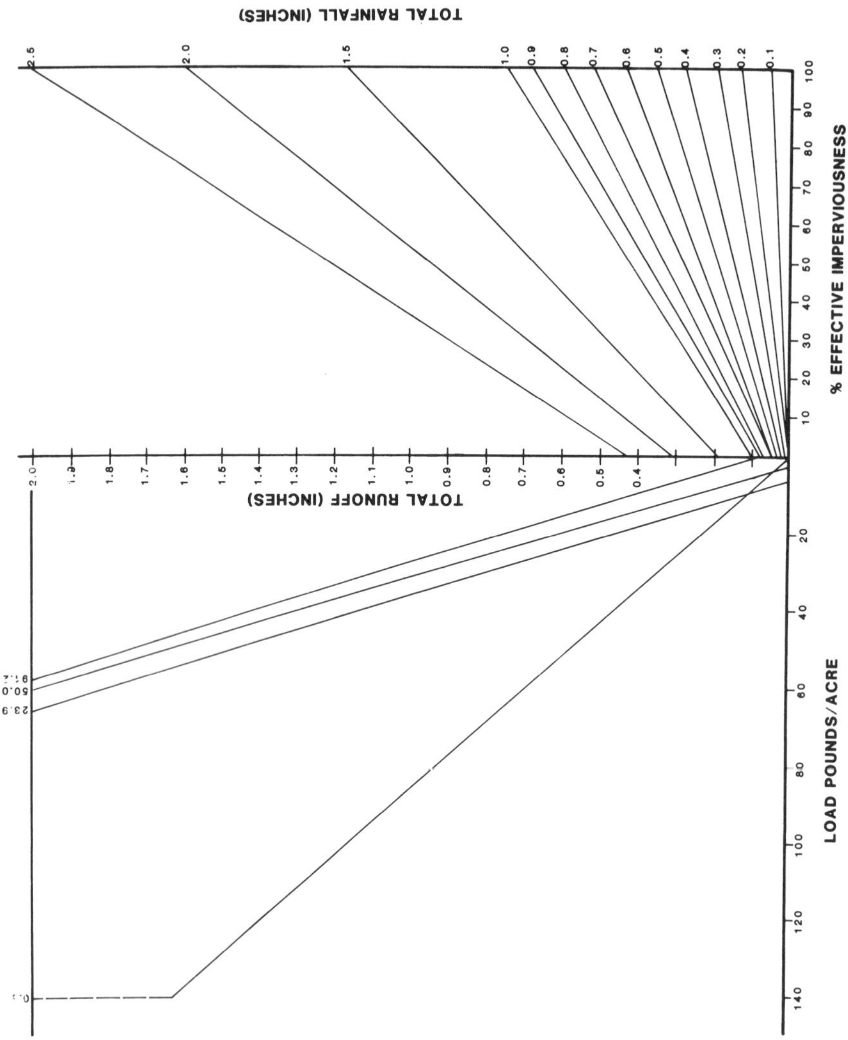

Figure 9.5 Nomograph relating total rainfall (inches), % effective imperviousness, total runoff (inches), and pollutant loads (lb/a) for suspended solids.

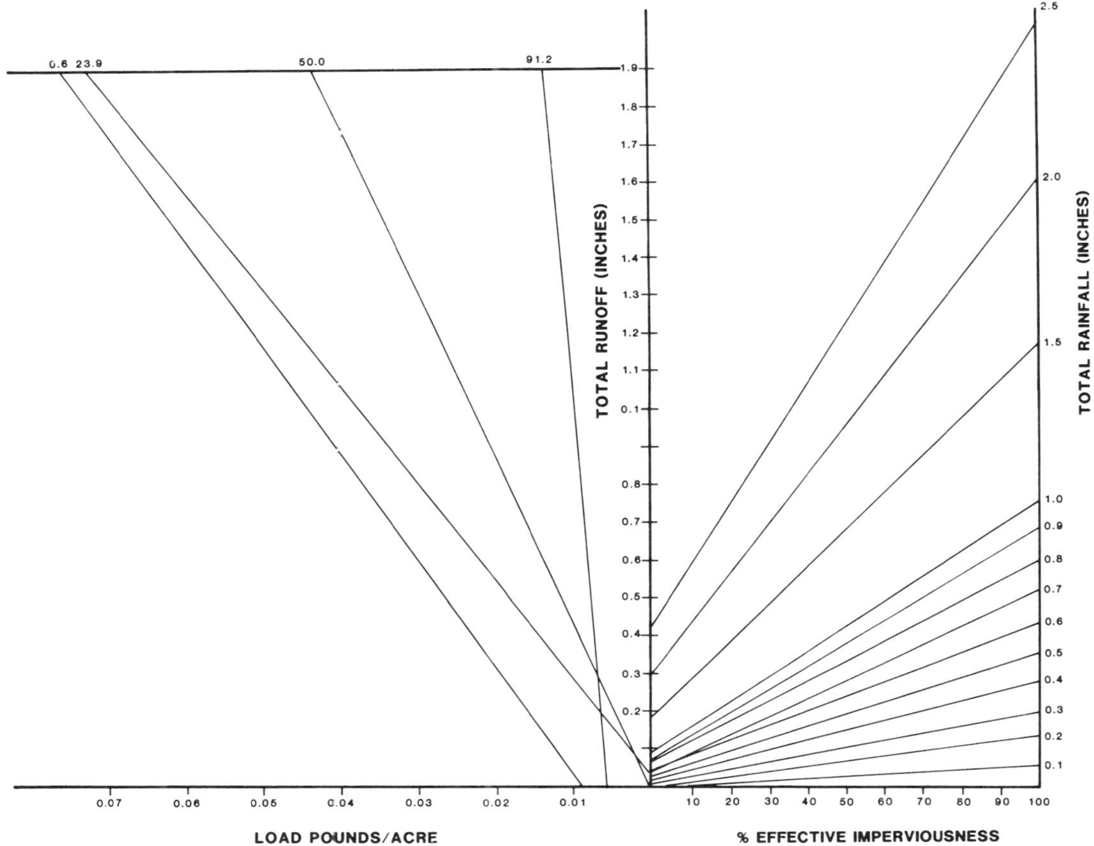

Figure 9.6 Nomograph relating total rainfall (inches), % effective imperviousness, total runoff (inches), and pollutant loads (lb/a) for orthophosphate.

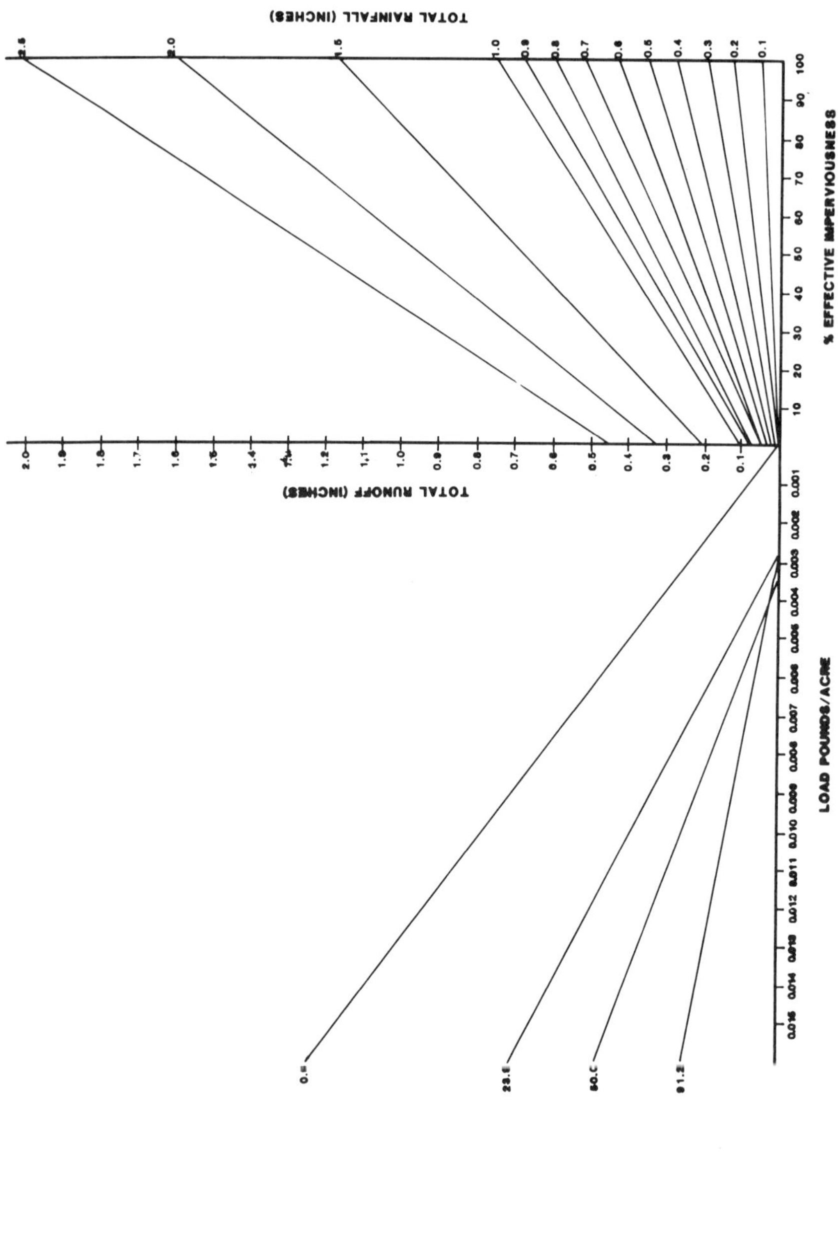

Figure 9.7 Nomograph relating total rainfall (inches), effective imperviousness, total runoff (inches), and pollutant loads (lb/a) for zinc.

Table 9.3 Water quality data for storm events sampled at in-stream South Platte River stations, 1980.

Date	*Station*	*Storm Discharge*	*Total P*[2]	*O-PO₄*	*TOC*	*Total Lead*	*Total Zinc*	*Total Copper*	*Total Manganese*	*Dissolved Ammonia*	*Dissolved Nitrate plus Nitrite*	*Total Suspended Solids*	*Chemical Oxygen Demand*	*Total Nitrogen*	*Total Kjeldahl Nitrogen*
7/11–12	8000[1]	2,440,800	345.5	33.8	4853.9	39.8	0.0	0.0	0.0	36.5	175.0	294487.0	10545.0	931.7	756.7
	3000	788,963	30.7	12.6	2257.1	22.7	4.8	0.48	4.2	27.0	9.5	31446.0	6291.6	121.1	111.5
	2000	8,805,150	1092.2	149.3	21262.8	21.1	0.0	21.7	135.6	412.5	165.9	801533.0	71792.5	2495.9	2330.1
	1000	10,154,700	882.6	358.2	26725.8	306.0	385.4	105.5	360.3	529.8	—	572755.1	94172.5	1890.2	1799.5
8/14–15	8000	2,957,400	134.9	22.5	9845.3	19.2	0.0	1.8	70.0	36.9	216.7	129535.2	12041.2	112.7	—
	7000	—[3]	144.2	18.8	8154.2	52.8	75.8	9.1	131.0	—	223.8	124499.9	25143.5	834.4	604.2
	6500	1,230,509	38.5	12.0	2370.4	24.6	19.7	2.0	44.8	—	21.0	32243.0	11407.9	305.4	285.1
	5500	1,358,293	173.4	11.9	6016.2	75.0	60.8	10.3	168.4	—	58.7	144372.0	26587.7	819.5	775.8
	4500	986,246	55.4	4.2	2983.8	30.8	30.1	5.1	78.5	—	40.9	52121.3	10198.1	323.1	281.9
	3500	683,732	78.9	4.6	3423.6	26.5	38.1	9.2	89.5	—	34.0	59200.7	14572.1	250.4	216.6
	3000	10,074,227	1332.4	163.1	57608.8	323.9	536.6	112.4	888.7	—	459.9	1.629 E 6	154511.0	4455.0	3995.2
	2000	4.209 E 7	8176.8	150.6	205221.7	1272.9	1990.8	428.9	4840.2	2766.8	2759.4	7.33 E 6	715092.5	26845.3	24085.9
	1000	4.56 E 7	8653.4	285.8	189237.5	1338.8	2193.5	439.5	4641.0	3415.5	3182.0	7.03 E 6	838032.6	20358.9	17176.8
9/8–9	8000	1,155,667	4.6	0.0	2174.6	5.1	21.9	11.9	18.4	24.6	84.6	—	3143.8	255.4	170.9
	3000	7,723,755	479.8	307.7	11236.6	19.5	42.1	17.1	79.6	265.6	1266.0	81500.3	33011.6	2086.1	1202.8
	2000	25,592,850	3009.9	1465.0	54639.1	171.9	376.9	65.9	779.5	3555.0	2006.3	1064352.1	158592.7	10706.9	8700.6
	1000	30,735,450	3074.0	1000.4	62902.5	249.6	629.1	181.4	1008.9	2788.7	1629.6	1.299 E 6	200953.7	10400.6	8771.1

[1] Station identification numbers are in decreasing order from upstream to downstream. See Figure 1 for a map of each station. Stations 7000, 6500, 5500, 4500, and 3500 are auxiliary tributary stations and were duly sampled once in 1980.

[2] Values for each constituent are total storm load in pounds determined by integrating the loadograph curve at 15 minute intervals.

[3] Discharge value was used in computation of load but was not printed and was lost.

To extrapolate from a monitored small site to the tributary basin and then to the South Platte River is complicated and risky, at best (McPherson 1979). Nevertheless, I propose a simple mathematical model to compare loads in a type of accounting format. This model is proposed in the absence of sufficient data to test or verify it. Data will become available at a later date and the model will be used to quantify load response relationships in the South Platte River in the Denver region. Figure 9.8 is a schematic illustration of the proposed model. A mathematical presentation and rationale follows.

Theoretically, all tributary loads are a function of accumulated small site

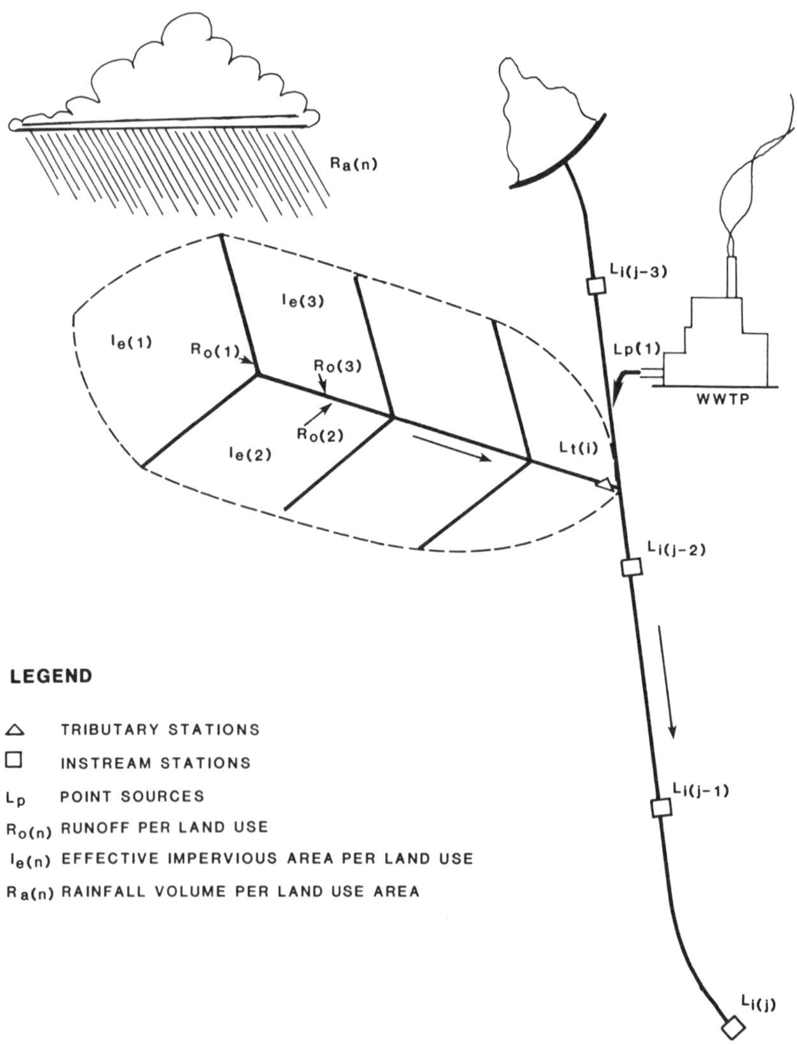

Figure 9.8 Urban runoff water quality conceptual model.

loads based on effective impervious area, rainfall, and runoff. Likewise, all in-stream loads are functions of the sum of the tributary and direct flow area loads. Therefore, equation 9.2 represents a simplistic accounting equation to predict loads at any given in-stream station j as a function of the upstream basin contributions.

$$LT_j = \Sigma L_t + \Sigma L_d + L_i + \Sigma L_p \tag{9.2}$$

where

L_t = tributary load (pounds)
L_d = direct flow area load (pounds)
L_i = in-stream load at station j-1 (pounds)
L_p = point source load (pounds)
LT_j = total in-stream load at station j (pounds)

L_t was measured directly and can be estimated from the area of the basin, % I_e, total storm runoff (R_o), and point source load. Therefore, L_t becomes

$$L_t = \Sigma\,(a_i \times R_{oi}) \times (Cf)(C) \times (T) \times (6.245 \times 10^{-5}) + (L_p) \tag{9.3}$$

where

L_t = tributary load (pounds)
a_i = area of basin$_i$ (acres)
R_{oi} = runoff volume of basin$_i$ (inches)
Cf = conversion factor from inches to cubic feet
C = concentration (ppm)
T = time (seconds)
L_p = point source load (pounds) measured during the event or estimated from NPDES permits

L_d, the direct flow area load, must be estimated because no actual measurements were made. L_d depends upon area, % I_e, and runoff. Therefore, L_d becomes

$$L_d = \Sigma\,(a_i \times R_{oi}) \times (Cf) \times (C) \times (T) \times (6.245 \times 10^{-5}) \tag{9.4}$$

where

L_d = direct flow area load (pounds)
a_i = area of basin$_i$ (acres)
R_{oi} = runoff of basin$_i$ based on effective impervious area relationships
Cf = conversion factor inches to cubic feet
C = concentration measured (ppm)
T = time (seconds)

L_i, the in-stream load at station j-1, the upstream mainstem station measured, is a function of L_t and L_d. L_i was measured in this study and is represented as a measured parameter.

$$L_i = (C \times Q) \times (6.245 \times 10^{-5}) \times (T) \tag{9.5}$$

where

L_i = instream load (pounds) measured at station j-1
C = concentration measured (ppm)
Q = instantaneous flow measured (cubic feet per second)
T = time (seconds)

L_p, the point source load, was estimated from NPDES discharge permitees' self-monitoring data. Most of these data were collected monthly; therefore, only rough approximations can be made. Equation 9.6 is an estimation of the point source load:

$$L_p = (EC) \times (EQ \times T_s \times 6.245 \times 10^{-5}) \tag{9.6}$$

where

L_p = point source load (pounds)
EC = effluent concentration measured or reported on NPDES permits
EQ = effluent discharge during storm period (cubic feet per second)
T_s = storm duration (seconds)

LT_j, therefore, is a combination of all these equations and can be represented by equation 9.7.

$$\begin{aligned} LT_j = {} & [\Sigma\,(a_i \times R_{oi}) \times (Cf \times (C)) \times (6.245 \times 10^{-5})] + [\Sigma\,(EC \times EQ \times T_s) \\ & \times (6.245 \times 10^{-5})] + [\Sigma\,(a_i\,(R_{oi})) \times (Cf\,(C)) \times (6.245 \times 10^{-5})] \\ & + \int_{to}^{ts} (c \times Q \times 6.245 \times 10^{-5}\,(T)) \\ & + [\Sigma\,((EC)\,(EQ)\,(TS)\,(6.245 \times 10^{-5}))] \end{aligned} \tag{9.7}$$

where

$\int_{to}^{ts}$ = duration of some storm event based on the hydrograph at station j

This equation represents a lump sum parameter model of water quality. It assumes that no changes occurred in constituents as they passed downstream as a function of either travel-time, biological or chemical reactions, or other factors known to influence chemical behavior. Obviously, these types of changes do occur. However, storm event pollution is transient in nature, in that slugs of pollutants move through the system. Lee (1981) addressed this problem in relation to toxicity values for fishes. Lee concluded that the duration of exposure of aquatic organisms to storm-generated pollutants is short and intermittent. This requires a different type of analysis than extrapolation of documented LC-50 values down to the duration of the storm event. Additionally, he questioned the availability of these toxins, especially metals, for biological uptake. The rationale behind his argument was that stream systems are constantly flowing and do not allow acute toxic conditions to occur if the stream concentrations fall below toxic values during ambient conditions. In short, Lee felt that the data collected by the National Urban Runoff Program were useless in determining toxicity. Therefore, this analysis will not attempt to delineate the effects of these other factors.

This model used storm runoff as an input rather than storm rainfall. While runoff is dependent upon rainfall, basin slope, channel length, ratio of pervious to impervious area, antecedent dry-days and other miscellaneous factors, I showed that runoff can be accurately estimated solely as a function of effective imperviousness (Judy 1981). This vastly simplified the dynamic process except that it required the use of the nomograph relating rainfall, runoff, and effective impervious area (see Figure 9.2). Rainfall, per se, is not an input into this model due to the increased difficulty in using the model.

Model Testing

The lack of certain pieces of data precludes an actual test of the model at this time. The model will be used as the basis for comparing predicted loads with the actual measurements made during the 1980 and 1981 sampling seasons. I have presented the model, primarily, for discussion and as a vehicle for testing based on the assumptions previously presented. The model will be tested in the following manner.

Average subbasin precipitation will be calculated based on a network of forty-two rain gages installed for the project. In conjunction with the rain gages a radar reflectivity analysis system, developed by Geophysical Research and Development Corporation, was used to predict rainfall in 1-sq.-mi.-sections of the study area. The predictions of subbasin rainfall were compared with actual 15-minute rainfall records and were best described by an equation of the form

$$y = a + b \ln x \tag{9.8}$$

where

y = basin rainfall per time interval predicted
a = intercept
b = slope of line fit
x = radar reflectivity value

This information will be used to calculate subbasin runoff. The relationships of unit-load to unit-volume runoff will then be calculated, and the model will be executed.

All predicted values will be compared as a ratio of predicted to measured pollutant at each tributary and in-stream station. The mean, deviation, and standard error of estimate will be calculated for all comparisons combined. The resulting statistics should allow a valid test of the proposed model.

BEST MANAGEMENT PRACTICES

The model will then be used in conjunction with Best Management Practice (BMP) pollutant removal efficiencies. The result will not tell where to implement control measures but will give a relative estimate of effectiveness of measures singly and in the aggregate. Best Management Practices are only effective in controlling small storms (U.S. EPA 1977; Malcolm 1980). Therefore, based on Colt (1980), only storms falling in the 0.1–0.7 inch total rainfall categories will be used for predictive

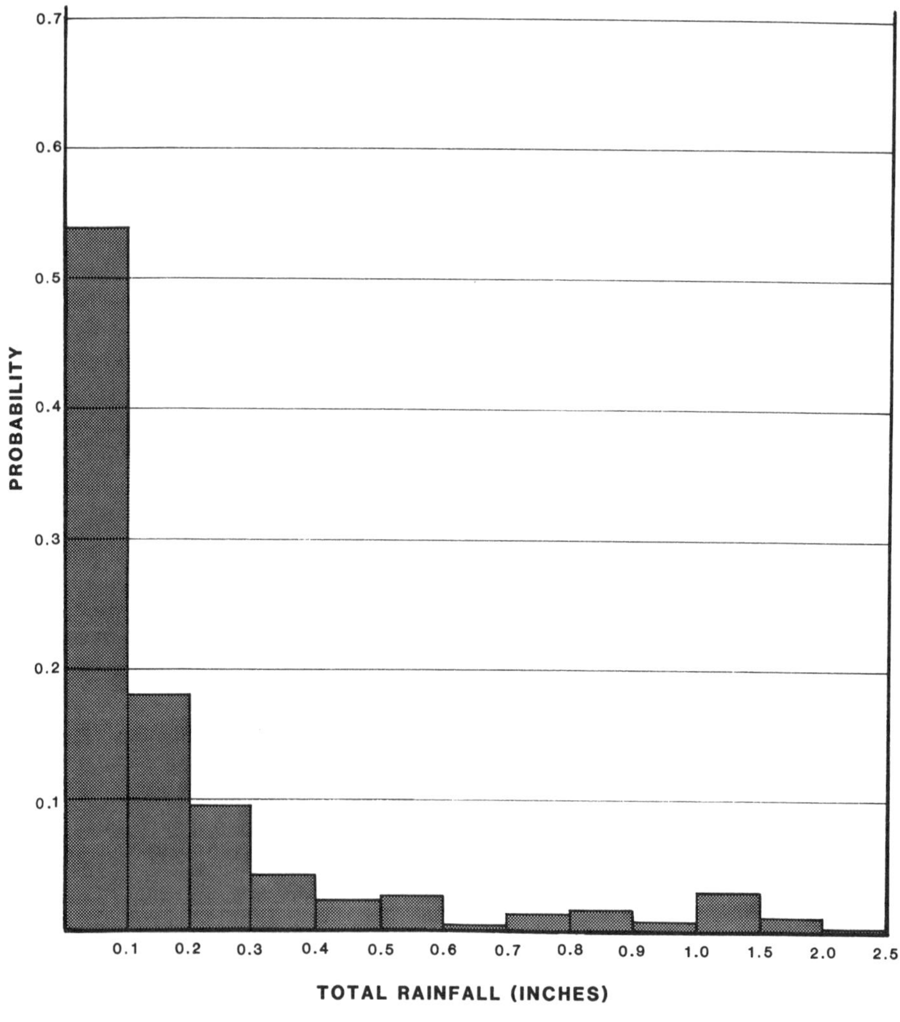

Figure 9.9 Probability distribution of total rainfall volumes at Stapleton Weather Center (NWS), Denver, CO. Courtesy of J. Colt, personal communication.

purposes. Figure 9.9 summarizes Colt's findings of total rainfall probability in the study area. An analysis of the total rainfall classes in Figure 9.9 shows that 90% of storms yield total storm volumes between 0.1 in. and 0.7 in. Therefore, control measures should focus on storms most likely to occur. Control measures should not impede flood waters to the degree that they greatly increase the likelihood of damage to property. The Best Management Practices should, however, have the capability of controlling small storms and providing cost-effective treatment of storm water without requiring extensive operation and maintenance procedures.

Runoff Control and Treatment

BMPs that hold promise for reduction of storm water contamination were briefly mentioned in an earlier section of this chapter. This section of the chapter will discuss the storm water control measure investigation and the respective pollutant removal efficiencies, benefits, and drawbacks.

Temporary and Permanent Construction Controls

Many Best Management Practices have been described in the literature for temporary and permanent control of runoff from developed, silvicultural, and agricultural areas (U.S. EPA 1977; DRCOG 1980; LWCOG 1977; USDA-SCS 1978). Effectiveness of Best Management Practices varies widely based upon upstream contributing areas, slope, soil types, degree of detention/retention, and maintenance. Best Management Practices selected for application in the Denver region were judged on the criteria that they: (1) could be accommodated within Colorado Water Law (Doctrine of Prior Appropriation); (2) required little overall maintenance; (3) were aesthetically pleasing; (4) had high potential removal rates; and (5) could be incorporated into local flood control ordinances.

Areas of the Denver region known to have problems with sedimentation of floodways and drainage channels were selected for possible involvement. DRCOG (1980) compiled construction-related Best Management Practices into a regional guide for nonpoint source and erosion control. This document compiled methods of erosion and storm water controls into a single guide available to local developers. Colt (1980) developed a set of guidelines for maximum permissible soil loss in Adams County, Colorado, due to construction of urban developments. His recommendations were based on the maximum acceptable soil loss for continued soil productivity, 5 tons/acre/year (USDA-SCS 1975). Drainageway capacity losses have been documented in Adams County and Aurora, Colorado, due to high soil losses. The transition period between the nondeveloped and stable states is thought to contribute amounts of sediment and sediment-related pollutants. Estimates of soil loss range from 1,000 to 100,000 tons/sq. mi./year or 20,000 to 40,000 times greater than that obtained from adjacent undeveloped areas in an equivalent period of time (USDA-SCS 1975). Unfortunately, no actual monitoring data of developed areas with and without control measures implemented during the construction phase are available. This is due, in part, to the opposition to erosion control ordinances by local developers and their reluctance to give permission to monitor storm runoff.

Stable Development Control Methods

The control methods selected for use in stable developments include percolation pits, flood control detention ponds, with and without water quality improvements outlets, and wetlands. *Street sweeping* was not investigated because the literature values showed no decrease in loading unless continuous cleaning methods were employed (Sartor and Boyd 1972; Pitt and Shawley 1981; Bender et al. 1981).

Colt (1980) mapped the soils of Adams County for permeability and calculated design criteria for the construction of *percolation pits* based on the ten-year recurrence interval. Colt (personal communication) has implemented a water quality protection zone around Barr Lake, Adams County, Colorado. Developments in this zone will be required to use effective runoff control measures. Contributing basins to the Barr Lake drainage area include the areas studied by the DRURP.

Questions arise, however, concerning the danger of contaminating the alluvial ground water systems in the region by runoff-generated pollutants. Table 9.2 shows concentrations of nitrate (NO_3), zinc (Zn), copper (Cu), lead (Pb), and other pollutants in runoff. Seahurn and Aronson (1974) measured the concentrations of pollutants entering percolation pits in Suffolk County, New York. They found similarly high concentrations of these pollutants. Measurements of pollutants in the ground water showed no measurable effect, however. They speculated that the contaminants were adsorbed onto the soil particles and organic matter at the contact surfaces of each pit.

Koppelman (1978) presented additional data on these structures showing no negligible effect six years after the initial investigation concluded. Based on these data, we have assumed no long-term contamination of the ground water due to these pits. Removal efficiencies for these structures are essentially 100% up to the percolation capacity of the pit. Percolation pits do require periodic maintenance but no data are currently available as to the frequency of maintenance required.

Flood control detention ponds were investigated at two sites representing single-family housing (Asbury Park) and a mixed multifamily and commercial development (Denver Federal Center). These flood control ponds were constructed to detain extremely large storms such as the 1-in-100-years flood. No data are yet available from these tests. Data will be evaluated as to effectiveness as they become available.

Relationship between Particle Size and Pollutants and Total/Dissolved Fractions

A special study was conducted to ascertain the fraction of pollutants associated with several particle size classes commonly found in urban runoff. Table 9.4 presents

Table 9.4 Concentrations of constituents in total sample and in filtrate form 8/31/81 storm at DPC

USGS Station	*Sample Type*	*Whatman Filter No.*	*Filter Size (microns)*	*CD μg/l*	*Cu μg/l*	*Fe μg/l*	*Pb μg/l*	*Mn μg/l*	*Zn μg/l*	*NH_4 mg/l*	*N Org mg/l*	*NH_4^+ Org mg/l*	*NO_2^+ NO_3 mg/l*	*Total N mg/l*	*Ortho P mg/l*	*Total P mg/l*
06711635	composite	—	—	3	100	21,000	420	680	790	—	—	9.2	2.6	—	.22	.90
06711635	composite	43	16	0	50	140	100	110	90	1.8	2.5	4.3	2.3	6.6	.09	.18
06711635	composite	40	8	0	0	380	200	120	170	1.7	2.7	4.4	2.4	6.8	.11	.20
06711635	composite	44	3	0	50	90	200	100	90	1.7	2.2	3.9	2.4	6.3	.08	.17
06711635	composite	42	2.5	0	0	110	100	110	150	1.8	2.4	4.2	2.4	6.6	.09	.19
06711635	composite	—	0.45	—	—	—	—	—	—	1.7	2.1	3.8	2.4	6.2	—	.07
06711637	discrete at 1745	—	—	2	60	9,400	280	370	420	—	—	7.8	3.1	—	.19	.52
06711637	discrete at 1750	—	—	2	100	19,000	320	570	640	—	—	7.8	2.8	—	.25	.59
06711637	composite 1745–1755	43	16	0	50	800	100	100	170	1.4	2.5	3.9	2.8	6.7	.10	.20
06711637	composite " "	40	8	0	0	360	100	90	170	1.5	2.5	4.0	2.8	6.8	.11	.21
06711637	composite " "	44	3	0	0	90	100	80	70	1.4	2.1	3.5	2.7	6.2	.08	.16
06711637	composite " "	42	2.5	0	50	80	200	80	60	1.5	2.2	3.7	2.8	6.5	.09	.17
06711637	discrete at 1745	—	0.45	—	—	—	—	—	—	1.0	3.3	4.3	2.9	7.2	—	.03
06711637	discrete at 1750	—	0.45	—	—	—	—	—	—	1.5	2.8	4.3	2.6	6.9	—	.13
06711637	discrete at 1805	—	—	1	53	12,000	280	330	340	—	—	6.1	2.7	—	.31	.47
06711637	composite 1800–1810	43	16	0	0	110	200	90	180	1.5	2.2	3.7	2.3	6.0	.11	.20
06711637	composite " "	40	8	0	50	700	0	80	90	1.6	2.4	4.0	2.1	6.1	.12	.23
06711637	composite " "	44	3	0	50	260	0	80	150	1.6	2.0	3.6	2.2	5.8	.10	.21
06711637	composite " "	42	2.5	0	50	140	100	80	160	1.6	2.1	3.7	2.2	5.9	.11	.20
06711637	discrete at 1805	—	0.45	—	—	—	—	—	—	1.7	1.7	3.4	2.5	5.9	—	.13
06711637	discrete at 1820	—	—	2	50	11,000	220	290	280	—	—	4.3	2.3	—	.18	.37
06711637	composite 1815–1820	43	16	0	0	290	200	90	140	1.3	2.0	3.3	2.1	5.4	.10	.19
06711637	composite " "	40	8	0	0	590	300	90	160	1.5	1.9	3.4	1.9	5.3	.11	.21
06711637	composite " "	44	3	0	0	60	100	80	130	1.4	1.8	3.2	1.8	5.0	.09	.18
06711637	composite " "	42	2.5	0	50	130	0	70	40	1.4	1.5	2.9	1.9	4.8	.09	.19
06711637	discrete at 1805	—	0.45	—	—	—	—	—	—	1.3	3.1	4.4	2.1	6.5	—	.32

Table 9.5 1980 Data dissolved and total metals concentrations in μg/l

		Cadmium		*Copper*		*Iron*		*Lead*		*Manganese*		*Zinc*	
Date	*Time*	*DIS*	*TOT*	*DIS*	*TOT*	*DIS*	*TOT*	*DIS*	*TOT*	*DIS*	*TOT*	*DIS*	*TOT*
06711635–Upper DFC													
9/8/80	@ 2155	0	3	50	66	170	16,000	40	500	240	510	280	970
6711637–Lower DFC													
9/8/80	@ 2205	0	2	0	48	40	10,000	3	230	−30	380	170	580
9/9/80	@ 0900	0	2	0	27	20	10,000	0	190	10	210	50	240
394236105042400–Villa Italia													
7/30/80	Composite	10	6	50	75	430	9,400	27	500	540	760	650	1060
9/8/80	@ 2145	0	3	0	34	40	2,200	7	170	120	240	200	360
9/9/80	@ 0210	2	1	10	9	20	320	0	18	60	70	100	120
06711635–Upper DFC													
2/21	0145	0	3	31	46	20	17,000	3	80	140	490	190	920
2/21	0215	0	6	31	39	40	12,000	8	30	110	380	170	640
2/21	0315	0	2	9	25	20	8,000	4	90	100	280	150	650
2/21	1005	7	7	31	46	20	16,000	0	10	170	530	230	750
2/21	1050	0	3	31	45	10	17,000	12	60	100	470	120	700
2/21	1150	0	3	0	47	20	18,000	9	70	80	470	120	650
2/21	1320	0	2	11	32	90	14,000	11	90	100	350	120	700
2/21	Comp.	1	2	31	35	50	12,000	10	60	120	320	180	530
3/3	Comp.	6	6	29	50	80	17,000	5	50	90	440	140	720
3/4	1035	0	2	0	40	50	12,000	4	20	70	310	60	420
3/4	1105	6	6	57	70	5700	21,000	6	70	80	580	60	730
3/4	1205	0	4	29	60	50	21,000	3	30	80	520	60	680
3/4	1335	6	6	29	70	30	24,000	3	80	50	590	40	650
3/4	1605	0	1	29	30	60	9,400	4	70	60	220	80	320
3/4	1805	1	1	29	30	30	8,400	2	80	60	190	90	240
3/5	Comp.	13	13	4	22	40	8,800	3	0	50	200	50	250
06711637–Lower DFC													
2/21	Comp.	0	3	31	41	40	13,000	13	90	110	360	160	630
3/3	Comp.	0	3	29	70	80	23,000	4	80	60	550	70	810
3/4	1125	6	6	29	130	90	54,000	3	80	40	1200	30	1300
3/4	1230	0	5	0	20	40	33,000	4	90	40	850	20	880
3/4	1400	6	6	29	70	90	28,000	3	50	40	700	30	760
3/4	1605	6	6	57	90	3400	15,000	6	70	300	360	690	710
3/4	1700	6	6	29	30	10	11,000	1	70	50	240	60	280
3/5	Comp.	0	1	29	29	20	9,300	0	50	10	210	30	240
06710225–Southglenn													
3/4	1305	6	6	0	100	160	67,000	3	490	20	1900	20	800
3/4	1335	6	6	29	40	100	23,000	3	380	10	710	10	320
3/20–21	Comp.	0	1	0	30	0	12,000	0	190	20	300	20	220
5/3	Comp.	0	1	48	80	90	16,000	8	460	10	440	10	360

Table 9.5 (*Continued*)

		Cadmium		*Copper*		*Iron*		*Lead*		*Manganese*		*Zinc*	
Date	*Time*	*DIS*	*TOT*	*DIS*	*TOT*	*DIS*	*TOT*	*DIS*	*TOT*	*DIS*	*TOT*	*DIS*	*TOT*
06710610–Rooney													
5/16	1415	0	0	0	13	50	2,900	2	98	40	110	50	150
5/17	0815	1	1	10	8	10	2,500	3	13	470	790	10	40
5/17	0915	1	1	10	7	20	2,400	1	4	300	440	10	20
5/17	1045	0	0	10	9	10	2,900	2	8	260	400	20	120
5/17	Comp.	0	0	0	10	20	2,600	2	8	290	430	30	60
06711585–Upper Asbury Park													
3/4	1300	6	6	29	40	40	10,000	2	540	90	300	90	290
3/4	1315	6	6	29	30	30	10,000	4	480	50	350	20	270
3/4	1330	0	2	0	30	70	8,000	2	460	60	250	30	230
3/4	1430	6	6	29	29	30	5,600	2	330	40	160	30	160
3/4	1500	6	6	0	20	30	4,700	2	310	50	130	30	140
5/3	2100	0	2	48	55	40	17,000	16	470	20	420	20	360
5/3	2110	0	2	0	70	40	21,000	14	560	20	520	10	410
5/3–4	Comp.	0	1	48	90	60	30,000	14	430	30	660	20	380
06711586–Lower Asbury Park													
3/4	1300	6	6	0	40	20	13,000	2	670	90	390	30	380
3/4	1315	6	6	0	30	50	9,400	5	480	70	280	30	270
3/4	1330	6	6	29	30	40	8,900	3	440	70	260	40	260
3/4	1400	0	1	0	30	50	6,800	2	400	50	200	30	200
3/4	Comp.	0	12	0	30	40	7,700	4	440	60	240	40	230
5/28	Comp.	0	0	0	120	70	5,200	1	670	40	1300	10	700
06713010–Cherry Knolls													
4/3	Comp.	0	1	0	11	60	2.500	4	110	30	80	50	100
5/12–13	Comp.	0	0	0	0	40	1,100	1	39	20	50	40	60
5/17	0530	1	1	10	4	10	230	2	20	200	220	20	20
5/17	0640	1	1	10	9	40	1,500	2	53	10	40	30	70
5/17	0830	0	0	10	3	30	1,000	4	29	10	30	30	50
06720420–Northglenn													
3/4	1525	0	0	0	30	0	11,000	1	300	10	240	0	170
3/4	1540	0	1	0	40	40	14,000	1	390	10	300	10	200
3/4	1555	6	6	0	40	20	13,000	2	320	0	290	0	170
4/3	0910	0	1	0	12	40	3,600	1	120	10	80	10	80
5/3	1950	0	4	0	110	100	36,000	26	1500	10	730	20	870
5/3	1955	0	3	0	80	70	25,000	16	1000	0	540	20	650
5/17–18	Comp.	0	0	0	9	40	1,800	17	63	10	50	10	60
6/3	1450	0	0	50	70	730	44,000	10	310	30	720	20	430
394236105042400–Villa Italia													
3/20	Comp.	0	1	0	29	40	6,300	3	320	100	240	100	290
3/21	Comp.	0	1	0	18	90	5,100	0	140	60	170	70	200

this information. Control of runoff pollution is compounded by the fact that these pollutants occur in both the total and dissolved forms. Lee (1981) has pointed to the difficulty in interpreting toxic effects of pollutants in storm water to aquatic life. While quantification of the differences between total, soluble, and available forms of toxic compounds has not been attempted, it is assumed, as a worst case, that the soluble fraction may be toxic under the proper combinations of pH, temperature, and hardness. Table 9.5 presents the results of the total and dissolved metals analysis. The dissolved form is defined as that fraction of the pollutant that passes through a 0.45 μm filter.

Table 9.5 illustrates that the majority of heavy metal pollutants are adsorbed onto the fine particles. The results of the particle size pollutant and total dissolved comparisons imply design criteria for any structure used in controlling storm-generated runoff.

CONCLUSIONS

Few hard conclusions can be presented at this time. Obviously the data show that the relationships between storm rainfall, runoff, effective impervious area, and pollutants are quantifiable. Urban runoff pollution is probably predictable and these predictions will be based on the model presented here. The model undoubtedly requires careful testing and evaluation. It may, in fact, require revision for other than conservative constituents. The results of the actual testing as well as testing by others in different geographical areas is necessary to determine its applicability to a wide variety of conditions.

Predictive results of the model utilizing Best Management Practices data and data from the special studies may have serious implications for the future of water quality management. These implications include implementation of a Best Management Practices program in conjunction with the NPDES program. Also, wet weather quality criteria that are in developmental stages (Athayde, U.S. Environmental Protection Agency, 1982 personal communication) may greatly impact state water quality standards and practice. This study is a step in determining the answers to these and other possible questions.

REFERENCES

Bender, G.M., M.L. Terstriep, and D.C. Noel. 1981. *Nationwide Urban Runoff Project, Champaign, Illinois: Evaluation of the Effectiveness of Municipal Street Sweeping in the Control of Urban Storm Runoff Pollution.* Champaign, IL: Illinois Institute of Natural Resources State Water Survey Division.

Colt, J.L. 1980. Determination of local annual sediment loss design limit, Adams County,

Colorado. Report to Denver Regional Council of Governments. Denver, Colorado.

______. 1981. Percolation pits: Their design, construction use, and maintenance for stormwater disposal, ground water recharge, and surface water quality protection in Adams County, Colorado. Adams County Planning Department.

DRCOG. 1977. *Denver Regional Council of Governments Clean Water Plan.* Denver Regional Council of Governments. Denver, Colorado.

______. 1980. *Managing Erosion and Sedimentation from Construction Activities: Interim Best Management Practices.* Denver Regional Council of Governments. Denver, Colorado.

Ellis, S.R., and W.M. Alley. 1979. *Quantity and Quality of Urban Runoff from Three Localities in the Denver Metropolitan Area, Colorado.* Reston, VA: USGS Water Resources Investigations No. 79–64.

Heaney, J.P., W.C. Huber, and M.E. Lehman, 1981. *National Assessment of Receiving Water Impacts from Urban Stormwater Pollution.* EPA-600/52-81-025. Washington, D.C.: U.S. Environmental Protection Agency.

Hutchinson, G.E. 1975. *A Treatise on Limnology.* Vol. 1, Part 2, *Chemistry of Lakes.* New York: John Wiley and Sons. P. 727.

Hynes, H.B.N. 1970. *The Ecology of Running Waters.* Toronto: University of Toronto Press. P. 543.

Judy, R.D., Jr. 1981. *Denver Regional Urban Runoff Program: Second Year Progress Report.* Denver Regional Council of Governments. Denver, Colorado.

Judy, R.D., Jr., and P.H. Davies. 1979. Effects of calcium addition as Ca $(NO_3)_2$ on zinc toxicity to fathead minnows, *Pimephales promelas,* Rafinesque. *Bull. Environm. Contam. Toxicol.* 22:88–94.

Koppelman, L.E. 1978. The Long Island comprehensive waste water treatment management plan. Volume II: Summary documentation. Nassau-Suffolk Regional Planning Board. Long Island, N.Y. P. 64.

Lee, G.F. 1981. Will EPA's nationwide urban runoff study achieve meaningful results? *ASCE Journal* (September 1981):86–88.

LWCOG. 1977. Larimer-Weld Council of Governments 108 areawide water quality management plan: Nonpoint source pollution control.

McPherson, M.B. 1979. *Overview of Urban Runoff Tools of Analysis.* EP79-R-20. Quebec: Ecole Polytechnique de Montréal, University of Montreal.

Malcolm, H.R. 1980. *A Study of Detention in Urban Stormwater Management.* UNC-WRRI-80-156. Raleigh, N.C.: University of North Carolina Water Resources Research Institute.

Pitt, R., and G. Shawley. 1981. San Francisco Bay area nationwide urban runoff project: A demonstration of nonpoint source pollution. Castro Valley, CA: Association of Bay Area Governments.

Sartor, J.D., and G.B. Boyd. 1972. *Water Pollution Aspects of Street Surface Contaminants.* EPA-R2-72-081. Washington, D.C.: U.S. Environmental Protection Agency.

Seahurn, G.E., and D.A. Aronson. 1974. *Influence of Recharge Basins on the Hydrology of Nassau and Suffolk Counties, Long Island, New York.* Reston, VA: U.S. Geological Survey Water-Supply Paper 2031.

Shaheen, D.G. 1975. *Contributions of Urban Roadway Usage to Water Pollution.* EPA-600/2-75-004. Washington, D.C.: U.S. Environmental Protection Agency.

U.S. Dept. of Agriculture. Soil Conservation Service. (USDA-SCS). 1975. *Standards and Specifications for Soil Erosion and Sediment Control in Developing Areas.* Washington, D.C.

———. 1978. *Water Management and Sediment Control for Urbanizing Areas.* Washington, D.C.

U.S. Environmental Protection Agency (U.S. EPA). 1977. *Preventative Approaches to Stormwater Management.* EPA-440/9-77-011. Washington, D.C.

———. 1980. *Zinc Water Quality Standards Criteria Digest: A Compilation of State and Federal Criteria.* Washington, D.C.: U.S. EPA Criteria and Standards Division.

INDEX